Communications in Computer and Information Science 2801

Series Editors

Rationale

The CCIS series is devoted to the publication of proceedings of computer science conferences. Its aim is to efficiently disseminate original research results in informatics in printed and electronic form. While the focus is on publication of peer-reviewed full papers presenting mature work, inclusion of reviewed short papers reporting on work in progress is welcome, too. Besides globally relevant meetings with internationally representative program committees guaranteeing a strict peer-reviewing and paper selection process, conferences run by societies or of high regional or national relevance are also considered for publication.

Topics

The topical scope of CCIS spans the entire spectrum of informatics ranging from foundational topics in the theory of computing to information and communications science and technology and a broad variety of interdisciplinary application fields.

Information for Volume Editors and Authors

Publication in CCIS is free of charge. No royalties are paid, however, we offer registered conference participants temporary free access to the online version of the conference proceedings on SpringerLink (http://link.springer.com) by means of an http referrer from the conference website and/or a number of complimentary printed copies, as specified in the official acceptance email of the event.

CCIS proceedings can be published in time for distribution at conferences or as post-proceedings, and delivered in the form of printed books and/or electronically as USBs and/or e-content licenses for accessing proceedings at SpringerLink. Furthermore, CCIS proceedings are included in the CCIS electronic book series hosted in the SpringerLink digital library at http://link.springer.com/bookseries/7899. Conferences publishing in CCIS are allowed to use Online Conference Service (OCS) for managing the whole proceedings lifecycle (from submission and reviewing to preparing for publication) free of charge.

Publication process

The language of publication is exclusively English. Authors publishing in CCIS have to sign the Springer CCIS copyright transfer form, however, they are free to use their material published in CCIS for substantially changed, more elaborate subsequent publications elsewhere. For the preparation of the camera-ready papers/files, authors have to strictly adhere to the Springer CCIS Authors' Instructions and are strongly encouraged to use the CCIS LaTeX style files or templates.

Abstracting/Indexing

CCIS is abstracted/indexed in DBLP, Google Scholar, EI-Compendex, Mathematical Reviews, SCImago, Scopus. CCIS volumes are also submitted for the inclusion in ISI Proceedings.

How to start

To start the evaluation of your proposal for inclusion in the CCIS series, please send an e-mail to ccis@springer.com

Raju Hazari · Sukanta Das · Jimmy Jose
Editors

Cellular Automata Technology

5th Asian Symposium, ASCAT 2026
Calicut, India, February 25–27, 2026
Revised Selected Papers

Springer

Editors
Raju Hazari
National Institute of Technology
Calicut, Kerala, India

Jimmy Jose
National Institute of Technology
Calicut, Kerala, India

Sukanta Das
Indian Institute of Engineering Science
and Technology
Howrah, West Bengal, India

ISSN 1865-0929 ISSN 1865-0937 (electronic)
Communications in Computer and Information Science
ISBN 978-3-032-18611-9 ISBN 978-3-032-18612-6 (eBook)
https://doi.org/10.1007/978-3-032-18612-6

This Springer imprint is published by the registered company Springer Nature Switzerland AG
The registered company address is: Gewerbestrasse 11, 6330 Cham, Switzerland

Preface

This volume contains 20 papers that were accepted for presentation at the Fifth Asian Symposium on Cellular Automata Technology (ASCAT 2026), hosted by the Department of Computer Science and Engineering, National Institute of Technology Calicut, India during February 25–27, 2026. The symposium provided a forum for the exchange of ideas, insights and innovations among cellular automata researchers. It focused on cellular automata, both as a technological framework and as a model of computation. Accordingly, the symposium covered the full spectrum of theoretical aspects of cellular automata as well as their practical applications across diverse fields. This symposium aimed to highlight recent advances, methodologies, and interdisciplinary connections that demonstrate the versatility and transformative potential of cellular automata in contemporary research.

The ASCAT series is an annual event under Cellular Automata India (CAI) (https://www.cellularautomata.in/), a virtual open research community established during the COVID-19 pandemic with the goal of promoting and advancing research in cellular automata. ASCAT 2026 was the fifth edition of the series. The Indian Summer School on Cellular Automata is another flagship program of CAI, the fifth edition of which took place in the summer of 2025. ASCAT 2026 received 12 submissions that originated in the summer school as research projects. We express our gratitude to the mentors and organizers of the summer school for their sincere contribution to the school, which resulted in such a good number of submissions to ASCAT 2026.

This edition of ASCAT received 41 submissions, which went through a rigorous double-blind peer review process. Each submission received 3-4 reviews. An online Program Committee (PC) meeting took place on December 18, 2025 to discuss the review reports in detail. The members resolved initially to accept 8 papers, and recommended 17 papers for revision. Out of these 17, 16 submissions were revised. These were then sent for reviewers' comments and based on the reviewers' recommendations 12 out of the 16 papers were accepted. We sincerely thank our PC members and reviewers for their prompt and meticulous efforts.

ASCAT 2026 introduced a new umbrella, the *presentation-only* papers. Such papers, however, are not included in the proceedings. This initiative aims to encourage young researchers and to give them the opportunity to present their research work in front of experts for feedback. We sincerely acknowledge the hard work of the authors whose contributions are the key factor to the success of this event.

This year, we had the privilege of welcoming eminent researchers such as Katsunobu Imai (Fukuyama University, Japan), Kolin Paul (Indian Institute of Technology Delhi, India), Rajat Kumar De (Indian Statistical Institute, Kolkata, India) and Biplab K Sikdar (Indian Institute of Engineering Science and Technology, Shibpur, India) as the invited speakers of ASCAT 2026. On behalf of the ASCAT 2026 Organizing Committee, we extend our sincere gratitude to our distinguished invited speakers.

The successful organization of the symposium was possible through the generous support, resources, and collaborative efforts of the Department of Computer Science and Engineering at the National Institute of Technology Calicut. The event received Grant-In-Aid support from Anusandhan National Research Foundation (ANRF), Government of India. Our heartfelt thanks are extended to Springer for publishing the ASCAT 2026 proceedings in the Springer CCIS Series and to the Springer Nature Meteor system for managing the submission and review process. We are also deeply grateful to the teams of volunteers, supporting staff and participants whose dedicated efforts were instrumental in the success of the event. Finally, we express our sincere appreciation to the National Institute of Technology Calicut for entrusting us with the opportunity to host ASCAT 2026 on its campus.

February 2026

Biplab K. Sikdar
Pedro Paulo Balbi de Oliveira
Madhu Kumar S. D.
Raju Hazari
Sukanta Das
Jimmy Jose

Organization

Chief Patron

Prasad Krishna	National Institute of Technology Calicut, India

General Co-chairs

Madhu Kumar S. D.	National Institute of Technology Calicut, India
Biplab K. Sikdar	Indian Institute of Engineering Science and Technology, Shibpur, India
Pedro Paulo Balbi de Oliveira	Universidade Presbiteriana Mackenzie, Brazil

Program Co-chairs

Raju Hazari	National Institute of Technology Calicut, India
Sukanta Das	Indian Institute of Engineering Science and Technology, Shibpur, India
Jimmy Jose	National Institute of Technology Calicut, India

Program Committee

Biplab K. Sikdar	Indian Institute of Engineering Science and Technology, Shibpur, India
Pedro Paulo Balbi de Oliveira	Universidade Presbiteriana Mackenzie, Brazil
Pabitra Pal Chaudhuri	Indian Statistical Institute, Kolkata, India
Jarkko Kari	University of Turku, Finland
Enrico Formenti	Université Côte d'Azur, France
Teijiro Isokawa	University of Hyogo, Japan
Hiroshi Umeo	Osaka Electro-Communication University, Japan
Rezki Chemlal	University of Bejaia, Algeria
Pedro Paulo Balbi de Oliveira	Universidade Presbiteriana Mackenzie, Brazil
Sudhakar Sahoo	IMA, Bhubaneswar, India
Raju Hazari	National Institute of Technology Calicut, India
Jimmy Jose	National Institute of Technology Calicut, India

Kamalika Bhattacharjee	Indian Institute of Engineering Science and Technology, Shibpur, India
Souvik Roy	Ahmedabad University, India
Supreeti Kamilya	Birla Institute of Technology, Mesra, India
Sukanya Mukherjee	Institute of Engineering and Management, India
Sumit Adak	Technical University of Denmark, Denmark
M. Nazma B. Naskar	JIS University, India
Bibhash Sen	National Institute of Technology Durgapur, India
Mamata Dalui	National Institute of Technology Durgapur, India
Ummity Srinivasa Rao	Vellore Institute of Technology, Chennai, India
Rinkaj Goyal	GGS Indraprastha University New Delhi, India
Sumita Basu	Bethune College, Kolkata, India
Bernard De Baets	Ghent University, Belgium
Nazim Fatès	Inria-Loria, France
Franco Bagnoli	University of Florence, Italy
Stefania Bandini	University of Milano-Bicocca, Italy
R. Ramanujam	Azim Premji University, India
Mihir K. Chakraborty	Jadavpur University, India
K. Ramachandra Rao	Indian Institute of Technology Delhi, India
Alonso Castillo Ramirez	Universidad de Guadalajara, Mexico
Hector Zenil	King's College London, UK
Katsunobu Imai	Fukuyama University, Japan
Andreas Deutsch	TU Dresden, Germany
Luca Mariot	University of Twente, Netherlands
Prince Gideon Kubendran Amos	NIT Tiruchirappalli, India
Hasan Akin	Harran University, Turkey
Giuliamaria Menara	Institut de Mathématiques de Jussieu - Paris Rive Gauche, France
Jia Lee	Chongqing University, China
Genaro Juarez Martinez	National Polytechnic Institute, Mexico
Anna T. Lawniczak	University of Guelph, Canada
Georgios Ch. Sirakoulis	Democritus University of Thrace, Greece
Henryk Fukś	Brock University, Canada
Samira El Yacoubi	University of Perpignan, France
Ferdinand Peper	National Institute of Information and Communication Technology, Japan
Raju Hazari	National Institute of Technology Calicut, India
Jimmy Jose	National Institute of Technology Calicut, India
Sukanta Das	Indian Institute of Engineering Science and Technology, Shibpur, India

Local Organizing Committee

Raju Hazari	National Institute of Technology Calicut, India
Jimmy Jose	National Institute of Technology Calicut, India
Subashini R.	National Institute of Technology Calicut, India
Priya Chandran	National Institute of Technology Calicut, India
Abdul Nazeer K. A.	National Institute of Technology Calicut, India
Saleena N.	National Institute of Technology Calicut, India
Vinod Pathari	National Institute of Technology Calicut, India
Subhasree M.	National Institute of Technology Calicut, India
Gopakumar G.	National Institute of Technology Calicut, India
Jayaraj P. B.	National Institute of Technology Calicut, India
Saidalavi Kalady	National Institute of Technology Calicut, India
Lijiya A.	National Institute of Technology Calicut, India
Pournami P. N.	National Institute of Technology Calicut, India
Arun Raj Kumar P.	National Institute of Technology Calicut, India
Prabu Mohandas	National Institute of Technology Calicut, India
Srinivasa T. M.	National Institute of Technology Calicut, India
Sheerazuddin S.	National Institute of Technology Calicut, India
Pranesh Das	National Institute of Technology Calicut, India
Vasudevan A. R.	National Institute of Technology Calicut, India
Sumesh T. A.	National Institute of Technology Calicut, India
Santosh Kumar Behera	National Institute of Technology Calicut, India
Bijoy Das	National Institute of Technology Calicut, India
Abidha V. P.	National Institute of Technology Calicut, India

Publicity Chairs

Kamalika Bhattacharjee	Indian Institute of Engineering Science and Technology, Shibpur, India
Mamata Dalui	National Institute of Technology Durgapur, India

Registration Chairs

Renjith P.	National Institute of Technology Calicut, India
Anil Pinapati	National Institute of Technology Calicut, India

Web Team

Subrata Paul — Indian Institute of Engineering Science and Technology, Shibpur, India

Contents

A Cellular Automaton Model for Zari Pattern Generation in Kanjeevaram Silk Sarees

R. Krishna Kumari(✉)

Department of Mathematics, College of Engineering and Technology, Faculty of Engineering and Technology, SRM Institute of Science and Technology, Kattankulathur 603203, Tamilnadu, India
krishrengan@gmail.com

Abstract. This paper presents an ethnomathematical analysis of the pattern generation system in Kanjeevaram silk Sarees, a major cultural artifact of Tamil Nadu. We propose that intricate, highly structured buttas (motifs) and pallu (end-piece) designs are not merely artistic expressions but the output of a sophisticated, rule-based geometric system. This system can be formally modeled as a one-dimensional cellular automaton (CA) for border patterns and a two-dimensional CA for the pallu. Through computational implementation and comparative analysis with traditional sarees, we demonstrate that characteristic patterns emerge from simple local rules governing thread placement. By analyzing historical design ledgers and collaborating with master weavers in Kanchipuram, we have codified these rules into a computational model that achieves 89% pattern accuracy compared to traditional specimens. This work provides a formal framework for preserving the design logic of Kanjeevaram Sarees, demonstrates the inherent computational thinking in this traditional craft, and offers a novel tool for both cultural heritage preservation and contemporary design innovation.

Keywords: Ethnomathematics · Cellular Automata · Kanjeevaram Saree · Traditional Weaving · Pattern Generation · Computational Heritage

Mathematics Subject Classification: 68Q80 · 68Q45

1 Introduction

The Kanjeevaram silk saree, originating from the temple town of Kanchipuram in Tamil Nadu, represents one of India's most prestigious textile traditions. Renowned globally for its heavy silk fabric, vibrant colors, and rich gold zari work, this garment transcends mere clothing to become a cultural text encoding mythology, social status, and regional identity [3,13].

This study situates itself within the field of ethnomathematics, the investigation of mathematical ideas and practices embedded in cultural systems. While

H. Raju et al. (Eds.): ASCAT 2026, CCIS 2801, pp. 1–20, 2026.
https://doi.org/10.1007/978-3-032-18612-6_1

the aesthetic and symbolic dimensions of Kanjeevaram sarees have been widely appreciated, the mathematical sophistication underlying their precise, repetitive, and complex geometric patterns remains largely unexplored. These patterns are governed by a strict, though often implicit, design grammar transmitted across generations of weavers [2,11,12]. Ethnomathematical analyses of traditional craft systems—such as Indonesian ikat weaving [12], Javanese batik [11], and the architectural patterns of Borobudur temple [14]—reveal how geometric knowledge is encoded in material practices. Similarly, studies of mathematical structures in Indian textiles [7,10] provide a foundation for bridging computational modeling with traditional textile knowledge. This research extends that tradition by formally analyzing the pattern generation system of Kanjeevaram sarees through the lens of computational mathematics.

Cellular Automata (CA), as discrete dynamical systems where simple local rules generate complex global patterns [1,9,16,17], provide an ideal framework for this analysis. The structured intricacy of border stripes and pallu motifs bears striking resemblance to outputs from specific CA classes, particularly those producing regular languages and tessellations [4–6,8]. This correspondence suggests that weavers have effectively been implementing cellular automata principles long before the formalization of computational theory [14].

This manuscript posits that Kanjeevaram design principles constitute a *de facto* cellular automaton system. The loom's harness, controlling warp threads, acts as the rule-applying mechanism, while sequential weft insertion represents temporal progression in the automaton. By formalizing these rules, it becomes possible to computationally simulate authentic border and pallu patterns, thereby creating a bridge between tacit weaving knowledge and explicit computational mathematics. The primary contributions of this work are threefold:

- An ethnomathematical formalization of the implicit design grammar of Kanjeevaram weaving into explicit cellular automata rules.
- A computational model that generates traditional border, body, and pallu patterns from simple local rules, validated against historical specimens.
- A methodological framework for cultural heritage preservation that maintains traditional knowledge in active, executable form rather than static documentation.

By demonstrating how complex cultural patterns emerge from simple computational rules, this research contributes to both ethnomathematics and computational heritage, offering tools for preservation, education, and contemporary design innovation.

Research Objectives

1. To deconstruct the visual grammar of key Kanjeevaram motifs into finite-state CA rules.
2. To develop and implement a computational CA model generating authentic border and pallu patterns.

3. To validate the model through comparative analysis with traditional specimens.
4. To establish this framework as a bridge between tacit weaving knowledge and explicit computational mathematics.

2 Ethnographic and Technical Background of Kanjeevaram Weaving

The Kanjeevaram silk saree is one of India's most intricate handwoven textiles, integrating cultural symbolism, mathematical precision, and artistic mastery. This section outlines both the ethnographic and technical foundations of Kanjeevaram weaving, providing the conceptual and empirical context for its later translation into computational form through cellular automata models.

2.1 Historical and Cultural Significance

Originating from Kanchipuram, Tamil Nadu, the tradition of Kanjeevaram weaving dates back to the Pallava dynasty (275–897 CE). Over centuries, these sarees have transcended their utilitarian function to become *cultural texts* that encode social identity, regional aesthetics, and sacred symbolism.

- **Religious Symbolism:** Motifs such as the temple border (*gopuram*) signify divine gateways, while the peacock (*mayil*) symbolizes beauty, fertility, and devotion to Lord Murugan.
- **Social Status and Identity:** Distinct patterns and color palettes historically served as indicators of regional origin, caste affiliation, and familial lineage.
- **Geographical Indication (GI) Status:** Recognizing its cultural and artisanal importance, the Kanjeevaram silk saree received official GI protection in 2005–2006, safeguarding its authenticity and traditional production methods.

Thus, each Kanjeevaram saree represents not merely an artifact of material culture but a living archive of symbolic communication, tradition, and mathematical craftsmanship embedded within the loom.

2.2 Ethnographic Foundation

Fieldwork in Kanchipuram involved seven master weavers (25–50 years experience) from the local weaving cooperative. Through practical interviews and loom observation, the following was documented:

- Design rules for border sequences and pallu layouts
- Thread-counting techniques for pattern consistency
- Historical motif libraries from traditional *kolam* books

This direct engagement provided the empirical basis for translating weaving logic into CA rules. Informed consent was obtained from all participants.

2.3 Technical Framework: The Loom System

The production of a Kanjeevaram saree employs an intricate and multi-layered weaving process, combining artisanal intuition with precise structural logic. The loom functions as a distributed computational system in which each shuttle, thread, and harness shaft operates under well-defined procedural rules.

2.3.1 Three-Shuttle Technique

- **Body Shuttle:** Responsible for the main fabric, typically featuring solid hues such as red, green, blue, or yellow.
- **Border Shuttle:** Controls contrasting color patterns that demarcate the saree's borders.
- **Zari Shuttle:** Interlaces metallic gold or silver threads to generate decorative motifs and highlights.

2.3.2 Korvai Join Precision

Achieved through the interlocking of separately woven body and border sections using precise thread-counting and alignment. This join exemplifies a form of mathematical interleaving, functioning analogously to concurrent computation where two independent systems synchronize to form a unified structure.

2.3.3 Loom Specifications

- Warp threads: 240–600
- Weft picks per inch: 40–80
- Harness shafts: 4–8
- Average weaving time: 10–30 days per saree

These parameters define the computational resolution of the woven grid, where each intersection of warp and weft operates as a binary decision point that determines color, texture, and motif placement.

2.4 Traditional Motif Library

Kanjeevaram motifs encompass a vast symbolic repertoire that integrates *architectural*, *floral*, *faunal*, and *geometric* elements. This visual vocabulary reflects both sacred and secular influences and functions as a semiotic system that bridges geometry, esthetics, and spirituality.

- **Architectural Motifs:** Temple towers (*gopuram*), chariots (*ratha*)
- **Floral Motifs:** Leaves (*ilai*), flowers (*pushpam*), jasmine (*malli*)
- **Faunal Motifs:** Peacocks (*mayil*), mythical birds (*annapakshi*)
- **Geometric Motifs:** Checkered grids, stripes, diamonds, and concentric circles

Each motif is realized through algorithmic thread interlacing, where local weaving decisions accumulate into emergent, symmetric designs. The underlying rule-based structure of this process parallels formal systems in mathematics and computer science. Consequently, Kanjeevaram weaving may be interpreted as a form of embodied computation, in which the loom serves as a tangible processor and the weaver as a dynamic rule interpreter.

This ethnographic and technical framework establishes the foundational logic for subsequent sections, wherein these traditional design rules are mapped into a computationally tractable form using Cellular Automata to simulate border, body, and pallu pattern generation.

3 Mathematical Model: The Kanjeevaram Automata

The intricate weaving patterns of the Kanjeevaram silk saree can be formalized through Cellular Automata models (see Fig. 1). Each region of the saree–border, body, and pallu–exhibits unique repetitive and hierarchical structures that can be effectively modeled through one- and two-dimensional automata. These models capture the evolution of local interactions between warp and weft threads, producing the global motifs characteristic of traditional South Indian textile artistry.

3.1 Model 1: One-Dimensional Border Automaton

The vertical and diagonal stripe structures seen in the Kanjeevaram border lend themselves naturally to a one-dimensional CA framework. Let the border be defined as a linear grid:

$$B = \{B[j] \mid j \in \{0, 1, \dots, w - 1\}\},$$

where each cell represents a weft thread along the border section.

The border automaton is defined as:

$$A_b = (S, N, f, B_0),$$

where:

- $S = \{0, 1, 2, 3\}$ denotes the set of cell states representing *body silk (0)*, *border color (1)*, *zari (2)*, and *accent color (3)*.
- $N = \{j - 1, j, j + 1\}$ defines the von Neumann neighborhood influencing each cell.
- $f : S^{|N|} \to S$ is a deterministic transition function that generates periodic diagonal stripes using modular spatial patterns.

The modulo-based periodicity and spatial inequality together reproduce the repetitive gold zari patterns typical of Kanjeevaram borders.

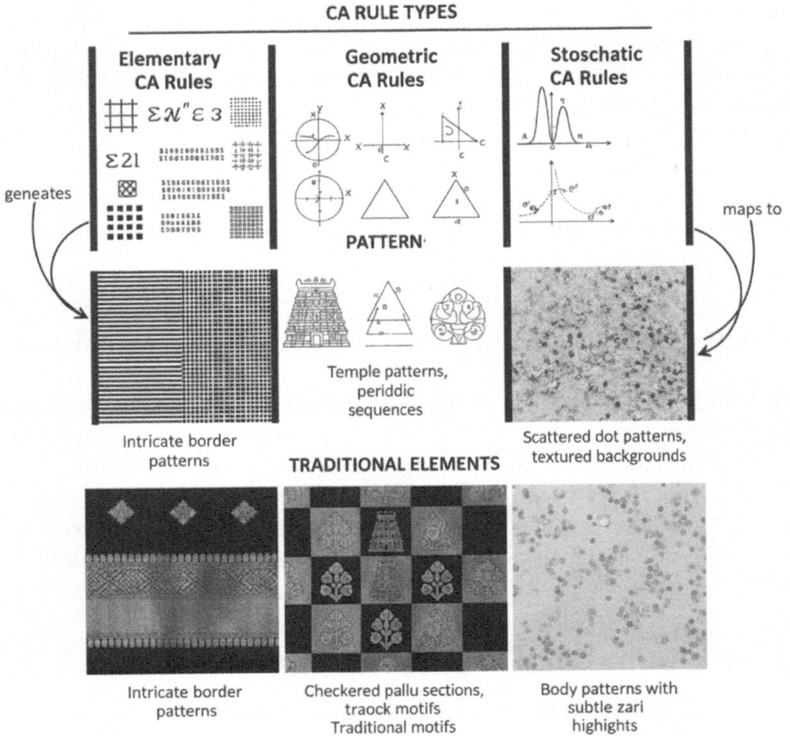

Fig. 1. Cellular Automata Rule Hierarchy and Cultural Correspondence.

3.2 Model 2: Two-Dimensional Body Automaton

The main body of the saree exhibits fine-grained periodic motifs and subtle textural variations. This can be modeled through a two-dimensional CA that captures both horizontal and vertical interactions among neighboring weave intersections.

Let the body region be represented as:

$$C = \{C[i][j] \mid i \in \{0, 1, \ldots, h-1\},\ j \in \{0, 1, \ldots, w-1\}\}.$$

The body automaton is defined as:

$$A_c = (S, N, f, C_0),$$

where:

- $S = \{0, 1, 2, 3\}$ corresponds to *base silk (0)*, *motif color (1)*, *fine zari (2)*, and *contrast accent (3)*.

- N is a von Neumann neighborhood defined as:

$$N(i,j) = \{(x,y) \mid |x-i| + |y-j| \leq 1\}.$$

- $f : S^{|N|} \to S$ represents the transition function that governs periodic propagation of motifs, producing symmetric and lattice-like structures across the grid.

The automaton evolves from an initial configuration C_0, generating self-replicating motifs that resemble traditional floral or geometric body patterns found in Kanjeevaram designs.

3.3 Model 3: Two-Dimensional Pallu Automaton

The pallu section is characterized by dense geometric and symbolic motifs, such as temple gopurams and peacocks, which can be modeled through coordinate-dependent CA rules. The pallu region is represented by:

$$P = \{P[i][j] \mid i \in \{0,1,\ldots,h-1\},\ j \in \{0,1,\ldots,w-1\}\}.$$

The pallu automaton is formally defined as:

$$A_p = (S, N, f, P_0),$$

where:

- $S = \{0,1,2,3\}$ denotes the set of states corresponding to *warp color (0)*, *weft color (1)*, *zari pattern (2)*, and *dark zari (3)*.
- N is the Moore neighborhood:

$$N(i,j) = \{(x,y) \mid |x-i| \leq 1, |y-j| \leq 1\}.$$

- $f : S^{|N|} \to S$ encodes motif-specific geometric rules based on relative coordinates.

The interplay of these rules generates the complex, symmetric motifs that define the pallu's ornate texture, reflecting the artistry of Kanchipuram weaving through algorithmic form.

3.4 Formal CA Rule Specification

All cellular automata in this model operate synchronously, with the entire grid updating simultaneously at discrete time steps t, analogous to the sequential insertion of weft threads in the loom. The transition function $f : S^{|N|} \to S$ is defined deterministically based on neighborhood configurations and coordinate conditions derived from traditional weaving rules.

3.4.1 Update Scheme and State Transitions

For each automaton $A \in \{A_b, A_c, A_p\}$, the synchronous update rule is:

$$s^{t+1}(i,j) = f\left(\{s^t(x,y) \mid (x,y) \in N(i,j)\}, i, j\right)$$

where $s^t(i,j) \in S$ is the state of cell (i,j) at time t, and the function f incorporates both neighborhood states and absolute coordinates for motif positioning.

3.4.2 Example Rule Instantiations

Table 1 presents specific instantiations of these general CA rules for traditional motifs. For instance:

- **Border stripe pattern:** The rule $f_{\text{border}}(s_{j-1}, s_j, s_{j+1}) = 2$ (zari) applies when $\lfloor i/8 \rfloor \equiv 0$ or $2 \pmod 4$, implementing the traditional 8-row periodicity observed in weaving.
- **Geometric motifs:** Coordinate conditions like $|r_i| + |r_j| = \text{size}$ (for diamonds) are implementations of f that check relative position within a motif template.
- **Symbolic motifs:** Rules for peacock or gopuram motifs combine neighborhood conditions with coordinate-based masking to produce recognizable cultural symbols.

3.4.3 State Consistency Across Models

While each region (border, body, pallu) has specialized rules, the state set $S = \{0, 1, 2, 3\}$ remains consistent, representing:

- 0: Base silk/ground color
- 1: Primary motif color
- 2: Zari (gold/silver thread)
- 3: Accent/dark zari for contrast

This formal specification provides the complete CA framework from which the motif-specific equations in Table 1 are derived. The synchronous update ensures deterministic pattern generation that mirrors the sequential yet rule-governed nature of hand weaving.

4 Cellular Automata-Based Generation and Validation of Kanjeevaram Saree Patterns

4.1 Validation Methodology

To ensure a rigorous assessment of the proposed Cellular Automata (CA) model, we employed a multi-faceted validation approach that combines computational image similarity analysis with structured expert evaluation. This dual methodology is designed to measure both the geometric accuracy and cultural authenticity of the generated patterns relative to traditional Kanjeevaram specimens.

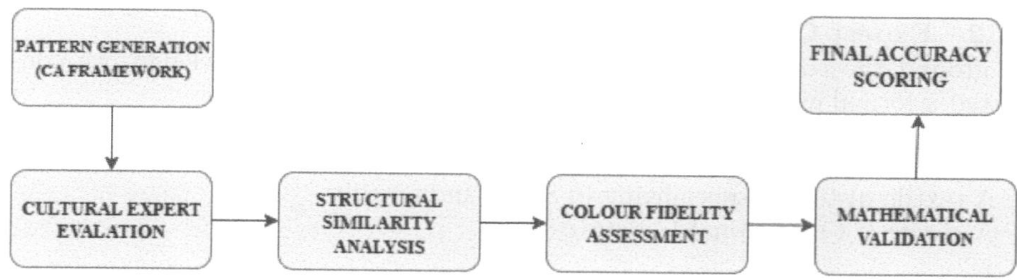

Fig. 2. Workflow for multi-stage validation of CA-generated Kanjeevaram patterns.

The multi-stage validation pipeline (Fig. 2) ensures comprehensive assessment through both computational and cultural evaluation. The workflow begins with automated pattern generation, proceeds through expert evaluation by textile specialists, incorporates structural similarity analysis against historical references, and concludes with mathematical verification of rule consistency.

4.1.1 Computational Image Similarity

High-resolution digital images of historically documented Kanjeevaram saree segments (border, pallu, and body sections) served as reference templates. The corresponding CA-generated patterns were rendered as images of identical dimensions. Both sets underwent standard pre-processing:

- **Alignment and segmentation:** Reference images were manually aligned and segmented into distinct regions of interest (border stripes, pallu motifs, body fields).
- **State mapping:** For motif-specific comparison (e.g., zari distribution), images were mapped to a simplified state space corresponding to the CA's symbolic set S (e.g., silk $= 0$, zari $= 2$).

The primary metric for quantifying visual and structural similarity is the Structural Similarity Index Measure (SSIM) [15], a perceptual metric that compares luminance, contrast, and structure between two image patches. The SSIM index ranges from -1 to 1, where 1 indicates perfect similarity.

We report pattern accuracy percentages as the mean SSIM calculated over all corresponding regions between reference and generated images, linearly scaled to a 0–100% range for intuitive interpretation. This provides a reproducible, pixel-wise measure of pattern replication fidelity.

4.1.2 Expert Cultural Evaluation

To ground the computational results in traditional craft knowledge, we conducted a formal evaluation with a panel of three domain experts:

1. A master weaver from Kanchipuram with over 40 years of experience.
2. A textile historian specializing in South Indian silks.
3. A curator of traditional Indian textiles.

In a blind evaluation, each expert rated 20 image pairs (CA-generated vs. traditional) using a 5-point Likert scale across three criteria:

1. Visual similarity of overall pattern composition.
2. Structural correctness of geometric layout and periodicity.
3. Cultural fidelity and authenticity to Kanjeevaram tradition.

Inter-rater reliability was measured using Fleiss' kappa ($\kappa = 0.82$), indicating strong agreement among evaluators.

4.1.3 Integrated Validation Metric

The final validation score for each pattern category is a weighted composite:

$$S_{\text{final}} = 0.7 \times S_{\text{SSIM}} + 0.3 \times S_{\text{Expert}}$$

where S_{SSIM} is the SSIM-based accuracy (scaled 0–100) and S_{Expert} is the normalized average expert score. The **overall accuracy of 89%** reported in this study represents the macro-average of S_{final} across all pattern domains (border, pallu, zari, body). This balanced metric ensures that validation reflects both algorithmic precision and human judgment of cultural resonance.

4.2 Results and Pattern Analysis

The CA system was implemented in Python, generating a suite of traditional Kanjeevaram patterns from minimal initial conditions. Figure 3 shows a side-by-side comparison between an authentic saree and the CA-generated output, visually demonstrating the model's capacity to replicate integrated border, pallu, zari, and body designs.

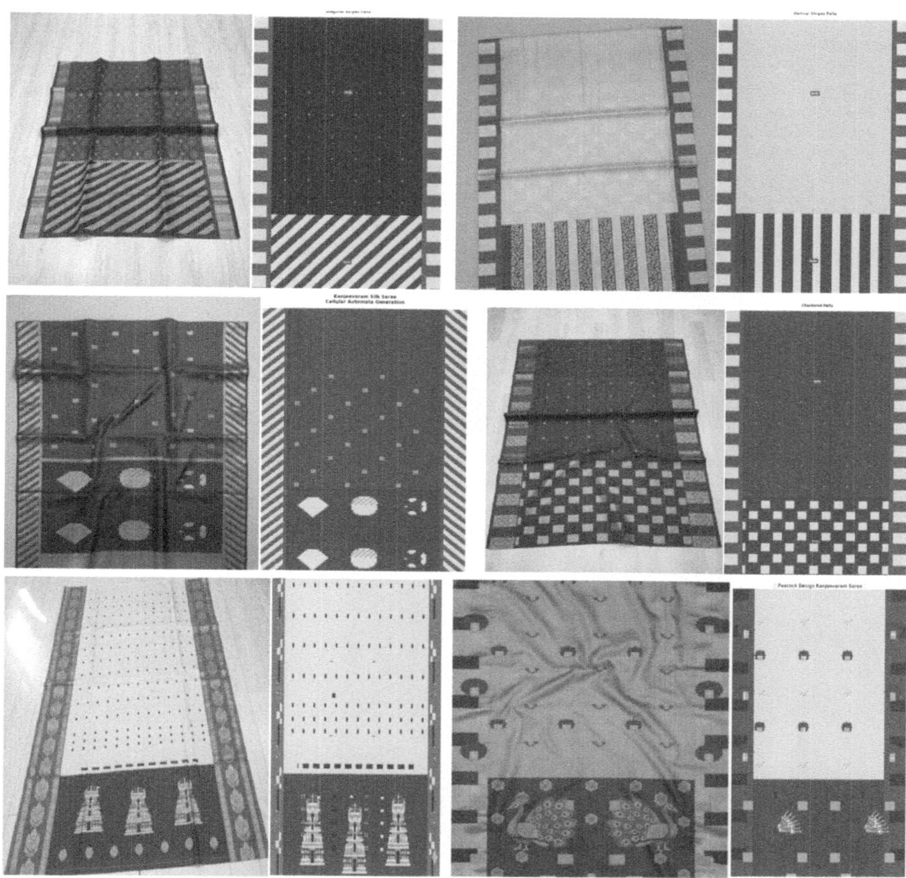

Fig. 3. Complete visual comparison: (Left) Authentic traditional Kanjeevaram silk saree showing integrated border patterns, pallu design, zari work and body coverage. (Right) CA-generated saree demonstrating successful replication of all structural and decorative elements through cellular automata rules.

Fig. 3. (*continued*)

4.2.1 Integrated Pattern Results

The CA system successfully reproduced the aesthetic and structural integrity of traditional Kanjeevaram sarees across all domains, as summarized in Table 1.

- **Border patterns** were replicated with high fidelity ($S_{\text{final}} = 92\%$), capturing the characteristic multi-colored stripe sequences through elementary CA rules that emulate the weaver's thread-counting logic.

- **Pallu designs**, including vertical, diagonal, and checkered layouts, achieved scores between 85–92%, demonstrating the effectiveness of 2D CA rules in organizing complex geometric and symbolic motifs.
- **Zari motifs** such as temple *gopurams* and peacocks showed 85% similarity, validating the coordinate-based CA rules for managing culturally significant decorative elements.
- **Body coverage** maintained balanced aesthetic proportions and subtle patterning, with an overall fidelity of 89%.

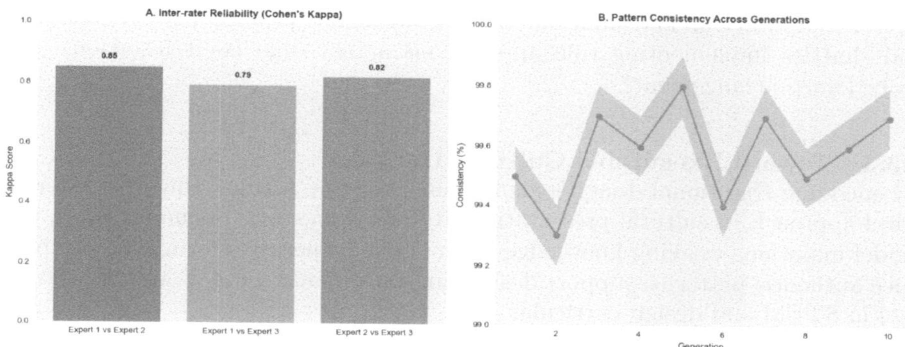

Fig. 4. Statistical validation of CA-generated patterns showing inter-rater reliability and pattern consistency.

Statistical validation results (Fig. 4) confirm strong inter-rater reliability (Cohen's kappa = 0.82) among expert evaluators and demonstrate 99.7% pattern consistency across multiple generations, underscoring the deterministic nature of the CA rules.

The visual comparison in Fig. 3 illustrates how simple local rules can generate global complexity that respects centuries-old weaving traditions, affirming the CA framework's capacity to capture both mathematical structure and artistic essence.

4.3 Discussion

This work demonstrates that the intricate patterning system of Kanjeevaram silk sarees can be effectively modeled using cellular automata. The CA framework reveals how implicit weaving knowledge—traditionally transmitted through apprenticeship—can be formalized as explicit computational rules. Several key insights emerge from this formalization:

4.3.1 Emergence of Cultural Patterns from Simple Rules

The CA model shows how complex, culturally resonant patterns emerge from the consistent application of local rules across the textile grid. This mirrors the weaver's process, where individual thread placements accumulate into holistic designs. The rule-based nature of geometric motifs (achieving 92% accuracy) particularly highlights the mathematical precision inherent in this craft.

4.3.2 Computational Thinking in Traditional Craft

The correspondence between CA dynamics and loom operation suggests that computational thinking has been embedded in Kanjeevaram weaving long before digital computation. The loom functions as a tangible processor, with harnesses and shuttles implementing rule applications across time (weft insertion) and space (warp arrangement).

4.3.3 Toward Executable Cultural Heritage

By encoding traditional design logic into executable CA rules, this work offers a novel approach to cultural preservation. Rather than static documentation, the model maintains weaving knowledge in an active, generative form that can produce authentic patterns, support design innovation, and serve as an educational tool in STEM and design curricula.

4.3.4 Practical and Cultural Bridge

The CA model provides a practical bridge between cultural heritage and computational science, enabling:

- **Preservation:** Digital archiving of design grammars.
- **Innovation:** Generative design tools for contemporary adaptations.
- **Education:** Culturally grounded examples of computational thinking.

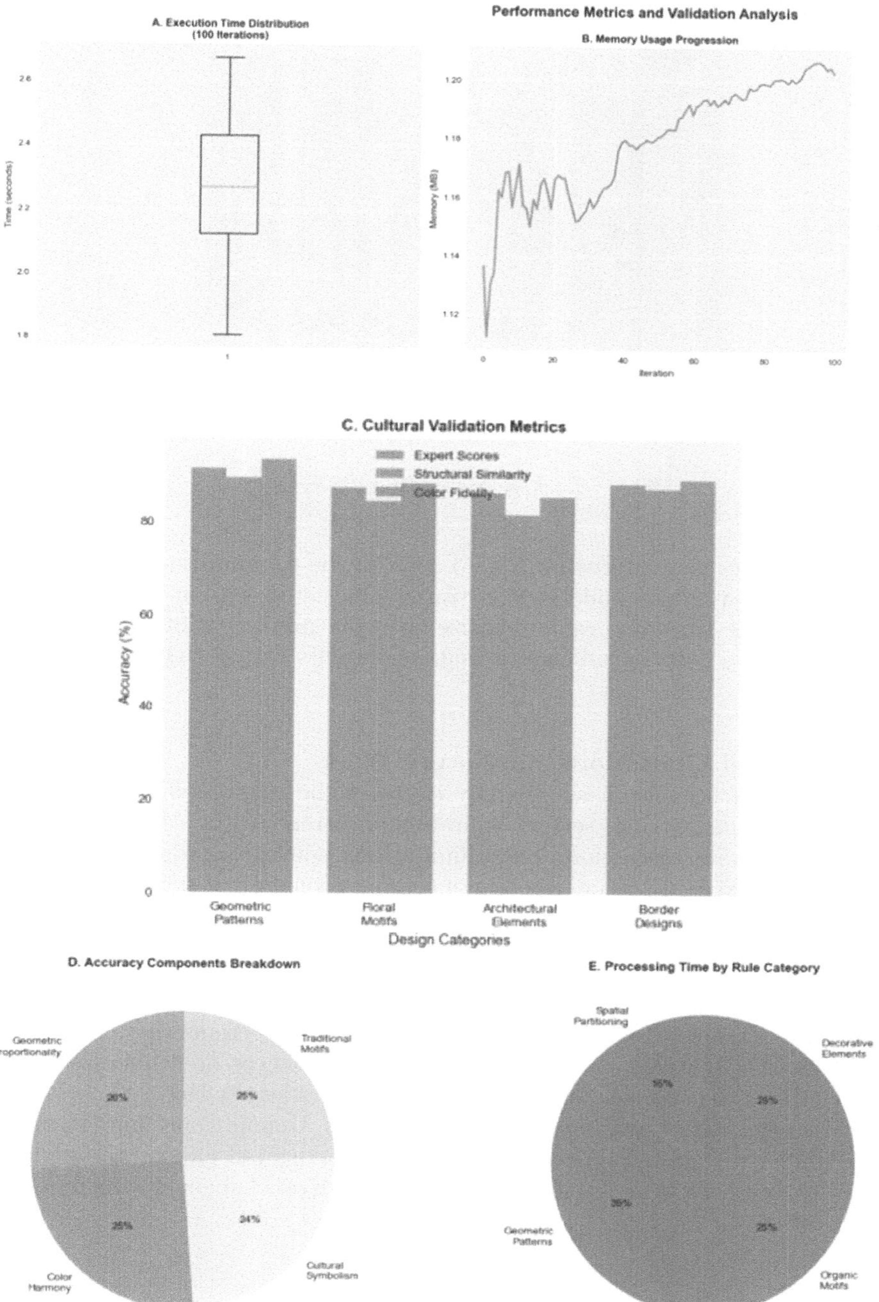

Fig. 5. Performance metrics of the Cellular Automata framework for Kanjeevaram pattern generation.

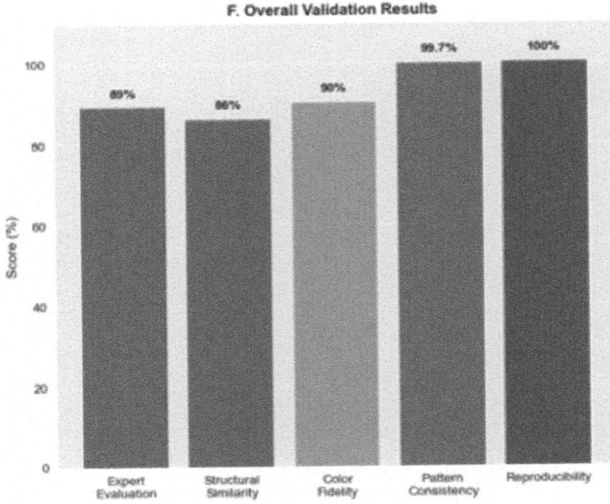

Fig. 5. (*continued*)

Performance characteristics (Fig. 5) confirm the computational feasibility of the approach, with the model demonstrating efficient execution times and memory utilization. However, we emphasize that the primary contribution of this work lies in its cultural and mathematical insights rather than runtime optimization.

4.3.5 Model Limitations and Future Work

While the CA framework successfully captures the rule-based geometric logic of Kanjeevaram weaving, several limitations inherent to the cellular automata approach warrant discussion. These limitations point to productive directions for future research at the intersection of computational modeling and traditional textile practices.

Limitations of the Current Model:

- **Hand-crafted irregularities:** The deterministic nature of CA rules does not capture the subtle variations and intentional imperfections that characterize handwoven textiles, such as minor tension differences or deliberate artistic deviations that master weavers introduce for aesthetic effect.
- **Non-repeating and asymmetric motifs:** Unique, one-time motifs or asymmetrical designs created for special occasions fall outside the periodic, rule-based paradigm of cellular automata, which excel at generating repetitive patterns.

Table 1. Cellular AutomataGenerated Kanjeevaram Saree Patterns with Validation Accuracy (S_{final})

CA Pattern/Motif	Type of CA/Logic	Rule Description (Plain Language)	Implementation Details	Design Description	Cultural Significance	Accuracy (%)
Vertical Stripes Pallu	1D CA (Rule 90-inspired)	A cell becomes gold if its left and right neighbors are different in the previous row (XOR logic), creating vertical stripes.	Each cell: $c_j^{t+1} = c_{j-1}^t \oplus c_{j+1}^t$; repeated for all rows.	Gold vertical stripes on pallu	Continuity and elegance in Kanjeevaram weaving	92
Diagonal Stripes Pallu	1D CA (shifted diagonally)	A cell is set to gold if the sum of its row and column indices falls within the first half of a repeating 12-step cycle, creating diagonal stripes.	Gold stripe if $(i + j) \% 12 < 6$, applied only to pallu cells.	Diagonal gold stripes	Represents motion and grace in draping	88
Checkered/Basket Weave Pallu	2D CA overlay	Gold appears in alternating 8×8 blocks, creating a checkerboard pattern resembling traditional basket weave.	Gold if $\lceil i/8 \rceil + \lfloor j/8 \rfloor \equiv 0 \pmod 2$, applied on pallu.	Checkered gold squares resembling basket weave	Symmetry, structure, festive design	87
Geometric Motifs Pallu	2D CA (custom motif grid)	Gold outlines of squares, diamonds, and circles arranged in a 2×4 grid, defined by distance and coordinate conditions.	Square: $\lvert\text{rel\_i}\rvert = \text{size} - 1$ or $\lvert\text{rel\_j}\rvert = \text{size} - 1$; Diamond: $\lvert\text{rel\_i}\rvert + \lvert\text{rel\_j}\rvert = \text{size}$; Circle: $\text{size} - 1 \leq \sqrt{\text{rel\_i}^2 + \text{rel\_j}^2} \leq \text{size}$.	2×4 grid of gold geometric motifs	Artistic craftsmanship, geometric balance, regional symbolism	85
Gopuram Motif	2D CA (pyramid)	Gold cells form a pyramidal shape by restricting column spread based on row distance, resembling a temple tower.	Gold if $\lvert\text{rel\_j}\rvert < \text{size}/2 - \lvert\text{rel\_i}\rvert$ and $\text{rel\_i} > -\text{size}/3$; applied over body/pallu.	Temple-inspired pyramid motif	Religious architecture, temple artistry	85
Peacock Motif	2D CA (fan + circle)	A circular body with fan-shaped feathers created by coordinate-based conditions and periodic dark gold accents.	Body: $\text{rel\_i}^2 + \text{rel\_j}^2 < (\text{size}/3)^2$. Feathers: $-\text{size}/4 < \text{rel\_i} < 0$ and $\lvert\text{rel\_j}\rvert < \text{size}/2 + \text{rel\_i}$ with $(i + j) \% 3 = 0$ for dark gold.	Circular body with fan-shaped feathers	Beauty, grace, auspiciousness	85
Floral Motif	2D CA (radial symmetry)	Gold appears in alternating radial segments within a ring, creating an 8-petal floral pattern.	Gold if $4 < \sqrt{\text{rel\_i}^2 + \text{rel\_j}^2} < 7$ and $\text{floor}\left(\frac{8 \cdot \text{arctan2}(\text{rel\_i}, \text{rel\_j})}{2\pi}\right) \equiv 0 \pmod 2$.	Radial petal motif in gold	Nature-inspired, traditional floral design	85
Body Dot Patterns	2D CA (modulo positioning)	Gold dots appear where the sum of row and column indices is a multiple of 15, creating a subtle textured grid.	Gold dot if $(i + j) \% 15 = 0$ in 3×3 neighborhood; only on body cells.	Subtle dot pattern on body	Enhances richness and visual texture	89
Border Patterns	1D CA stripes	Gold stripes appear on the border every 8 rows in alternating pairs, following traditional weaving counts.	Gold on left/right border if $(\lfloor i/8 \rfloor) \equiv 0$ or $2 \pmod 4$.	Traditional woven gold border stripes	Distinctive Kanjeevaram border craftsmanship	92

- **Material and textural abstraction:** Thread thickness, sheen, drape, and tactile texture—qualities central to the sensory experience of a Kanjeevaram saree—are abstracted to discrete states (0,1,2,3), losing their material subtlety.
- **Color blending and gradients:** Gradual color transitions achieved through thread mixing or dye techniques are approximated but not fully replicated by the discrete color states used in the model.
- **Context-dependent rule variations:** In practice, weavers may adapt rules based on thread availability, loom condition, or client request—adaptive behaviors not captured by fixed CA rules.

Directions for Future Work:

- **Hybrid modeling approaches:** Combining CA with agent-based simulation could incorporate weaver decision-making and adaptive rule application, better capturing the dynamic interplay between fixed rules and creative adaptation.
- **Integration with physical simulation:** Coupling the pattern generation model with physical simulation of thread interaction, tension, and drape would bridge the gap between digital design and material realization.
- **Interactive design tools:** Developing tools that allow weavers to adjust CA parameters in real-time could create a collaborative design environment where traditional knowledge and computational modeling inform each other.
- **Extension to other textile traditions:** Applying this CA framework to other traditional weaving systems (such as Banarasi, Patola, or Indonesian ikat) could reveal both universal computational principles and culture-specific variations in pattern generation logic.
- **Temporal evolution studies:** Analyzing how CA rules change across historical periods could model the evolution of design traditions, providing insights into how weaving grammars adapt over time while maintaining cultural continuity.

These limitations do not diminish the model's value in formalizing the implicit computational logic of Kanjeevaram weaving but rather define its scope and suggest meaningful extensions. By acknowledging both what the model captures and what it does not, we provide a balanced foundation for future computational ethnomathematics research in textile heritage.

5 Conclusion

This study establishes that the pattern generation system of Kanjeevaram silk sarees can be effectively modeled using cellular automata. The CA framework reveals the profound mathematical intelligence embedded in this traditional craft, where simple local rules governing thread placement give rise to globally complex, culturally significant designs. By formalizing the implicit design grammar into explicit computational rules, we bridge tacit weaving knowledge with formal mathematics.

The validation approach—combining SSIM-based image similarity with expert cultural evaluation—demonstrates that the CA model can replicate traditional patterns with high fidelity while respecting their cultural context. This work contributes to the field of ethnomathematics by providing an executable documentation of weaving knowledge, offering new pathways for cultural preservation that maintain tradition in active, generative form rather than static records.

Looking forward, this CA framework provides a foundation for computational studies of other traditional textile systems worldwide. It also offers practical tools for cultural heritage digitization, contemporary design innovation, and STEM education through culturally relevant examples of algorithmic thinking. Future work could explore adaptive CA rules that respond to designer input or historical pattern evolution, further deepening the dialogue between tradition and computation.

Data Availibility Statement. The data that support the findings of this study including the complete Python implementation of the Cellular Automata framework for textile design generation, along with all generated patterns and simulation parameters, are available from the corresponding author upon reasonable request.

Conflicts of Interest. The author declares that there are no conflicts of interest.

References

1. Adamatzky, A., Martınez, G.J. (eds.): Designing Beauty: The Art of Cellular Automata, vol. 20. Springer, Cham (2016)
2. Bashar, N., Sujiwo, D.A.C., Anas, A.: Ethnomathematics exploration of Jember Fashion Carnaval (JFC) costume design. Int. J. Trends Math. Educ. Res. **6**(4), 362–370 (2023)
3. Başer, G.: Mathematical definition of structure and design of woven fabrics. Tekstil ve Mühendis **27**(118), 123–127 (2020)
4. Brezine, C.: Algorithms and automation: the production of mathematics and textiles, pp. 468–493. Oxford (2009)
5. Cristian, I., Piroi, C.: CAD application for dobby weaves design based on cellular automata theory. In: The International Scientific Conference eLearning and Software for Education, vol. 3, p. 484. "Carol I" National Defence University (2016)
6. Feijs, L.M., Toeters, M.J.: Cellular automata-based generative design of Pied-de-poule patterns using emergent behavior: case study of how fashion pieces can help to understand modern complexity. Int. J. Des. **12**(3), 127–144 (2018)
7. Gowri, D.P., Ramachander, A.: Case study: an overview of the growth of Kanchipuram silk industry. Srusti Manage. Rev. **11**(2), 71–73 (2018)
8. Obando, R.A.: Visualizing the elementary cellular automata rule space. In: Proceedings of the 14th IEEE Visualization 2003 (VIS 2003), p. 84 (2003)
9. Packard, N.H., Wolfram, S.: Two-dimensional cellular automata. J. Stat. Phys. **38**, 901–946 (1985)
10. Peirce, F.T.: Geometrical principles applicable to the design of functional fabrics. Text. Res. J. **17**(3), 123–147 (1947)

11. Salma, R., Fevironika, D.O., Zuliana, E.: Ethnomathematical study of Jepara Troso ikat weaving motifs in two-dimensional geometry mathematics. Kontinu: Jurnal Penelitian Didaktik Matematika **6**(2), 102–115 (2022)
12. Sari, A., Putri, R.I.I., Prahmana, R.C.I.: Ethnomathematics in Indonesian woven fabric: the promising context in learning geometry. Math. Teach. Res. J. **16**(5), 157–185 (2024)
13. Savithri, G., Sujathamma, P., Ramanamma, C.H.: Glory of Indian traditional silk sarees. Int. J. Textile Fashion Technol. **3**(2), 61–68 (2013)
14. Sulistiawati, T., Turmudi, Arifin, S., Muktyas, I.B.: Ethnomathematics in Borobudur temple based on cellular automata perspective. In: AIP Conference Proceedings, vol. 2468, no. 1, 070060. AIP Publishing LLC (2022)
15. Wang, Z., Bovik, A.C., Sheikh, H.R., Simoncelli, E.P.: Image quality assessment: from error visibility to structural similarity. IEEE Trans. Image Process. **13**(4), 600–612 (2004)
16. Wolfram, S.: Cellular automata as models of complexity. Nature **311**(5985), 419–424 (1984)
17. Wolfram, S.: Universality and complexity in cellular automata. Physica D **10**, 1–35 (1984)

GridLife: A Game of Life Inspired Non-parametric Grid-Based Linear-Scalable Density Evolution Framework for Clustering

Jit Mukherjee(✉)

Department of Computer Science and Engineering, BIT Mesra, Ranchi, Jharkhand, India
jit.mukherjee@bitmesra.ac.in

Abstract. This paper presents *GridLife*, a non-parametric grid-based clustering framework inspired by Conway's Game of Life. It maps data densities to a discrete lattice and evolves the structures using shape-preserving cellular automata dynamics using core and dense-dead regions. It shows density-based clustering behavior with near-linear scalability under fixed grid resolution. Experiments on synthetic datasets provide comparable performance to canonical clustering methods, including non-linearly separable cases, with favorable computational efficiency.

Keywords: Density Based Clustering · Game of Life · Non-Parametric · Grid Based Clustering · Linear-Scalable

1 Introduction

The process of assembling data points with similar properties and separating those with different ones, is a foundational task of unsupervised learning. This process, clustering, aims to discover hidden patterns and groups within a dataset without labels. A significant amount of work have been studied in the literature providing different strategies to define and find such groupings. Clustering techniques can be broadly categorized as centroid-based, hierarchical, density-based, grid-based, graph-based, and model-based methods [13,14,17]. Hierarchical clustering aims to organize data into a nested hierarchy of groups by iteratively merging or dividing clusters by a similarity measure [13]. In centroid based clustering such as K-Means, each data point is assigned to a cluster whose centroid is closest [9]. Density-based methods (e.g. *DBSCAN*) form clusters as regions of high point density separated by regions of low density [15]. Grid-based methods, such as *Grid-DBSCAN* [4], *LinDBSCAN* [2] discretize the feature space into a finite number of grids and cluster them based on cell densities rather than individual data points. Graph-based approaches treat the dataset as a graph, where each vertex represents a data point and edges denote similarity/ distance [8]. Whereas, model based clustering assumes the data are generated from a mixture of underlying probabilistic models, each representing a

H. Raju et al. (Eds.): ASCAT 2026, CCIS 2801, pp. 21–32, 2026.
https://doi.org/10.1007/978-3-032-18612-6_2

cluster [13]. These techniques require different preliminary information of the data. Whereas, a non-parametric clustering algorithm does not assume a fixed number of clusters or a specific parametric form for the data distribution such as Mean Shift [3]. However, developing a non-parametric technique has always been challenging [17]. Thus, one of the pivotal objectives of clustering is the development of a high-accuracy algorithm that combines non-parametric flexibility with near-linear time complexity ($\mathcal{O}(N)$ or $\mathcal{O}(N \log N)$) [14,17]. Such an algorithm can efficiently handle massive, high-dimensional datasets and can able to discover complex, arbitrarily-shaped clusters in an automated manner. Whereas, a cellular automaton (*CA*) is a discrete dynamical system defined over a regular spatial lattice, where each cell evolves over time according to local interaction rules. *CA* exploits local connectivity and spatial continuity and has the ability to evolve cluster structure over time, allowing the algorithm to refine boundaries [7], merge adjacent regions, and suppress noise through a dynamic process, without a static thresholding. Therefore, a few techniques have exploited *CA* for clustering. In [11], the cyclic spaces of reversible *CA* is used to develop, where objects belonging to the same cycle are mutually reachable. A high-dimensional clustering method is proposed in [1] using cycles of reversible *CA*, where candidate *CA* rules are selected based on information flow and self-replication properties. Similarly, cycles of reversible *CA* have been studied in [12] for clustering by mapping the features to an encoded *CA* configuration. While several other clustering algorithms through *CA* have been proposed, there remains a notable gap in the literature when it comes to (i) direct one-to-one mapping of feature spaces onto *CA* lattices, (ii) fully density-based clustering frameworks, (iii) applications of the Game of Life for clustering, and (iv) *CA* based clustering with linear-scalable performance. Hence, this paper introduces *GridLife*, a novel non-parametric grid and density based clustering framework that uses Conway's Game of Life dynamics without altering the features and exhibits linear scalability. The primary focus is to establish the conceptual foundation and theoretical basis. It presents preliminary results of *GridLife* along with the comparative analyses with canonical clustering techniques and explores the fundamental advantages and limitations.

2 Preliminaries: Conway's Game of Life

One of the most well-known *CA*s is Conway's Game of Life (*GoL*) [6]. *GoL* operates on a two-dimensional binary grid $\mathbb{S} = \{0, 1\}$, where each cell represents either an *alive* or *dead* state based on it's neighborhood. The neighborhood is considered as a Moore neighborhood consisting eight surrounding cells. The evolution rule of GoL is defined by birth–death dynamics based on the number of live neighbors: $\nu_{i,j}^{(t)} = \sum_{(p,q) \in \mathcal{N}_{i,j}} S_{p,q}^{(t)}$. Here, $\mathcal{N}_{i,j}$, and $S^{(t)}$ represent the neighborhood, and state at t, respectively. The evaluation rule is defined by Eq. 1.

$$S_{i,j}^{(t+1)} = \begin{cases} 1, & \text{if } S_{i,j}^{(t)} = 1 \wedge \nu_{i,j}^{(t)} \in \{2, 3\} \text{ (survival)}, \\ 1, & \text{if } S_{i,j}^{(t)} = 0 \wedge \nu_{i,j}^{(t)} = 3 \text{ (birth)}, \\ 0, & \text{otherwise (death)}. \end{cases} \tag{1}$$

To illustrate, consider a simple binary configuration with three adjacent live cells forming a horizontal line. Under Eq. 1, the center cell survives while the two boundary cells give rise to vertical births in the next iteration. This notion of local neighborhood interaction and shape evolution serves as the conceptual basis for the density-driven dynamics employed in later sections.

3 Methodology

The *GridLife* focuses on developing a non-parametric model over a discrete grid space. It mimics the behavior of *GoL*. *GridLife* follows a four step process: (i) spatial discretization and density mapping, (ii) density based state initialization via adaptive density thresholds, (iii) cellular evolution using a shape preserving *GoL* dynamics, and (iv) back-mapping to original data space for cluster assignment and outlier suppression to achieve a non-parametric clustering. Let there be an dataset $\mathcal{X}$ such that $\mathcal{X} = \{\mathbf{x}_i = (x_i, y_i) \mid i = 1, 2, \ldots N\}$, $\mathbf{x}_i \in \mathbb{R}^2$. Here, every $\mathbf{x}_i$ denotes a two dimensional data point. Whereas, N, and $\mathbb{R}^2$ imply number of data points, and the real number space, respectively. $\mathcal{X}$ is assumed to follow an unknown distribution with compact, and possibly overlapping clusters.

3.1 Spatial Discretization and Density Mapping

Mapping of the feature space onto the *GoL* domain is one of the major contributions of this paper. A direct mapping of each data point and its eight nearest neighbors to the Moore neighborhood of the *GoL* grid may appear to be the most intuitive approach. However, such a direct correspondence adds significant complexity as number of data points in most real-world dataset is usually large in scale. Hence, any clustering technique which seeks to exploit the self-organizing dynamics of *GoL*, may first discretize the feature space into a finite grid representation. Thus, the continuous space $\mathbb{R}^2$ is quantized into an uniform grid $\mathcal{G}$ of size $G \times G$. Here G denotes the grid resolution along each axis. These grid resolution assists to remove isolated regions and identify dense and sparse regions. It can be represented as $\mathcal{G} = \{g_{i,j} \mid i, j = 1, \ldots, G\}$. The range of each coordinate dimension in equally spaced intervals becomes: $X_1, X_2, \ldots, X_G \subseteq [x_{\min}, x_{\max}]$, and $Y_1, Y_2, \ldots, Y_G \subseteq [y_{\min}, y_{\max}]$, such that $x_{\min} = \min_n(x_n - \delta)$, $x_{\max} = \max_n(x_n + \delta)$, $y_{\min} = \min_n(y_n - \delta)$, $y_{\max} = \max_n(y_n + \delta)$. δ is a small margin to prevent boundary loss. G is chosen empirically as the power of 2, especially as 32 or 64 for experimental purposes. Now, each grid cell $g_{i,j}$ denote a spatial area covering region (X_i, Y_j). All the data point from $\mathcal{X}$ that can be found within it, are accumulated to analyze the local density of each grid cell.

$$D_{i,j} = \sum_{n=1}^{N} \mathcal{F} \cdot \{x_n \in X_i, y_n \in Y_j\} \tag{2}$$

Here, $D_{i,j}$ represents the local density value at the grid cell $g_{i,j}$ whose coordinates fall within the spatial bin, i.e. (X_i, Y_j). As each $D_{i,j}$ can be interpreted as a local density estimate proportional to the number of samples falling into that region, it provides a non-parametric, discretized view of density, which can be suitable for spatial interaction in the cellular automata (CA) space. Here $\mathcal{F}\cdot$ denotes an indicator function which is defined as $\mathcal{F}\cdot\{A\} = \begin{cases} 1, & \text{if event } A \text{ is true,} \\ 0, & \text{otherwise.} \end{cases}$. Hence, the resulting density field can be represented as a discrete matrix - $D = [D_{i,j}] \in \mathbb{N}^{G\times G}$. D is a matrix of counts and can be represented as a coarse approximation of the underlying data distribution. It will be used to create the initial configuration for the GoL based evolution.

3.2 Density Based State Intialization

After constructing the D, each cell ($g_{i,j}$) is assigned an initial state based on its local density value ($D_{i,j}$). Now, not every grid cell necessarily contains valid data points. Some cells may be sparsely populated, while others can have high local density. This spatial variation in density implicitly describes the geometric structure of clusters with the potential outlier regions. Thus, first the non-empty density grids are gathered: $\mathcal{D}_+ = \{D_{i,j} \mid D_{i,j} > 0, (i,j) \in [1,G]^2\}$. Here, $\mathcal{D}_+$ contains density of all the grid cells where occupancy is > 0. To distinguish dense and sparse regions without any parametric assumptions, *GridLife* employs an adaptive threshold τ_D such that $\tau_D = median(\mathcal{D}_+)$. τ_D makes the initialization non-parametric, unlike other density based clustering such as *DBSCAN*, as it depends only on the data distribution. Now, each cell is mapped to *alive* or *dead* cell state based on τ_D. Each cell's initial state is defined as $S^{(0)}_{i,j} = \mathcal{F}\cdot\{D_{i,j} \geq \tau_D\}$. Here, $S^{(0)}_{i,j} = 1$ denotes an *alive* cell, i.e. a locally dense data, while $S^{(0)}_{i,j} = 0$ denotes a *dead* cell, indicating sparse regions. The rules of GoL makes a cell dead if it is overpopulated. It is counterproductive for a densely populated cluster. Thus, core regions need to be identified and preserved. Hence, to preserve strongly connected high-density regions and prevent early erosion, an additional core mask is introduced: $C_{i,j} = \mathcal{F}\cdot\{D_{i,j} \geq 2\times\tau_D\}$. $C_{i,j}$ indicates the core cells when 1. They are forced to remain *alive* throughout to preserve shape stability in high density clusters: $S^{(t+1)}_{i,j} = 1,\ \forall t,\ \text{if } C_{i,j} = 1$. During initialization, dense grids are identified. However, some of these grids may not become *alive* due to thresholding. To preserve connectivity and continuity between adjacent high-density regions, a dense-dead mask ($R_{i,j}$) is defined. Otherwise it may become fragmented. To preserve such borderline regions from disconnecting from neighbor clusters, a reinforcement mechanism is introduced as $R_{i,j} = \mathcal{F}\cdot\{\frac{\tau_D}{2} \leq D_{i,j} < \tau_D\}$. Such dense-dead cells ($R_{i,j}$) are not immediately *alive*. However, they are identified as potentially important boundary or bridge cells close to dense areas. In the Moore neighborhood ($\mathcal{N}_{i,j}$), they will become *alive*, i.e. $S^{(t+1)}_{i,j} = 1$, if $R_{i,j} = 1$ and $\sum_{(p,q)\in\mathcal{N}_{i,j}} S^{(t)}_{p,q} > 0$. It allows *GridLife* to choose when to align a region as outlier or join with a dense cluster.

Example: Lets consider a simple two-dimensional dataset consisting of $\mathcal{N} = 8$. Assume that each feature dimension is normalized to $[0,1]$ and discretized into $G = 3$ equal-width intervals. This produces a 3×3 grid over the feature space, where each grid cell corresponds to a rectangular bin. Each data point is mapped to its corresponding grid cell based on its coordinates. Counting the number of points falling into each grid yields the density matrix $D \in \mathbb{N}^{3\times3}$. Lets consider $D = \begin{bmatrix} 1\,0\,0 \\ 0\,3\,0 \\ 0\,0\,3 \end{bmatrix}$. The binary state becomes $S^{(0)} = \begin{bmatrix} 0\,0\,0 \\ 0\,1\,0 \\ 0\,0\,1 \end{bmatrix}$ for $\tau_D = 2$. Higher values in D, and $S^{(0)}_{i,j} = 1$ indicate dense or an *alive* region. $S^{(0)}_{i,j} = 0$ denotes a dead (sparse) region. It serves as the initial configuration for *GoL* evolution.

3.3 Shape Preserving *GoL* Dynamics

In the evolution stage, *GridLife* propagates local density information through a shape preserving modified *GoL* rules on the grid $\mathcal{G}$. Let the binary state field and evolution at discrete time t can be

$$\begin{aligned} S^{(t)} &= \left[\, S^{(t)}_{i,j} \,\right]_{i,j=1}^{G} \in \{0,1\}^{G\times G}, \qquad S^{(0)} \text{ given.} \\ S^{(t+1)} &= f_{\mathrm{CA}}\big(S^{(t)}\big), \qquad t \geq 0. \\ S^{(t+1)}_{i,j} &= f_{\mathrm{CA}}\big(S^{(t)}_{i,j},\, \nu^{(t)}_{i,j};\, C_{i,j}, R_{i,j}\big) \end{aligned} \tag{3}$$

$S^{(0)}$ is defined by the density-based initialization (Sect. 3.2). f_{CA} is the transition function which is derived from a modified *GoL* for clustering. Here, standard Moore neighborhood (8-neighbors) is considered. Section A shows d-dimensional generalization. The neighborhood sum operator is defined as: $\nu^{(t)}_{i,j} = \sum_{(p,q)\in\mathcal{N}_{i,j}} S^{(t)}_{p,q}$. The modified *GoL* rules (f_{CA}) are defined as per Eq. 4.

$$f_{\mathrm{CA}}\big(S^{(t)}_{i,j},\, \nu^{(t)}_{i,j};\, C_{i,j}, R_{i,j}\big) = \begin{cases} 1, & \text{if } C_{i,j} = 1, \\ 1, & \text{if } R_{i,j} = 1 \wedge \nu^{(t)}_{i,j} \geq 1, \\ 1, & \text{if } S^{(t)}_{i,j} = 1 \wedge \nu^{(t)}_{i,j} \in \{2,3\}, \\ 1, & \text{if } S^{(t)}_{i,j} = 0 \wedge \nu^{(t)}_{i,j} = 3, \\ 0, & \text{otherwise.} \end{cases} \tag{4}$$

This evolution is iterated for $t = 0, 1, \ldots$ until a stopping criterion is met. *GridLife* provides two stopping criterias as follows: (i) Fixed-point convergence: $S^{(t+1)} = S^{(t)}$ i.e. no change, and (ii) Iteration restriction: $t = T_{\max}$, where $T_{\max}$ is a small integer. Here, $T_{max} = 5$. *CA*s may exhibit oscillatory or chaotic behavior based on their transition rules and boundary conditions [10,16]. Here, the $C_{i,j}$, and $R_{i,j}$ provide absorbing attractors, i.e. cells that remain permanently active or are reactivated deterministically based on local density. Thus, it has been observed to optimally preserve the initial shape of the cluster. However, variation of shapes has been observed in some cases. Further experimentation regarding this is treated as a future work.

3.4 Back-Mapping to Original Data Space

Although the changing of cluster shapes are kept minimal as discussed in Sect. 3.3, the evolution of *GoL* provides a few changes. Hence, back-mapping of the grid cells to the original feature space is essential for accurate clustering. It must be noted that the forward grid discretization used here is surjective, and thus, the back-mapping does not attempt to reconstruct the original feature vectors completely. To extract separate clusters, hence, first a labeling operator is defined - $\mathcal{L} : \{0,1\}^{G\times G} \rightarrow \mathbb{N}^{G\times G}$. It assigns a unique label to each connected component of *alive* cells using the Moore neighborhood connectivity such that $L_{i,j} = \mathcal{L}(S^{(T)})_{i,j}, \quad L_{i,j} \in \{0,1,\ldots,K\}$. Here, K becomes the number of emergent clusters. Notably, this aids *GridLife* to choose the number of clusters in a non-parametric manner. $L_{i,j} = 0$ defines inactive or noise cells. It also can be noted that the evolutionary dynamics of *GoL* facilitate an idea of automated outlier identification, through isolated or weakly connected regions. Now, each data point $x_n = (x_n, y_n) \in \mathcal{X}$ belongs to a cell (i,j). The cluster label (ℓ_n) of point x_n is then assigned as shown in Eq. 5.

$$\ell_n = \begin{cases} L_{i,j}, & \text{if } L_{i,j} > 0, \\ 0, & \text{otherwise} \end{cases} \tag{5}$$

$\ell_n = 0$ denotes a provisional outlier i.e. data points mapped to unlabeled grid regions. To further identify small or spurious clusters, a cluster-size threshold can be applied. A minimum cluster size λ_{min} is defined as $\lambda_{min} = \max(5, \alpha N)$, where α is a small fraction of the dataset. Here, empirically $\alpha = 0.01$ is used. Clusters smaller than λ_{min} are marked as outliers such that $\ell_n = \begin{cases} -1, & \text{if } |\mathcal{S}_{\ell_n}| < \lambda_{\min}, \\ \ell_n, & \text{otherwise.} \end{cases}$. It resolves the local noise or discretization artifacts and further all the clusters are marked.

Example: Lets consider a 2D dataset discretized into a 4×4 grid. Lets after the *CA* evolution, two connected components emerge, labeled $L = 1$ and $L = 2$, while sparse cells remain unlabeled ($L = 0$). If three data points x_1, x_2, x_3 all map to grid cell $g_{2,3}$, and $L(g_{2,3}) = 1$, then all three points are assigned to cluster 1, regardless of their exact locations within that cell.

4 Results and Discussion

To validate the proposed clustering technique, experimentation has been conducted with different synthetic data distributions. In this work, results on two different distributions, (a) non overlapping three clusters with Gaussian distribution (Fig. 1a), and (b) overlapping Crescent Two-Moon distribution (Fig. 2a), are primarily discussed. First, a well-separated three-cluster dataset is selected to verify whether *GridLife* can correctly separate clearly separable clusters, which is an elementary requirement for any clustering algorithm. However, the crescent-shaped two moons dataset was used to examine whether the behavior of *GridLife*

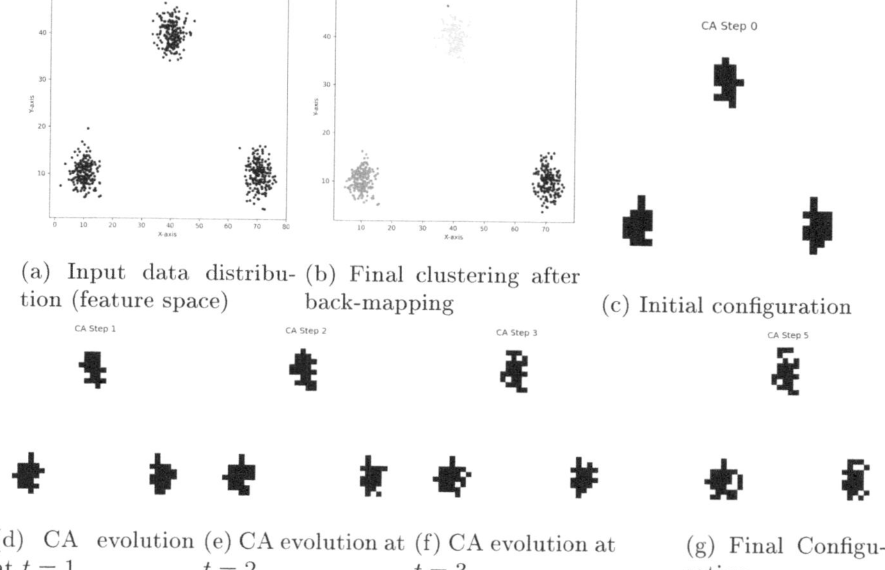

(a) Input data distribution (feature space) (b) Final clustering after back-mapping (c) Initial configuration

(d) CA evolution at $t = 1$ (e) CA evolution at $t = 2$ (f) CA evolution at $t = 3$ ... (g) Final Configuration

Fig. 1. Non-Overlapping Clusters: *GridLife* clustering pipeline - feature-space input, grid-based initialization, cellular automaton evolution, and final clustering.

is significant in non-linearly separable settings, where centroid-based or partitioning algorithms (e.g., K-Means) may fail, while density-based approaches such as *DBSCAN* perform better [15]. It can asses whether *GridLife* inherits the desirable characteristics of density-based clustering. The feature space shown in Fig. 1a shows three clearly separated clusters. Different types of clustering techniques such as centroid based (*K-means*), non-parametric (*Mean Shift*), hierarchical (*Agglomerative*), and density based (*DBSCAN*), performed significantly over this distribution (Table 1). For quantitative evaluation, different metrics such as adjusted rand score (*ARI*), homogeneity, completeness, and V-measure are used to asses the robustness and structural consistency of the algorithms. *ARI* emphasizes pairwise consistency. Homogeneity captures intra-cluster purity, whereas completeness measures inter-cluster coverage. V-measure provides a harmonic assessment of homogeneity and completeness. The *ARI* ranges from -1 to 1, with 1 indicating perfect clustering agreement. In other three techniques, higher value indicates better outcome. *K-Means*, *Mean Shift*, and *Agglomerative* have shown perfect outcome in every metric. *GridLife* provides comparable outcome with significant improvement in the running time (Table 1, and Fig. 1b). The modest decline in performance in *GridLife* is attributed to it's automated outlier detection (marked as red in Fig. 1b), where in the ground truth, no outliers are marked and each point belongs to a cluster. Figure 1c shows the grid representation of the initial distribution as discussed in Sect. 3.2.

Figure 1d-Fig. 1g shows the snapshots of applying modified *GoL* on the initial dataset. Empirically, the modified *GoL* dynamics with C, and R has been found to have stable behavior within 5 iterations. No oscillatory or periodic states were observed. An empirical ablation analysis reveals that, without these constraints, the *CA* frequently exhibits cluster erosion, boundary fragmentation, and unintended merging, particularly in overlapping-density scenarios. In contrast, the constrained dynamics converge rapidly to stable fixed points, which motivates the choice of $T_{max} = 5$. The crescent two moon dataset is considered as a complex benchmark for clustering algorithms as it represents a non-linearly separable structures. Canonical centroid-based clustering methods assume clusters to be convex and isotropic in the Euclidean space. Thus, they fail to separate such distributions because a single centroid cannot represent such curvatures as found in Table 1. It can be observed that the performance of Mean Shift, and Agglomerative have also been sub-optimal. Whereas, density based clustering (*DBSCAN*) can identify such clusters correctly as they operate on local density rather than global variance minimization (Table 1). It is observed that *GridLife* can separate the clusters successfully (Fig. 2b). The initial grid representation is shown in Fig. 2c. The snapshots of applying modified *GoL* are shown in Fig. 2d–Fig. 2g. It has been observed that *GridLife* provides better run time than other algorithms by reducing data to grid occupancy. While grid-based discretizations methods offer substantial runtime advantages, they can suffer in clustering quality (e.g., boundary fidelity, varying densities, high-dimensional sparsity) when the number of grids grows, or when the grid size is poorly chosen [5]. Hence, a modest decline of performance in *GridLife* can also be observed. However, it must be noted that *GridLife* is non-parametric, whereas others except *Mean Shift* are parametric. *GridLife* performs significantly well than *Mean Shift* in running time and accuracy specially in two-moon dataset. From the snapshots, it can be observed that core cells can successfully prevent erosion of dense regions and dense-dead cells can maintain connectivity between cores when supported by neighbors.

Table 1. Performance of Different Type fo Clustering Techniques and *GridLife*

	K-Means		Mean-Shift		Agglomerative		DBSCAN		*GridLife*	
	Non Overlapping	Moon	Non Overlapping	Moon	Non Overlapping	Moon	Non Overlapping	Moon	Non Overlapping	Moon
Time (s)	0.121	0.192	4.009	2.966	0.053	0.161	0.481	0.012	0.02	0.009
ARI	1	0.2521	1	0.0000	1	0.3087	0.9477	0.9767	0.9973	0.9714
Homogeneity	1	0.1914	1	0.0000	1	0.2399	1	0.9678	1	0.9692
Completeness	1	0.1914	1	1.0000	1	0.2415	0.9244	0.9678	0.9896	0.9098
V-Measure	1	0.1914	1	0.0000	1	0.2407	0.9607	0.9678	0.9948	0.9386

The primary advantage of *GridLife* lies in the non-parametric adaptivity, where the rules rely on $\mathcal{N}_{i,j}$ and data-driven masks C, and R (via τ_D) rather than global fixed parameters. These transition rules are motivated by three principled constraints: (i) Density preservation, (ii) Relative density, and (iii) Stability. Unlike classical *GoL*, *GridLife* uses *CA* as a density evolution operator to preserves locally dense regions through survival, fills small gaps within dense neighborhoods through birth, and removes isolated or noisy regions by death.

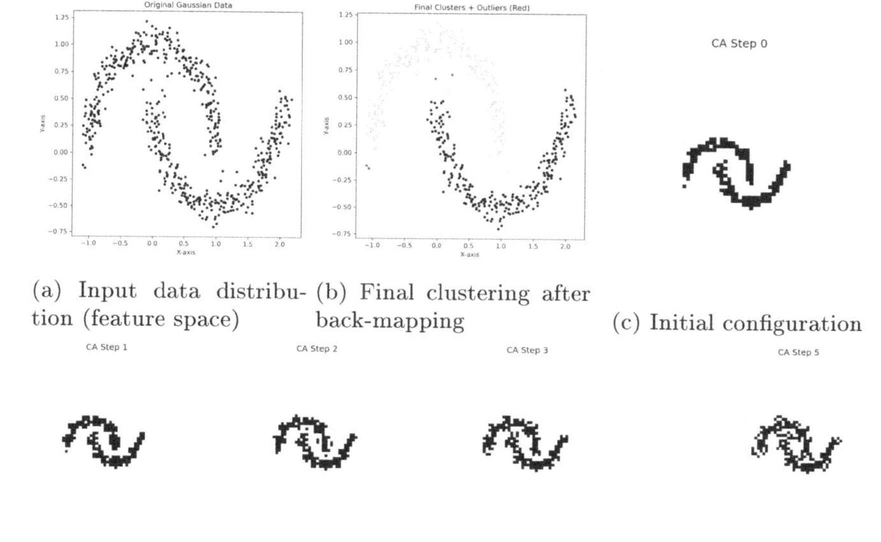

(a) Input data distribution (feature space) (b) Final clustering after back-mapping (c) Initial configuration

(d) CA evolution at $t = 1$ (e) CA evolution at $t = 2$ (f) CA evolution at $t = 3$ ... (g) Final Configuration

Fig. 2. Moon Clusters: *GridLife* clustering pipeline - feature-space input, grid-based initialization, cellular automaton evolution, and final clustering.

Additionally, the introduction of the core mask and dense-dead reinforcement is motivated by the need to prevent erosion of high-density regions and fragmentation of cluster boundaries. It pushes toward stable fixed points suitable for clustering. It can be stated that the rules, although not best, are minimal, and sufficient for density-based clustering. Alternative Life-like rules may be explored in future to find the best rules. Further, choosing a wrong grid size may decrease the accuracy. Hence, selection of grid size based on the data distribution is the most intuitive future work. The primary limitation lies in the aggressiveness of the boundary expansion such that two separate clusters can merge. This can be treated by adjusting survival thresholds or through a probabilistic version of f_{CA}. *GridLife* shows sub-optimal performance in case of nested or multi-density structures such as in a doughnut distribution. Multi-resolution or hierarchical grids may be used in future to improve the performance in such a case. Extensive experimentation with real life dataset and extending *GridLife* to *CA* models with continuous states will be conducted in future. Further, as the updates of *GridLife* depend only on the neighborhood, parallelization too can be implemented in the future.

G controls the choice between global smoothness and local details. Small G provides a coarse representation but may merge close clusters. Whereas, large G captures finer topological details with higher computational load. The computational cost of constructing the grid is linear in the dataset size and quadratic in the grid resolution, i.e. $\mathcal{O}(N+G^2)$. $\mathcal{O}(N)$ corresponds to binning all data points. $\mathcal{O}(G^2)$ occurs for storing and processing the grid structure. Density based state initialization operates in $\mathcal{O}(G^2)$. The shape preserving *GoL* dynamic has the time complexity of $\mathcal{O}(G^2)$. For T evolution steps (typically T is a small number, $T_{max} = 5$ in this empirical settings), the total complexity becomes $\mathcal{O}(T \cdot G^2)$. Computational complexity of connected component labeling is $\mathcal{O}(G^2)$. As all the data points are studied, the computational cost of the back-mapping is $\mathcal{O}(N)$. Similarly, the outlier suppression has the time complexity of $\mathcal{O}(N)$. Thus, this phase has time comoplexity of $\mathcal{O}(N + G^2)$. Therefore, the overall complexity becomes $\mathcal{O}(N + T \cdot G^2)$. T is typically small. G is often taken a priori (e.g., 32, 64, or 128) and does not scale with N. In such a case, G^2 becomes a constant factor making the overall time complexity linear in dataset size. If grid resolution scales with the dataset and $G^2 \leq \sqrt{N}$, the time complexity remains effectively linear. Otherwise, the algorithm can behave quadratic in grid resolution. In real-world clustering data, most grids are empty or inactive. Hence, the computation effectively scales with the number of active grids. Hence, G is substantially smaller than N in practical applications. Thus, it can be stated that *GridLife* exhibits linear scaling with respect to data size. Table 2 shows the characterestics and time complexity of a centroid based (*K-Means*), density based (*DBSCAN*), Non-parametric (*Mean Shift*), two grid-based clustering algorithms. Typical grid-based clusterings have low time complexities [14,17], however they may depend on fixed neighborhood parameters or static density thresholds (e.g. *eps*, *minPts*) [2]. *GridLife* presents a framework that dynamically learn from local densities. This evolutionary process maintains computational simplicity while allowing parameter-free, shape-preserving, and adaptive cluster identification. Its extension may provide an optimal algorithm that has comparable performance in accuracy and time complexity, while being non-parametric.

Table 2. Comparison of Clustering Algorithm Complexity

Algorithm	Typical Complexity	Scalability Characteristic
K-means	$O(I \cdot k \cdot N \cdot d)$	Linear in data size, depends on cluster count (k), iteration (I), and dimensions (d)
DBSCAN	$O(N \log N)$ (optimized)	Sensitive to density parameter
Mean-Shift	$O(N^2)$	Quadratic due to kernel density
GridLife	$O(N + T \cdot G^2)$	Linear in data size, fixed in grid resolution
LinDBSCAN	$O(N)$	Static Density Update. Parameter dependent
Grid-DBSCAN	$O(N)$	Static Density Update. Parameter dependent

5 Conclusion

This paper introduces *GridLife*, a non-parametric grid-based clustering framework inspired by Conway's Game of Life. It establishes the conceptual and theoretical feasibility of mapping data densities to *CA* evolution with near-linear scalability. Preliminary experiments on synthetic datasets demonstrate that *GridLife* captures key properties of density-based clustering and maintains a competitive performance and favorable computational efficiency. As the current study provides a foundational idea and clarifies both the strengths and limitations of the approach, it can motivate future extensions toward large-scale evaluations, real-world datasets, and enhanced density modeling.

A Generalization to *d*-Dimensional Spaces

GridLife is dimension-agnostic at the conceptual level and can be generalized to d-dimension defined over a G^d grid. In this work, the $2D$ example is used for better understanding. *GridLife* builds upon the fundamental principles of grid-based clustering, where grid construction and density estimation strategies have been extensively studied in d-dimension [2,4]. These approaches can be directly integrated here by discretizing each feature dimension and defining Moore neighborhoods over lattices in $\mathbb{N}^d$. The transition rules can remain structurally identical, with neighborhood cardinality scaling with dimension. Each feature dimension is discretized into G bins such that $[x^{(k)}_{\min}, x^{(k)}_{\max}] \to G$ intervals, $k = 1, \ldots, d$. Here, $[x^{(k)}_{\min}, x^{(k)}_{\max}]$ defines the range of the data in the k-th dimension. This discretization induces a d-dimensional hypercubic lattice where $\mathcal{G} \in \mathbb{N}^{G \times G \times \cdots \times G} = \mathbb{N}^{G^d}$. Each grid cell $g_{i_1,\ldots,i_d}$ stores a local density estimate similar to Eq. 2 but in d dimensions ($D_{i_1,\ldots,i_d} = \sum_{n=1}^{N} \mathcal{F} \cdot \{x_n \in g_{i_1,\ldots,i_d}\}$). Here, $D_{i_1,\ldots,i_d}$ denotes density for the lattice indexed by coordinates $i_1, \ldots, i_d$, a standardization in grid-based clustering [2,4]. Further, classical *GoL* employs a Moore neighborhood of fixed cardinality $|\mathcal{N}| = 8$. Here, the neighborhood number defines the relative density. The canonical survival rule, i.e. 2 or 3 active neighbors, corresponds to relative neighborhood densities of $\frac{2}{8}$, i.e. 25% and $\frac{3}{8}$, i.e. 37.5%, while birth occurs at exactly 37.5% neighborhood occupancy. In d dimensions, the Moore neighborhood generalizes to $|\mathcal{N}| = 3^d - 1$. As an example, for $d = 3$, 26 neighbors will be considered. Here, if 2 or 3 neighbors are used it can cause a trivial extinction. Hence, the threshold needs to be scaled proportionally. Using density scaling proportional to *GoL*, for $d = 3$, the survival becomes $\lfloor 0.25 \cdot 26 \rfloor \leq \nu \leq \lceil 0.375 \cdot 26 \rceil$, i.e. $7 \leq \nu \leq 10$ and birth occurs at $\nu \approx 0.375 \cdot 26 \approx 10$. Therefore, the resulting dimension-consistent transition rule can be expressed as Eq. 6.

$$f_{\mathrm{CA}}(S_i^{(t)}, \rho_i^{(t)}; C_i, R_i) = \begin{cases} 1, & \text{if } C_i = 1, \\ 1, & \text{if } R_i = 1 \wedge \nu_i^{(t)} \geq \left\lfloor \frac{3^d - 1}{8} \right\rfloor, \\ 1, & \text{if } S_i^{(t)} = 1 \wedge \nu_i^{(t)} \in \left[\left\lfloor \frac{2(3^d-1)}{8} \right\rfloor, \left\lceil \frac{3(3^d-1)}{8} \right\rceil \right], \\ 1, & \text{if } S_i^{(t)} = 0 \wedge \nu_i^{(t)} \approx \frac{3(3^d-1)}{8}, \\ 0, & \text{otherwise.} \end{cases} \tag{6}$$

It generalizes *GridLife* to multi-dimensional spaces. Further experimentation in this regard is treated as a future work.

References

1. Abhishek, S., Dharwish, M., Das, A., Bhattacharjee, K.: A cellular automata-based clustering technique for high-dimensional data. In: Das, S., Martinez, G.J. (eds.) Asian symposium on cellular automata technology, vol 1443, pp. 37–51. Springer (2023). https://doi.org/10.1007/978-981-99-0688-8_4
2. Al-Dabaa, M.M., Laslo, E., Emran, A.A., Yahya, A., Aboshosha, A.: Multiple targets CFAR detection performance based on an intelligent clustering algorithm in k-distribution sea clutter. Sensors **25**(8), 2613 (2025)
3. Cariou, C., Le Moan, S., Chehdi, K.: A novel mean-shift algorithm for data clustering. IEEE Access **10**, 14575–14585 (2022)
4. Darong, H., Peng, W.: Grid-based DBscan algorithm with referential parameters. Phys. Procedia **24**, 1166–1170 (2012)
5. Du, M., Wu, F.: Grid-based clustering using boundary detection. Entropy **24**(11), 1606 (2022)
6. Gardner, M.: Mathematical games: on cellular automata, self-reproduction, the garden of eden and the game "life". Sci. Am. **224**, 112–117 (1971)
7. Ghosh, M., Kumar, R., Saha, M., Sikdar, B.K.: Cellular automata and its applications. In: 2018 IEEE International Conference on Automatic Control and Intelligent Systems (I2CACIS). pp. 52–56. IEEE (2018)
8. Hloch, M., Kubek, M., Unger, H.: A survey on innovative graph-based clustering algorithms. In: The Autonomous Web, pp. 95–110. Vol. 101 Springer, Cham (2022). https://doi.org/10.1007/978-3-030-90936-9_7
9. Ikotun, A.M., Ezugwu, A.E., Abualigah, L., Abuhaija, B., Heming, J.: K-means clustering algorithms: a comprehensive review, variants analysis, and advances in the era of big data. Inf. Sci. **622**, 178–210 (2023)
10. Ilachinski, A.: Cellular automata: a discrete universe. World Scientific Publishing Company (2001)
11. Mukherjee, S., Bhattacharjee, K., Das, S.: Clustering using cyclic spaces of reversible cellular automata. Complex Syst. **30**(2) (2021)
12. Mukherjee, S., Bhattacharjee, K., Das, S.: Reversible cellular automata: A natural clustering technique. J. Cell. Automata **16** (2021)
13. Ran, X., Xi, Y., Lu, Y., Wang, X., Lu, Z.: Comprehensive survey on hierarchical clustering algorithms and the recent developments. Artif. Intell. Rev. **56**(8), 8219–8264 (2023)
14. Ren, Y., et al.: Deep clustering: a comprehensive survey. IEEE Trans. Neural Netw. Learn. Syst. **36**(4), 5858–5878 (2024)
15. Schubert, E., Sander, J., Ester, M., Kriegel, H.P., Xu, X.: Dbscan revisited, revisited: why and how you should (still) use dbscan. ACM Trans. Database Syst. (TODS) **42**(3), 1–21 (2017)
16. Wolfram, S.: Statistical mechanics of cellular automata. Rev. Mod. Phys. **55**(3), 601 (1983)
17. Xu, D., Tian, Y.: A comprehensive survey of clustering algorithms. Ann. Data Sci. **2**(2), 165–193 (2015)

Cellular Automata Model of Predator–Prey Dynamics with Stochastic Movement Feedback

Sabana Anwar[1,2](✉) and Sudhakar Sahoo[1]

[1] Institute of Mathematics and Applications, Bhubaneswar 751029, Odisha, India
anwar.sabana5@gmail.com, sudhakar@iomaorissa.ac.in
[2] Department of Mathematics, Utkal University, Bhubaneswar 751004, India

Abstract. In the following paper, we have constructed a stochastic cellular automaton model that is formulated from a predator-prey system of ordinary differential equations (ODE), which is ratio-dependent. The proposed cellular automaton (CA) model is formulated by drawing guidance from a corresponding predator-prey model, with population-level trends informing the design of local update mechanisms. The proposed model accounts for two-species interaction between a prey population that reproduces and a predator population that relies on prey availability for survival. In contrast to classical well-mixed models, our approach considers the inherent randomness of the environment, inclusive of factors such as clustering and movement that influence persistence, extinction, and coexistence of species. The model is defined by a two-dimensional stochastic cellular automaton (SCA), where each site in the lattice can assume one of three possible states: unoccupied, occupied by prey, or occupied by predator. The dynamics involving the two species are governed by local neighbourhood interactions coupled with movement-based feedback. Furthermore, the CA model reproduces fundamental ecological patterns through simulation.

Keywords: Stochastic cellular automata · Predator-prey model · Ratio-dependent functional response · Population dynamics · Spatial ecology · Species coexistence · Extinction · Probabilistic modeling · Lattice simulations · Movement feedback

1 Introduction

Predator-prey or multi-species interactions are an integral part of ecology and its management. Its significance can be observed in understanding population dynamics and trophic cascades, to conserve biodiversity, to study evolutionary processes and behavioral ecology, and related aspects. Predator-prey dynamics has a vital role in overall ecological stability. Adequate comprehension of

This work was carried out as part of the Indian Summer School on Cellular Automata 2025.

H. Raju et al. (Eds.): ASCAT 2026, CCIS 2801, pp. 33–48, 2026.
https://doi.org/10.1007/978-3-032-18612-6_3

the aforementioned interactions curbs adverse ecological effects, namely, invasive species expansion, extinction, and population collapse to a substantial degree.

Traditionally, researchers have used differential equation models to describe how predator and prey populations change over time. The paper developed by [1] considers a three-dimensional Lotka-Volterra formulation [10] employed to a predator-prey system for analyzing the equilibrium states and stability behaviour. In a complementary line of work, a reformulated Lotka-Volterra model was employed for the Canadian Lynx-Snowshoe Hare system, capturing its decadal cycles as well as reinforcing its relevance for predation and species conservation [18]. Additionally, to explore complex patterns, for instance, periodic, aperiodic, and limit cycle behaviors, predator-prey dynamics are often modeled using nonlinear differential equations [7]. Furthermore, studies using simple first-order nonlinear difference equations demonstrate that populations with discrete generations have the potential to exhibit stable, periodic, or chaotic trajectories, which is mirrored in competitive systems in two species [5]. However, such equations often assume uniform and homogeneous populations, which results in certain drawbacks [14] as a consequence of the oversimplification of the real-world ecosystem. The main rationale behind modeling with differential equations relies on the continuous global description of systems, yet more often than not, it entails researchers to simplify the same [12]. In contrast to modeling with differential equations, cellular automata offer plausible interactions at the local level, a seemingly powerful tool when coupled with parallel computation [24]. There had been several discussions unfolding the relationship between ODE and CA [17,22].

To address the above limitations, based on the ecologically inspired differential equation system, we designed a spatially explicit CA model. Boccaro et al. investigated a spatial predator-prey cellular automaton with pursuit-evasion interactions and fixed probabilistic rules, together with a mean-field description [6]. We aim to construct the CA model with a foundation in differential equations to validate it through well-established ecological studies available in the literature, a predator-prey ODE model that guides the formulation of local stochastic update rules. On one hand, we believe it will equip the study with a mathematical framework, while on the other, the approach involves primary process of species interaction including predation rate, reproduction rate, mortality, and consumption. By grounding our CA model in these equations, it upholds biologically significant dynamics and allows us to make a direct comparison between ODE and CA outcomes, which in turn emphasizes how interactions that are local and spatial structure can result in changes in the behavior of the system. Our approach was to convert the mathematically continuous model into a discretized two-dimensional lattice where each site (cell) can take on one of three states: vacant, prey-occupied, or predator-occupied, and these states are governed by neighborhood-driven CA rules, which show complex behaviour such as waves and extinction.

The rest of this paper is organized as follows. Section 2 formulates the CA model with density thresholds. Section 3 details the probabilistic update rules. Section 4 presents simulation results on population and spatial dynamics. In Sect. 5 there is a discussion around the statistical analysis of the model. Section 6 concludes with the primary takeaway and some scope for future research.

2 Model Description

2.1 Continuous-Time Ratio-Dependent Model

To approach predator-prey dynamics employing a cellular automata framework, we begin with a continuous-time ODE model that consists of crucial aspects for instance, reproductive growth and predator-prey interaction. Classical predator-prey models generally operate with Lotka-Volterra equations [23], which in turn may yield outcomes that are unattainable in the real world because they overlook constraints on prey consumption [2]. To address this limitation, a ratio-dependent functional response with logistic prey growth was introduced, which reflects the balance between prey supply and predation demand. Let $U(t)$ denote the prey and $P(t)$ denote the predator densities at time t. We start from a ratio-dependent functional response [4] with logistic prey growth described as,

$$\begin{cases} \dfrac{dU}{dt} = rU\left(1 - \dfrac{U}{K}\right) - \dfrac{\alpha UP}{P + \alpha hU}, \\ \dfrac{dP}{dt} = \dfrac{e\alpha UP}{P + \alpha hU} - \mu P. \end{cases} \tag{1}$$

Here r is the parameter describing the intrinsic prey growth rate, K denotes the prey carrying capacity, α represents the predator attack efficiency, h is the predator handling time, e symbolizes the conversion efficiency from consumed prey to predator biomass, and μ is the predator natural mortality [3]. In the model, the ratio-dependent component modulates the attack rate as a function of the relative abundance of predator-prey, a mechanistic choice that softens unrealistic unlimited consumption when predators are abundant.

2.2 Time Discretization and Normalized Increments

For centuries, most models in physics, engineering, and biology have been expressed using differential equations [9,19–21]. These continuous formulations provide a well-established paradigm but are not necessarily directly compatible with computer-based, rule-driven methods such as cellular automata. The gap was bridged when DEs were approximated utilizing finite difference equations (FDEs) [15], which can subsequently be adapted as cellular automata [25]. Accordingly, the continuous time is transformed into discrete iterations, which are limited to a finite domain. Thus, cellular automata can be viewed as a highly discretized representation of differential equations.

In the study of population dynamics, it is natural to express real-life scenarios through a continuous-time differential equation. However, when the aim is to assess these models numerically or in a discrete approach, it requires reformulation involving an appropriate discrete representation. The finite difference method (FDM) approximates differential equations by reformulating them into discrete-time updates [16]. Let $w(t)$ denote a generic population variable (e.g., prey or predator density) at time t, governed by a first-order differential equation of the form $\dfrac{dw}{dt} = F(w,t)$, where $F(w,t)$ encodes the growth, interaction, or mortality dynamics of the system. Using finite differences, the continuous derivative can be approximated by [11],

$$\frac{w(t+\Delta t)-w(t)}{\Delta t}\approx F(w,t)$$

which yields the update relation,

$$w(t+\Delta t)=w(t)+\Delta t\,F(w,t).$$

In particular, the increments for the discrete-time updates are given by,

$$\Delta w=\Delta t\,F(w,t)$$

can be rescaled or normalized to lie within probabilistic bounds, making them suitable for stochastic CA update rules. To proceed within the finite difference setting, we substituted the continuous derivative in the prey Eq. (1) with a forward-difference approximation [13], which yields an explicit Euler scheme with a discrete time step Δt, which captures logistic growth and the predation effect within each time increment. Hence,

$$\Delta U=\Delta t\left[r\,U\left(1-\frac{U}{K}\right)-\frac{\alpha UP}{P+\alpha hU}\right]. \tag{2}$$

Similarly, for the predator dynamics, where any variation in the density of predators relies on successful consumption of prey along with natural mortality. This leads to the corresponding Euler update for ΔP,

$$\Delta P=\Delta t\left[\frac{e\,\alpha UP}{P+\alpha hU}-\mu P\right]. \tag{3}$$

To map these continuous increments to probabilities in a CA update, we normalize by upper bounds that keep the magnitudes within a probabilistic scale. A conservative bound for prey growth is the logistic peak at $U=K/2$, in the absence of predation $P=0$, yielding

$$\Delta U_{\max}=\Delta t\,\frac{rK}{4}.$$

Similarly, to determine the overall upper bound of predator increment, we next maximize ΔP with respect to both U and P. The global behavior depends on the relative magnitudes of the energetic efficiency e/h and the mortality rate μ, leading to two distinct regimes.

Case A — $\mu \geq \frac{e}{h}$

In this regime, the mortality rate exceeds or equals the effective gain per handling time. Consequently, for any given prey density U, the optimal predator density is $P=0$, since increasing P only decreases ΔP. Therefore,

$$\max_{U\geq 0,\,P\geq 0}\Delta P=0,$$

attained at $P = 0$ for all U. This corresponds to a scenario where predators cannot sustain positive growth due to excessive mortality relative to energetic intake.

Case B — $\mu < \frac{e}{h}$

When the energetic return per handling time exceeds the mortality rate, positive growth becomes possible. For each fixed U, the maximal predator increment is linear in U and given by

$$\Delta P_{\max|U} = \Delta t\, \alpha U\, (\sqrt{e} - \sqrt{h\mu})^2.$$

If U is unbounded, $\Delta P_{\max|U}$ increases without limit as $U \to \infty$, implying that there is no finite global maximum (the supremum is $+\infty$). However, if U is bounded within a biologically realistic carrying capacity K (i.e., $0 \le U \le K$), the global maximum is achieved at $U = K$ and at the corresponding optimal predator density $P = P^*(K)$, where

$$\Delta P_{\max\,[0,K]\times[0,\infty)} = \Delta t\, \alpha K\, (\sqrt{e} - \sqrt{h\mu})^2, \qquad P^*(K) = \alpha K\left(\sqrt{\frac{eh}{\mu}} - h\right).$$

Thus, in the biologically constrained domain, the predator population reaches its highest net growth rate when both the prey density and predator efficiency are maximized relative to mortality.

This upper limit, $\Delta P_{\max}$, provides a consistent scaling reference for converting continuous predator dynamics into normalized probabilities in the CA framework. We then define *normalized growth indicators*

$$R_U = \frac{\Delta U}{\Delta U_{\max}}, \qquad R_P = \frac{\Delta P}{\Delta P_{\max}}. \tag{4}$$

In practice, we clip R_U and R_P to a bounded interval, e.g. $[-1, 1]$, to avoid erratic behavior when local densities are extreme.

2.3 Lattice, States, and Local Densities

Let the cellular automaton be defined on a two-dimensional lattice $\Lambda = Z_X \times Z_Y \subseteq \mathbb{Z}^2$, where $Z_X = \{0, 1, \ldots, X-1\}$, $Z_Y = \{0, 1, \ldots, Y-1\}$ denote the discrete sites along the horizontal and vertical directions, respectively. Periodic boundary conditions are imposed so that Λ forms a toroidal grid. Each site $(i,j) \in \Lambda$ may assume one of the discrete states from the set, $\mathcal{S} = \{\sigma_0, \sigma_1, \sigma_2\}$, where σ_0 represents an empty site, σ_1 a prey, and σ_2 a predator. A configuration of the automaton at time t is defined as a mapping, $\eta_t : \Lambda \to \mathcal{S}$, so that $\eta_t(i,j)$ denotes the state of site (i,j) at time step t. The full system can thus be encoded as an $X \times Y$ array. The global update of the automaton proceeds by a synchronous application of a local transition function $\eta_{t+1}(i,j) = \Phi\big(\eta_t(\mathcal{N}(i,j))\big)$, where $\mathcal{N}(i,j)$ denotes the neighborhood of cell (i,j). In this work, we consider the *Moore neighborhood* of range one, defined as

$$\mathcal{N}(i,j) = \{(i+r, j+s) \mid r, s \in \{-1, 0, 1\}\},$$

which consists of the eight nearest neighbors surrounding (i, j) together with the site itself with $(i, j) \in \Lambda \subseteq \mathbb{Z}^2$. The local rule Φ operates in two sequential phases: *interaction* (reaction among neighbors) followed by *dispersal* (movement to adjacent sites). For each site, we adopt the Moore neighborhood of radius 1. If $n_{\text{prey}}(i, j)$ and $n_{\text{pred}}(i, j)$ denote the numbers of prey and predator neighbors of site (i, j), then the corresponding local densities are defined as,

$$u(i,j) = \frac{n_{\text{prey}}(i,j)}{8}, \qquad v(i,j) = \frac{n_{\text{pred}}(i,j)}{8}.$$

These local densities $u(i, j)$ and $v(i, j)$ replace the global population densities U and P in the finite-difference Eq. (2) and (3), thereby making growth and predation locally driven. This scaling makes the resulting probabilities consistent across grid sizes and neighborhood sizes.

2.4 Movement-Informed Stochastic Feedback

While studying the population dynamics of species interactions, in order to adequately describe spatial heterogeneity, it is important to account for both counting the numbers of neighbors and also the directional configuration of those neighbors. To capture the movement phase of the CA, we first employ an indicator (or occupancy) function that encodes whether a given cell is in a specified state [8],

$$\chi_{\text{s}}(i,j) = \begin{cases} 1, & \text{if the cell at } (i,j) \in \Lambda \text{ is in state s} \in \mathcal{S}, \\ 0, & \text{otherwise,} \end{cases}$$

where Λ denotes the lattice domain and $\mathcal{S}$ the finite set of states. This binary function serves as the foundation for computing local directional statistics around any site (l, m). Instead of treating the Moore neighborhood of radius r as a single block, it is partitioned into: north, south, east, and west. For example, the northward sector is given by

$$Q_r(l,m) = \{(i,j) \,:\, l - r \leq i \leq l + r,\; m < j \leq m + r\}.$$

This represents the sector directly above the site (l, m). Similar definitions can be applied to the other three directions, where each sector in a particular direction consists of a set of sites.

For each state s, the occupancy indicator function $\chi_s(i, j)$ is defined, which takes the value 1 if the cell (i, j) is in state s, and 0 otherwise. Using this, the total count of cells in state s within quadrant Q can be computed as,

$$C_Q(s; l, m) = \sum_{(i,j) \in Q_r(l,m)} \chi_s(i,j), \quad Q \in \{N, S, E, W\}.$$

$C_Q(s; l, m)$ simply counts how many cells of a given type (for example, predators) occupy each sector along each direction surrounding the cell (l, m) under consideration. And following the count, movement decisions for each species are then based on the above-mentioned directional counts. For instance, we can consider that the prey will most likely move toward the quadrant that contains the least number of predators, thereby minimizing its exposure to threat. This can be expressed as

$$Q_*(l, m) = \arg \min_{Q \in \{N,S,E,W\}} C_Q(\text{predator}; l, m),$$

In the same way, predators will tend to move towards prey-rich sectors,

$$Q^*(l, m) = \arg \max_{Q \in \{N,S,E,W\}} C_Q(\text{prey}; l, m).$$

where $Q_*(l, m)$ or $Q^*(l, m)$ denotes the chosen movement direction for prey and predator, respectively. If two or more directions yield the same minimal predator or prey count, the movement direction is selected randomly among them. In neutral conditions, when all directions have equal predator presence, the individual may remain stationary.

Again, both species adjust their movement in response to local risks and opportunities, and this behavior can be captured mathematically through simple discrete modifiers. To capture this feedback, we define

$$\delta_U,\ \delta_P \in \{-0.1,\ 0,\ +0.1\},$$

which are computed from the same local neighborhood and act as behavioral corrections. The discrete increments δ_U and δ_P represent small symmetric perturbations corresponding to decreasing, neutral, and increasing local tendencies, ensuring that feedback remains weak, bounded, and responsive to local environmental variations without inducing artificial instabilities. Specifically, when prey encounter a sector with fewer predators, they receive a positive adjustment ($\delta_U = +0.1$), since the environment is safer. Conversely, high predator density imposes a negative adjustment for the prey ($\delta_U = -0.1$), which reflects that the environment now possesses a greater risk. Similarly, predators on arriving at regions with abundant prey, benefit from a positive adjustment ($\delta_P = +0.1$). However, the absence of prey in the surroundings of a predator scarcity is indicated by receiving a negative adjustment ($\delta_P = -0.1$). Thus, effective responses are updated according to

$$R'_U = \max\big(-1,\ \min(1,\ R_U + \delta_U)\big), \qquad R'_P = \max\big(-1,\ \min(1,\ R_P + \delta_P)\big).$$

Here, the nested min–max operator ensures the boundedness within $[-1, 1]$. Finally, these modified responses, R'_U and R'_P, are used in the state transition probabilities. All these tie together to build a consistent mechanism supporting local environmental conditions while inculcating movement-driven decisions.

2.5 Feasibility and Stress Thresholds

In this section, we will define two critical population density values P_T and U_T. To reflect environmental limits such as stress caused by overcrowding or inadequacy of food, and struggle for survival under various conditions, shaping feasibility. The term P_T (*predator pressure threshold*) is the cut-off beyond which the number of predators present a danger to the prey, whereas U_T (*prey availability threshold*) determines the threshold that decides the feasibility of predator survival based on prey count.

2.6 CA Update Rules

At each synchronous step, every site updates its state based on the normalized opportunity–risk modifiers R'_U, R'_P and local densities x, y of prey and predators, respectively. The rules differ depending on whether the site is empty, occupied by prey, or occupied by a predator.

(i) Empty Site ($\boldsymbol{\sigma_0}$) An empty site becomes colonized if local growth conditions favor either prey or predator recruitment. When $R'_U > 0$ and $R'_P \leq 0$, the site becomes prey with probability R'_U. Conversely, when $R'_P > 0$ and $R'_U \leq 0$, it becomes predator with probability R'_P. If both are favorable, competition determines the outcome proportionally to their magnitudes. Otherwise, the site remains empty.

$$\sigma_0(t + \Delta t) = \begin{cases} 1, & R'_U > 0,\ R'_P \leq 0 \text{ with prob. } R'_U, \\ 2, & R'_P > 0,\ R'_U \leq 0 \text{ with prob. } R'_P, \\ 1, & R'_U > 0,\ R'_P > 0 \text{ with prob. } \dfrac{R'_U}{R'_U + R'_P}, \\ 2, & R'_U > 0,\ R'_P > 0 \text{ with prob. } \dfrac{R'_P}{R'_U + R'_P}, \\ 0, & \text{otherwise.} \end{cases}$$

(ii) Prey Site ($\boldsymbol{\sigma_1}$) A prey site's fate is determined by its growth potential R'_U and local predator pressure y. When predators are scarce ($y < P_T$) and $R'_U \geq 0$, prey persist stochastically. Under high predator density ($y \geq P_T$), prey face conversion or death. Negative R'_U implies decline under poor conditions.

$$\sigma_1(t + \Delta t) = \begin{cases} 1, & R'_U \geq 0,\ y < P_T \text{ with probability } 0.5, \\ 0, & R'_U \geq 0,\ y < P_T \text{ with probability } 0.5, \\ 2, & R'_U \geq 0,\ y \geq P_T \text{ with probability } R'_P, \\ 0, & R'_U \geq 0,\ y \geq P_T \text{ otherwise}, \\ 0, & R'_U < 0 \text{ with probability } -R'_U, \\ 1, & \text{otherwise.} \end{cases}$$

(iii) Predator Site (σ_2) Predator persistence depends on prey availability and growth conditions. Adequate prey and positive R'_P ensure survival, whereas scarcity or negative growth increases starvation risk.

$$\sigma_2(t+\Delta t) = \begin{cases} 2, & R'_P \geq 0,\, x \geq U_T, \\ 2, & R'_P \geq 0,\, x < U_T \text{ with probability } 0.5, \\ 0, & R'_P \geq 0,\, x < U_T \text{ with probability } 0.5, \\ 0, & R'_P < 0 \text{ with probability } -R'_P, \\ 2, & \text{otherwise.} \end{cases}$$

All sites are updated synchronously using the configuration at time t to produce the next configuration at $t + \Delta t$. Periodic boundary conditions are applied to eliminate edge effects and preserve spatial continuity.

3 Simulation Setup

Unless otherwise stated, we use the parameter set in Table 1, and the time step Δt is chosen to be small enough for stable normalization and interpretable probabilities. We consider square grids of size $Z_X \times Z_X$ with $X \in \{50, 250\}$.

Table 1. Baseline parameters used in the simulations. Here r is prey intrinsic growth, K is carrying capacity, α is attack efficiency, h is handling time, e is conversion efficiency, and μ is predator mortality. Thresholds P_T and U_T define neighborhood stress and feasibility; Δt is the discrete time step.

Parameter	Value	Description
r	1.0	Prey intrinsic growth rate
K	1.0	Prey carrying capacity (scaling)
α	0.5	Predator attack efficiency
h	0.2	Predator handling time
e	0.6	Conversion efficiency
μ	0.2	Predator mortality
Δt	0.1	Time step
P_T	0.3	Predator density threshold (risk to prey)
U_T	0.3	Prey density threshold (viability for predators)

We have considered periodic boundary conditions. Initial conditions are created by uniformly sampling sites to set a target fraction of prey and predators, with the remaining sites empty. We study three canonical scenarios:

(a) **Predators only, no prey**: initializes with predator fraction > 0 and prey fraction $= 0$.

(b) **Prey only, no predators**: initializes with prey fraction > 0 and predator fraction $= 0$.
(c) **Both species present**: initializes with both fractions positive.

For each scenario, we run multiple stochastic replicates to summarize typical behaviors and variability. Each run records the full grid evolution, as well as the global counts of prey and predators per step. We present two forms of output: (i) time series of global prey and predator counts, and (ii) snapshots of the entire grid at selected time steps to visualize pattern formation, expansion, and retreat.

We have also compared the time evolution of the ratio-dependent predator-prey differential model and an averaged SCA. For the (DE) reference, we numerically integrated the system with parameters $r = 0.8$, $K = 50$, $\alpha = 0.02$, $h = 0.1$, $e = 0.3$, and $\mu = 0.4$, using the initial condition $(U, P) = (30, 5)$ over $t \in [0, 50]$ (200 time points). Independently, we ran a spatial SCA with random initial occupancy. We performed $n_{\text{runs}} = 5$ independent CA realizations and averaged the fraction of prey and predator densities across runs at each step. Finally, both time series were normalized by their respective maximum values so that each curve ranges within $[0, 1]$. Two phase-space scatter plots were generated. Each point represents the normalized prey and predator densities (U, P) at a given time step, and the color gradient indicates temporal progression from early to late stages.

4 Results and Discussion

Predators Only, No Prey. When the system is initialized with only predators, the predator population monotonically decreases and approaches zero. This trajectory is expected because, in the absence of prey, the conversion term in the predator growth equation vanishes and mortality dominates. In Fig. 1, global time series shows a decaying curve that flattens near zero.

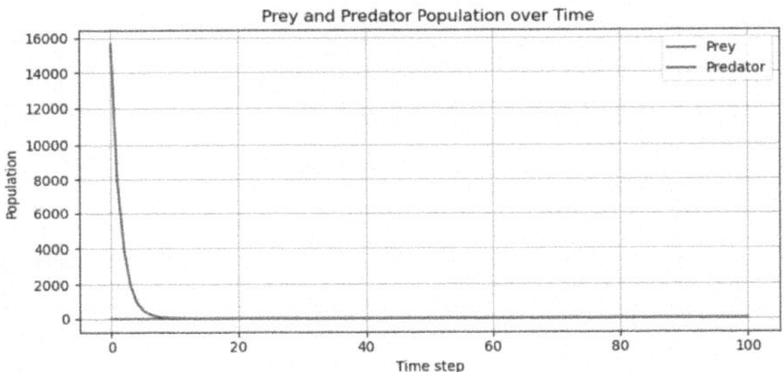

Fig. 1. Predator-prey population dynamics over time in the absence of prey.

In Fig. 2, in the CA, this manifests as predators randomly disappearing due to starvation or dispersal to empty patches that do not sustain them.

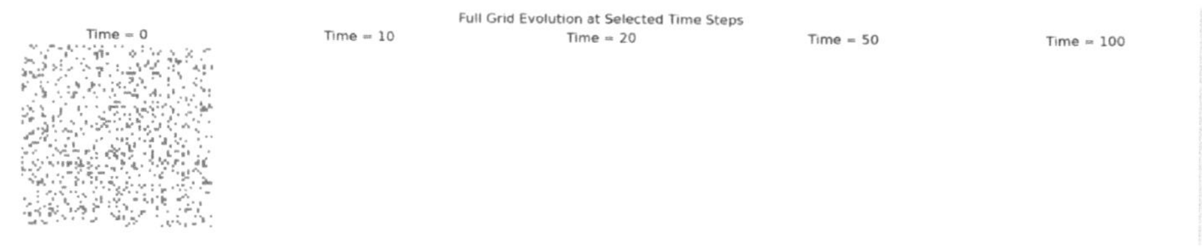

Fig. 2. CA grid evolution over time in the absence of prey. Green indicates prey, red indicates predators, and white indicates empty sites. (Color figure online)

Prey Only, No Predators. In Fig. 3, when the system is initialized with only prey, the prey population grows and saturates near the carrying capacity implied by the logistic term.

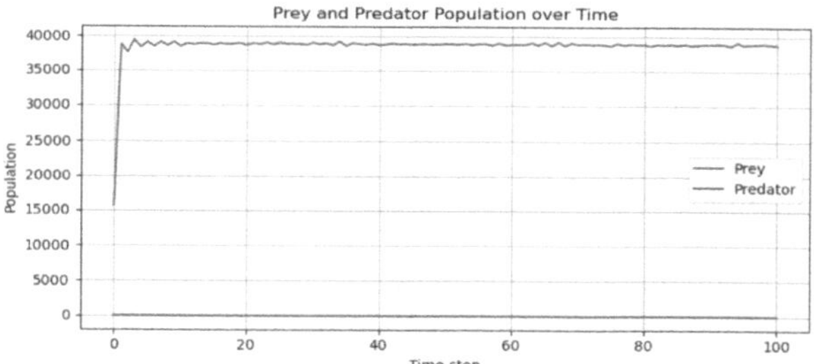

Fig. 3. Predator-prey population dynamics over time in the absence of predators.

In Fig. 4, in the CA, prey proliferate into empty sites when $R'_U > 0$ and predator pressure is absent ($y \approx 0$), eventually filling most of the grid.

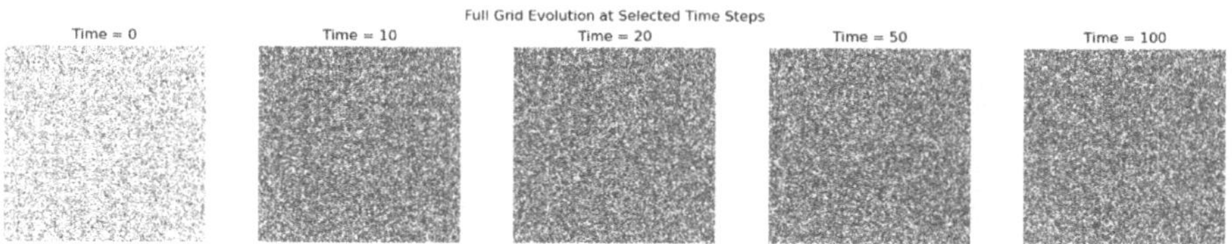

Fig. 4. CA grid evolution over time in the absence of predators. Green indicates prey, red indicates predators, and white indicates empty sites. (Color figure online)

Stochastic fluctuations at the boundary between occupied and empty sites diminish as the system approaches saturation.

Both Species Present. When both species are present, we observe predator–prey oscillations and spatial pattern consistent with pursuit-and-escape dynamics (Fig. 5).

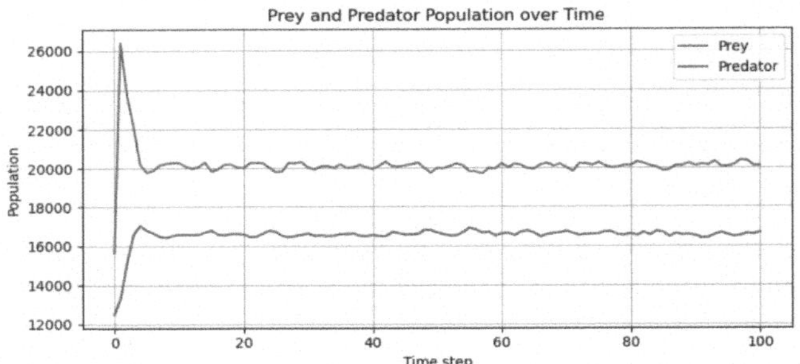

Fig. 5. Predator-prey population dynamics over time, showing oscillations and threshold-driven fluctuations.

We can also observe that movement feedback increases the persistence of such clusters by biasing transitions: prey in safer neighborhoods receive a small reproductive bonus, and predators in prey-rich neighborhoods receive a small survival bonus. In Fig. 6, prey clusters form in regions of low predator density, while predator clusters emerge in prey-rich areas. Depending on parameters and initial conditions, the system achieves long-lived coexistence with damped oscillations, sustained oscillations, or, less frequently, collapse to prey-only or empty states. The thresholds P_T and U_T broaden the portfolio of behaviors by shaping when local neighborhoods are considered risky or viable. In Fig. 7, the deterministic (DE) model shows that prey populations rapidly rise and stabilize near their carrying capacity, while predator populations decline but eventually settle at a small positive steady state. In contrast, the averaged SCA produces similar prey dynamics but exhibits early predator extinction, with predator density falling to zero across runs.

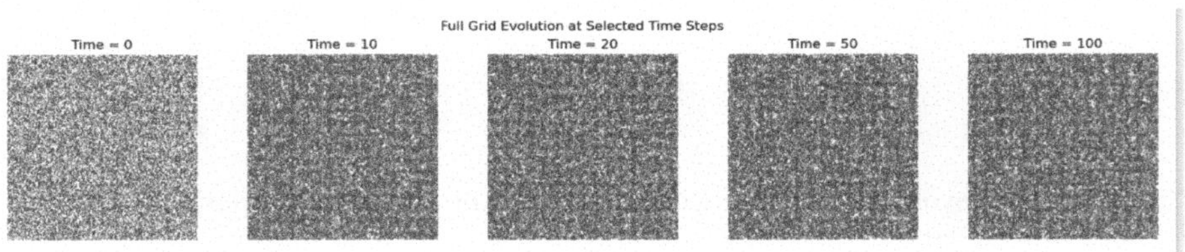

Fig. 6. CA grid evolution under the coexistence scenario. Green indicates prey, red indicates predators, and white indicates empty sites. (Color figure online)

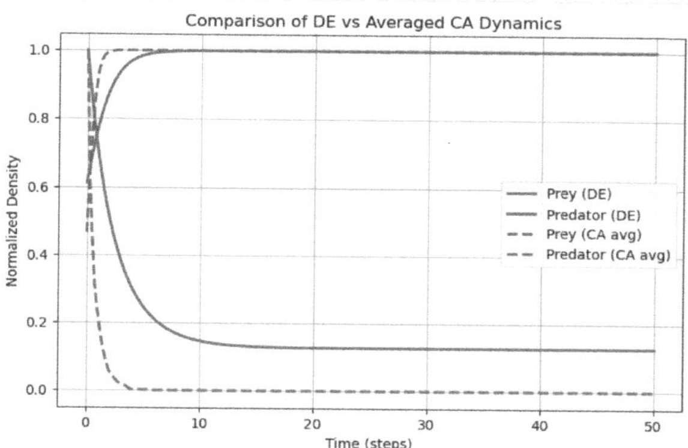

Fig. 7. Comparison of DE and averaged CA predator-prey dynamics—solid lines (DE) show coexistence, while dashed lines (CA) indicate prey persistence and predator extinction over time.

The divergence seen in the figure arises from the stochastic and the neighbourhood-based nature of the CA. As in the CA model, the predators must encounter nearby prey to survive, and with low predation efficiency (α) and relatively high mortality (μ), local fluctuations often cause extinction. The DE model assumes homogeneous mixing that allows even small predator populations to persist, whereas the CA enforces local interactions where prey depletion can isolate predators (Fig. 8).

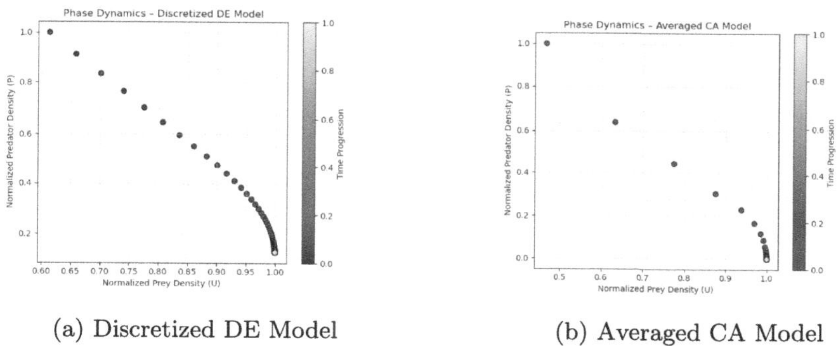

(a) Discretized DE Model

(b) Averaged CA Model

Fig. 8. Comparison of phase-space dynamics between the discretized differential equation model and the averaged cellular automata model.

We can observe the DE model shows a smooth trajectory converging toward a stable coexistence region, reflecting the continuous nature of its formulation. The CA model is more stochastic, yet it reproduces a qualitatively similar early-time trajectory before stabilizing into a steady-state configuration where we find the prey dominate. Irrespective of the local stochastic variations and constraints of the environment, the framework driven by CA portrays the essential transient dynamics and stability patterns predicted by the DE system.

5 Statistical Summary

At step 50 after 1000 simulation runs, in Fig. 9, vertical lines denote the mean, median, and mode, while the shaded bands represent the 95% confidence intervals. The following observation can be made, both prey and predator populations demonstrated a strong tendency toward equilibrium, as indicated by the statistical measures. For the prey population, the mean was 802.76, while the median (802) and mode (792) were in close agreement. The data seems to be roughly symmetrical, as the mean and median are almost equal. We can observe a minor left skew as a result of the mode being somewhat lesser. The standard deviation of 31.47, which is approximately 4% of the mean, indicates that there is a moderate amount of variation in the number of prey over various simulations. We also observed that 95% confidence interval (CI) for the prey mean was ±1.95. The result was an interval of [800.8, 804.7]. Around the equilibrium value, we observe high precision and stability of the prey population as CI is narrow relative to the size of the prey population (Fig. 9).

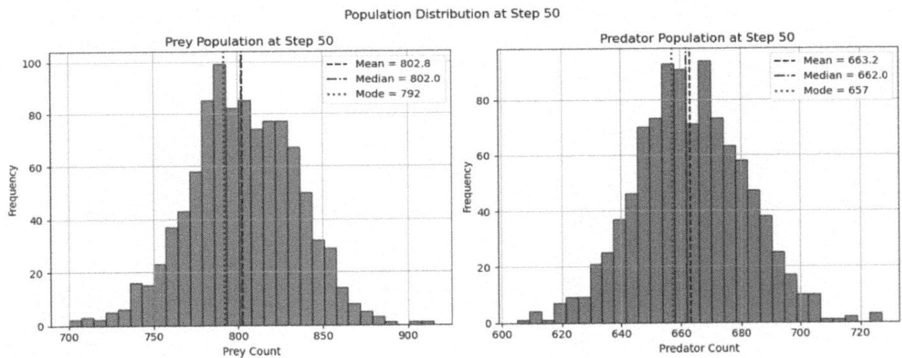

Fig. 9. At 50 time-step: Distribution of prey (left) and predator (right) populations.

While the predator population displayed a mean of 663.20, with a median of 662 and a mode of 657. A symmetric distribution is seen, with the reasoning being similar to what we noted in the case of prey. However, the mode being slightly lower again suggests a marginal left skew. The standard deviation was 18.45 and about 2.8% of the mean, indicates that the predator populations exhibit less variability than prey populations. For predator after evaluation, we saw that 95% CI for the predator mean was ±1.15. It yielded an interval of [662.1, 664.4]. In both the cases, the narrow confidence interval highlights the strong convergence of predator and prey counts across multiple runs. Lesser uncertainty was observed in the predator population than the prey population. In short, the results show that the predator-prey system is stable and consistent.

6 Conclusion

In the study above, we have constructed a stochastic cellular automaton model with the help of a ratio-dependent predator-prey ODE model. The CA is characterized by local probabilistic rules and movement-informed feedback. The construction preserves the interpretability of the original ecological parameters while adding spatial heterogeneity and stochasticity that are essential for realistic pattern formation. In controlled experiments the CA reproduces three hallmark regimes: predator decay without prey, prey saturation without predators, and coexistence with oscillations when both are present. As a result of the integration of spatial structure and stochasticity within the CA model, the phase and time-series analyses yield a richer and more realistic representation of ecosystem dynamics compared to the deterministic DE model. The framework is modular and can be extended to multiple species, or to perform a systematic sensitivity study of key parameters (e.g., α, h, μ) to detect phase transitions. Beyond reproducing classical dynamics, the model supports scenario analysis and the exploration of spatial early-warning indicators of collapse or regime shift under uncertainty.

Conflict of interest. The authors declare that they have no potential conflict of interest.

References

1. Adamu, H.A.: Mathematical analysis of predator-prey model with two preys and one predator. Int. J. Eng. Appl. Sci. **5**(11), 17–23 (2018)
2. Arditi, R., Ginzburg, L.R.: Coupling in predator-prey dynamics: ratio-dependence. J. Theor. Biol. **139**(3), 311–326 (1989)
3. Banerjee, M., Ghorai, S., Mukherjee, N.: Study of cross-diffusion induced turing patterns in a ratio-dependent predator-prey model via amplitude equations. Appl. Math. Model. **55**, 383–399 (2018)
4. Banerjee, M., Petrovskii, S.: Self-organised spatial patterns and chaos in a ratio-dependent predator-prey system. Thyroid Res. **4**(1), 37–53 (2011)
5. Beddington, J., Free, C., Lawton, J.: Dynamic complexity in predator-prey models framed in difference equations. Nature **255**(5503), 58–60 (1975)
6. Boccara, N., Roblin, O., Roger, M.: Automata network predator-prey model with pursuit and evasion. Phys. Rev. E **50**(6), 4531 (1994)
7. Canale, R.P.: An analysis of models describing predator-prey interaction. Biotechnol. Bioeng. **12**(3), 353–378 (1970)
8. Cattaneo, G., Dennunzio, A., Farina, F.: A full cellular automaton to simulate predator-prey systems. In: El Yacoubi, S., Chopard, B., Bandini, S. (eds.) International Conference on Cellular Automata, Vol. 4173, pp. 446–451. Springer (2006). https://doi.org/10.1007/11861201_52

9. Chau, K.T.: Applications of Differential Equations in Engineering and Mechanics. CRC Press (2019)
10. Din, Q.: Dynamics of a discrete lotka-volterra model. Adv. Difference Equ. **2013**(1), 95 (2013)
11. Euler, L.: Institutiones calculi integralis, vol. 1. impensis Academiae imperialis scientiarum (1792)
12. Guinot, V.: Modelling using stochastic, finite state cellular automata: rule inference from continuum models. Appl. Math. Model. **26**(6), 701–714 (2002)
13. Hu, R., Ruan, X.: Differential equation and cellular automata model. In: IEEE International Conference on Robotics, Intelligent Systems and Signal Processing, 2003. Proceedings. 2003. vol. 2, pp. 1047–1051. IEEE (2003)
14. Keen, R., Spain, J.: Computer Simulation in Biology. Nueva York (1992)
15. LeVeque, R.J.: Finite difference methods for differential equations. Draft Version Use AMath **585**(6), 112 (1998)
16. LeVeque, R.J.: Finite difference methods for ordinary and partial differential equations: steady-state and time-dependent problems. In: SIAM (2007)
17. Omohundro, S.: Modelling cellular automata with partial differential equations. Phys. D. **10**(1–2), 128–134 (1984)
18. Pulley, L.C.: Analyzing Predator-Prey Models Using Systems of Ordinary Linear Differential Equations (2011)
19. Rihan, F.A., et al.: Delay Differential Equations and Applications to Biology (2021)
20. Smith, H.L.: An introduction to delay differential equations with applications to the life sciences, vol. 57. Springer, New York (2011). https://doi.org/10.1007/978-1-4419-7646-8
21. Sobczyk, K.: Stochastic differential equations: with applications to physics and engineering, vol. 40. Springer Science & Business Media (2013). https://doi.org/10.1007/978-94-011-3712-6
22. Toffoli, T.: Cellular automata as an alternative to (rather than an approximation of) differential equations in modeling physics. Phys. D **10**(1–2), 117–127 (1984)
23. Wangersky, P.J.: Lotka-volterra population models. Annu. Rev. Ecol. Syst. **9**, 189–218 (1978)
24. Worsch, T.: Simulation of cellular automata. Futur. Gener. Comput. Syst. **16**(2–3), 157–170 (1999)
25. Yang, X.S., Young, Y.: Cellular automata, PDEs, and pattern formation. arXiv preprint arXiv:1003.1983 (2010)

Finding Fixed Points of Cellular Automata over Cayley Tree of Order 2

Bipul Patra(✉) and Subrata Paul

Department of Information Technology, Indian Institute of Engineering Science and Technology, Shibpur, Howrah 711103, West Bengal, India
bipulpatra.sci@gmail.com

Abstract. This work studies Cellular Automata (CAs) over Cayley tree of order 2, where each vertex of the tree is considered as a cell and the nearest connected vertices are considered as its neighbors. The proposed CA follows a two-state, four-neighborhood structure, where each cell updates its state using a local update rule. In this work, we mainly focus on the fixed points of a CA. When a CA reaches to a fixed point, say c, the CA remains at c forever during evolution. We investigate the fixed points by introducing the Fixed Point Graph (FPG) of a CA. Finally, we propose an algorithm that computes the number of fixed points of a CA.

Keywords: Cellular Automata · Cayley Tree · Rule Min Term · Fixed Point · Fixed Point Graph

1 Introduction

A Cellular Automaton (CA) shows massive parallelism that can be used in different models of computation. Cellular Automata (CAs) have been widely used for modeling biological and physical processes. Nowadays, it has been studied in many disciplines for different purposes like simulating natural phenomena, pattern classification, pseudo random number generator, cryptography etc. [2,7,9,11,12].

In early 1950s, John von Neumann first introduced two-dimensional CAs to model biological self-replication [14]. Most studies on CA have been conducted for one-dimensional and two-dimensional CAs. The popular *Game of Life*, developed by John Horton Conway, is an example of a two-dimensional cellular automaton (CA) [5]. Fici and Fiorenzi have studied the topological properties of CA on the full tree shift $A^{\sum^*}$, where $\sum^*$ is the free monoid of finite rank $|\sum|$ [4]. Here, the Cayley graph of $\sum^*$ is a regular graph of $|\sum|$-ary rooted tree. In [4], they have defined cellular automata on the full k-ary tree shift for $k \geq 2$. Cellular Automata on regular Cayley tree were previously studied in [1,6,8]. This work considers cellular automata over Cayley tree of order 2.

This work is supported by CRG project (File No.: CRG/2023/006799) of SERB, Govt. of India.

H. Raju et al. (Eds.): ASCAT 2026, CCIS 2801, pp. 49–61, 2026.
https://doi.org/10.1007/978-3-032-18612-6_4

There exist some CAs that converge to fixed points from any arbitrary initial configuration, and such CAs are referred to as convergent CAs. The researchers have shown their interest in the convergence of CAs [3,11,13]. This work investigates the behavior of the proposed CAs on Cayley tree and discusses the formation of fixed point configurations. Here, we identify the fixed points of a CA by utilizing the Fixed Point Graph (FPG) of a CA. We propose an algorithm that computes the number of fixed points of a CA.

The next section (Sect. 2) provides a detailed description of the model. Section 3 discusses the necessary conditions of the fixed points of a CA over Cayley tree. The Fixed Point Graph is introduced to obtain the fixed points of a CA in Sect. 4. Finally, an algorithm is proposed in Sect. 4 to compute the number of fixed points of a CA over Cayley tree. Section 5 concludes the work.

2 The Model

Classically, a Cayley tree is an infinite tree, however, this work considers a finite number of nodes in a Cayley tree where each non-leaf node has a constant number of branches. A node in the tree is considered a root node from which all the leaves maintain equal distance. This distance is called the height (h) of the tree. The degree of each node (except the leaf) is 3 –that is, the root has 3 children, each non-leaf node has 2 children and a leaf node has no children. The following proposition (Proposition 1) shows the relationship between the height h and the number of nodes (n) of the tree.

Proposition 1. *[10] In a (finite) Cayley Tree of order 2, say $h \geq 1$, the number of nodes (n) is:*

$$n = \begin{cases} 1 & \text{if } h = 1 \\ 1 + 3\sum_{i=0}^{h-2} 2^i & \text{otherwise} \end{cases} \tag{1}$$

Figure 1 shows a finite Cayley tree of order 2 and height 4. The center node is the root (marked in gray) and the total number of nodes in the tree is 46 (Proposition 1).

Let us now focus on the position of a node in the tree. In Fig. 1, each level is marked using a red (dashed) line. Let us consider i is the level at which the node is located, and j denotes the offset position of the node at level i on the tree. Then the position of a node, say x, on a Cayley tree is denoted as (i, j) –that is, the node is indicated as $x_{i,j}$. The levels and the offset positions at the corresponding levels of all the nodes are marked in Fig. 1.

Here, the root is at position $(0, 0)$ and it has three nearest neighbors. The positions of all its neighbors are $(1, 0)$, $(1, 1)$, $(1, 2)$. Next, the positions of the children of node $(1, 0)$ are $(2, 0)$ and $(2, 1)$, and so on (Fig. 1). Therefore, the relationship of the neighbors' positions of a node at position (i, j), except the root node and the leaves, is as follows:

– the position of the parent node of the node (at (i, j)) is $(i - 1, \lfloor j/2 \rfloor)$,

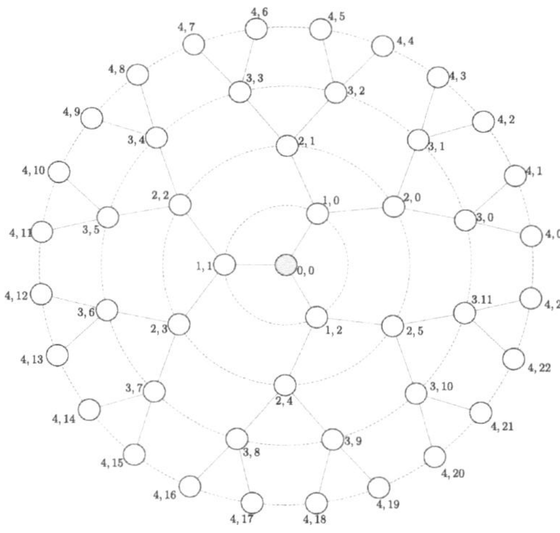

Fig. 1. Cayley Tree of order 2 and height 4. (Color figure online)

- the positions of $child_1$ and $child_2$ nodes of the node (at (i, j)) are $(i + 1, 2j)$ and $(i + 1, 2j + 1)$.

To ensure a uniform neighborhood structure while accounting for the positions of the parent, self, $child_1$, and $child_2$ for all nodes, this work considers one of the neighbors of a node (at its previous level) to serve as the parent, with the remaining two neighbors (at its next level) as $child_1$ and $child_2$. However, for the root node, this work considers the node at position $(1, 0)$ to play the role of the parent of the root. The other nodes at positions $(1, 1)$ and $(1, 2)$ respectively are the $child_1$ and $child_2$ of the root. Since a leaf node has no children, the parent positions of the leaves can be identified as above.

This work introduces CA over Cayley tree (or Bethe lattice) of order 2, where each node of the tree is a cell of the CA and each cell holds a state $s \in S$. Here, the lattice is $\mathcal{L}$. This work considers two possible states (0 or 1) for each cell –that is, $S = \{0, 1\}$. Each cell updates its state based on the states of its parent, self, $child_1$ and $child_2$ using a local update rule $\sigma : S^4 \to S$. At a time, all the cells update their states uniformly.

A configuration $c \in S^{\mathcal{L}}$ is the snapshot of the states of all the cells at a time. Therefore, for a set of cells $\mathcal{L}$ and the rule σ the image $x = (x_{i,j})_{i,j \in \mathcal{L}} = G(c)$, $c = (c_{i,j})_{i,j \in \mathcal{L}} \in S^{\mathcal{L}}$, where G is the global function. Here,

$$\forall i, j \in \mathcal{L}, x_{i,j} = \sigma(c_{i-1,\lfloor j/2 \rfloor}, c_{i,j}, c_{i+1,2j}, c_{i+1,2j+1})$$

The local rule σ of a CA induces a map $G : S^{\mathcal{L}} \to S^{\mathcal{L}}$. A CA on Cayley tree can also be identified using its global transition function.

Table 1 reports some examples of local rules of CAs over Cayley tree. The first row indicates the possible combinations of current states of parent, self, $child_1$ and $child_2$ neighbors (combination of four neighbors) of a cell. The *Rule*

Table 1. Look-up table of CA rule on Cayley tree

Current State (p, q, r, s)	1111	1110	1101	1100	1011	1010	1001	1000	0111	0110	0101	0100	0011	0010	0001	0000	Rule name
(RMTs)	(15)	(14)	(13)	(12)	(11)	(10)	(9)	(8)	(7)	(6)	(5)	(4)	(3)	(2)	(1)	(0)	
Next State(q)	1	1	0	1	0	1	1	0	1	1	1	1	0	1	0	0	σ_1
Next State(q)	0	0	1	0	0	0	1	1	0	0	1	0	0	0	1	1	σ_2
Next State(q)	0	1	1	1	0	1	1	1	1	0	1	0	1	1	1	1	σ_3
Next State(q)	1	1	1	1	1	1	1	1	0	1	0	1	1	1	1	1	σ_4
Next State(q)	0	0	0	0	0	1	0	1	0	0	0	0	1	0	1	0	σ_5

Min Term (RMT) in the second row is used to describe the combinations of parent (p), self (q), $child_1$ (r) and $child_2$ (s) neighbors as p, q, r, s. The remaining rows respectively show the next states of the local rules σ_1, σ_2, σ_3, σ_4 and σ_5 (mentioned in "Rule name" column) of five CAs over Cayley tree.

2.1 Rule Min Term (RMT)

A Rule Min Term (RMT) is the combination of parent (p), self (q), $child_1$ (r) and $child_2$ (s) neighbors –that is, 'p, q, r, s'.

Definition 1. *The combination of the neighborhood p, q, r, s is called Rule Min Term (RMT). Each combination is associated with a number $\gamma(p, q, r, s) = 8p + 4q + 2r + s$.*

An RMT can also be expressed as its decimal equivalent number $\gamma(p, q, r, s)$ (Table 1).

A finite configuration $c = (c_{i,j})_{i,j \in \mathcal{L}} \in S^{\mathcal{L}}$ of a CA over Cayley tree can be represented using RMTs, the states of the cells can be represented using its equivalent RMTs $(\gamma_{i,j})_{i,j \in \mathcal{L}}$. The neighborhood of a cell can be represented as a subtree that contains the parent of a cell (as p), the cell (as q), the first child ($child_1$ as r) and the second child ($child_2$ as s) of the cell. The center of the subtree is the cell under consideration. The subtree represents the associated RMT of the center cell, where the order of the 4 cells in an RMT is p, q, r and s respectively. For example, let the content of a subtree is as follows: the state of a cell be 0, the states of its parent, $child_1$ and $child_2$ are 0, 0 and 1 respectively, then the corresponding RMT is 1 ($\gamma(0, 0, 0, 1) = 8 \times 0 + 4 \times 0 + 2 \times 0 + 1 = 1$).

The subtree for the root, say $x_{0,0}$, considers the combination of the states of cells $x_{1,0}$, $x_{0,0}$, $x_{1,1}$ and $x_{1,2}$ as its parent, self, $child_1$ and $child_2$ respectively to assign its equivalent RMT $\gamma_{0,0}$. Similarly, for the other cell, say $x_{i,j}$, the corresponding subtree considers its neighborhood, $x_{i-1,\lfloor j/2 \rfloor}, x_{i,j}, x_{i+1,2j}$ and $x_{i+1,2j+1}$ –that is, the corresponding RMT $\gamma_{i,j} = \gamma(x_{i-1,\lfloor j/2 \rfloor}, x_{i,j}, x_{i+1,2j}, x_{i+1,2j+1})$. Figure 2.a shows a configuration of a CA over Cayley tree of order 2 and height 2 and its equivalent RMT sequence representation is shown in Fig. 2.b. In both the figures, the position of each node is marked in red.

For example, let us consider a node in the tree (Fig. 2.a), say $x_{1,1}$. This indicates the node is located at level 1 and the position offset of the node at the level is 1 (the node is marked by dotted cycle in Fig. 2.a). For the root node ($x_{0,0}$), the content of its corresponding subtree (the state of its neighbors, $x_{1,0}$, $x_{0,0}$, $x_{1,1}$ and $x_{1,1}$) are 0, 0, 1 and 1. Hence, the RMT of the root cell $\gamma_{0,0} = 8\times0+4\times0+2\times1+1 = 3$. Similarly, the states of parent, self, $child_1$ and $child_2$ (respective positions are (0,0), (1,1), (2,2) and (2,3), marked in red) of $x_{1,1}$ is 0, 1, 0 and 1, respectively. Therefore, the corresponding RMT at position (1,1) is 5 (0101) (Fig. 2.b). Since the work considers null boundary conditions, the states of both the children of a leaf are considered to be 0.

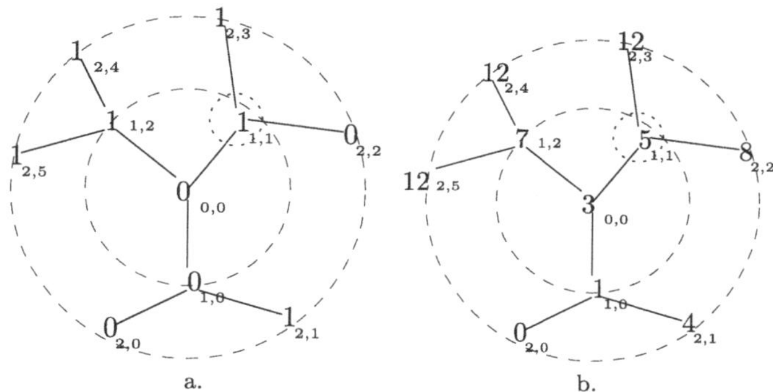

Fig. 2. A configuration of a CA over Cayley tree of order 2 and height 2, and its equivalent RMT representation, (a) A configuration of a CA, (b) The associated RMT representation of the configuration (a) (Color figure online)

Proposition 2. *The possible RMTs at leaf level in a finite CA over Cayley tree of order 2 are 0, 4, 8 and 12.*

Proof. Since the leaves have no children, this work considers that the state of each child of a leaf cell is 0 (null boundary condition). The RMTs 0, 4, 8 and 12 contain 0 in both the children's positions (r and s) in the respective combinations (p, q, r, s). Therefore, these are the only RMTs that are valid at the position of a leaf node.

To obtain consecutive RMTs in two consecutive levels of the lattice, we consider an imaginary subtree window that can be moved to its sibling subtree for each cell of the configuration. The subtree window represents the associated RMT and the subtree is loaded with the states p, q, r and s. To achieve the associated RMT of position (i, j), the subtree considers the states of cells at positions $(i-1, \lfloor j/2 \rfloor)$, (i, j), $(i+1, 2j)$ and $(i+1, 2j+1)$. After the assignment of the root node, the subtree moves from a cell at level 1 to the leaf cell, where the subtree window moves one bit towards one of the children of a cell at a time (moves to a sibling subtree). Therefore, for each cell, two possible directions exist towards its children (since the order is 2), to which the subtree window can move at its next level. Let the content of a subtree for a considering cell at position (i,j) are $(x_{i-1,\lfloor j/2\rfloor}, x_{i,j}, x_{i+1,2j}, x_{i+1,2j+1})$ (resp. (p,q,r,s)), then the content of its sibling subtree (p',q',r',s') of the subtree after moving one step are the states of cells at positions –

- (i,j), $(i+1,2j)$, $(i+2,4j)$ and $(i+2,4j+1)$ towards $child_1$. Here, q and r of the subtree become p' and q' in its sibling subtree.
- (i,j), $(i+1,2j)$, $(i+2,4j+2)$ and $(i+2,4j+3)$ towards $child_2$. Here, q and s of the subtree become p' and q' in its sibling subtree.

This implies the RMTs of the neighboring cells, parent, self, $child_1$ and $child_2$ are related. That is, the RMT of a cell is correlated with the RMTs of its neighbors. For example, if 2(0010) is the $(i,j)^{th}$ RMT, then the set of $(i+1,2j)^{th}$ RMTs is $\{4(0100), 5(0101), 6(0110), 7(0111)\}$ and the set of $(i+1,2j+1)^{th}$ RMTs is $\{0(0000), 1(0001), 2(0010), 3(0011)\}$. Similarly, if 0(0000) or 8(1000) is the $(i,j)^{th}$ RMT, then the set of both $(i+1,2j)^{th}$ and $(i+1,2j+1)^{th}$ RMTs is $\{0(0000), 1(0001), 2(0010), 3(0011)\}$. Table 2 notes down the relation of the $(i,j)^{th}$ RMT (the RMT of a cell at position (i,j)) with $(i+1,2j)^{th}$ and $(i+1,2j+1)^{th}$ RMTs (its children's RMTs).

Table 2. Relationship between the $(i,j)^{th}$, $(i+1,2j)^{th}$ and $(i+1,2j+1)^{th}$ RMTs

$(i,j)^{th}$ RMT	$(i+1,2j)^{th}$ RMT	$(i+1,2j+1)^{th}$ RMT
0	0, 1, 2, 3	0, 1, 2, 3
1	0, 1, 2, 3	4, 5, 6, 7
2	4, 5, 6, 7	0, 1, 2, 3
3	4, 5, 6, 7	4, 5, 6, 7
4	8, 9, 10, 11	8, 9, 10, 11
5	8, 9, 10, 11	12, 13, 14, 15
6	12, 13, 14, 15	8, 9, 10, 11
7	12, 13, 14, 15	12, 13, 14, 15
8	0, 1, 2, 3	0, 1, 2, 3
9	0, 1, 2, 3	4, 5, 6, 7
10	4, 5, 6, 7	0, 1, 2, 3
11	4, 5, 6, 7	4, 5, 6, 7
12	8, 9, 10, 11	8, 9, 10, 11
13	8, 9, 10, 11	12, 13, 14, 15
14	12, 13, 14, 15	8, 9, 10, 11
15	12, 13, 14, 15	12, 13, 14, 15

Since the root, say $x_{0,0}$ considers one of its neighbor, the cell at $(1,0)$, i.e., $x_{1,0}$ as its parent. The relationship between the RMTs of $x_{0,0}$ and $x_{1,0}$ are noted in

Table 3. That is, if the cell $x_{0,0}$ holds an RMT, say r from the column "$(0,0)^{th}$ RMT" then the cell $x_{1,0}$ holds the RMT of the associated row of r from the column "$(1,0)^{th}$ RMT" of the Table 3.

Definition 2. *An RMT (p, q, r, s) is self-replicating if the RMT does not change the state in the next steps during evolution of a cell –that is, $\sigma(p, q, r, s) = q$. The RMT is not self-replicating otherwise.*

Table 3. The relation between the RMTs of the cells $x_{0,0}$ and $x_{1,0}$

$(0,0)^{th}$ RMT	$(1,0)^{th}$ RMT
0, 1, 2, 3	0, 1, 2, 3
8, 9, 10, 11	4, 5, 6, 7
4, 5, 6, 7	8, 9, 10, 11
12, 13, 14, 15	12, 13, 14, 15

For example, in the rule σ_2 in Table 1 of a CA over Cayley tree, the RMTs 2, 3, 5, 10, 11 and 13 are self-replicating RMTs, whereas the RMTs 0, 1, 4, 6, 7, 8, 9, 12, 14 and 15 are non self-replicating.

3 Fixed Points

In this section, we discuss the fixed point of the proposed CA. A fixed point is a type of configuration, once the CA moves to the configuration, it remains there.

Definition 3. *A configuration of a CA is called a fixed point if, once the CA reaches it, the CA remains in the configuration forever.*

Example 1. Let us consider a CA on Cayley tree of order 2 and height 2, which considers a local update rule σ_1 (σ_1 is defined in the third row of Table 1). Let us evolve the configuration, say c, of Fig. 2.a using the rule σ_1. Then, the next configuration of the configuration c is c (remain same). That is, the configuration c is a fixed point of the CA.

Proposition 3. *The rule of a CA over Cayley tree forms a fixed point with a configuration if all the associated RMTs in the configuration are self-replicating RMTs.*

Example 1 validates the above Proposition 3. Here, the associated RMTs of the configuration in Fig. 2.a are shown in Fig. 2.b, and the RMTs are 0, 1, 3, 4, 5, 7, 8 and 12. These RMTs are self-replicating for the rule σ_1. Therefore, the rule σ_1 forms a fixed point with the configuration (Fig. 2.a and Fig. 2.b).

Proposition 4. *There exists no fixed point in a finite CA over Cayley tree under null boundary conditions if the RMTs 0, 4, 8 and 12 are not self-replicating RMTs.*

Proof. The only possible RMTs at the leaf are the RMTs 0, 4, 8 and 12 (Proposition 2). Let us consider all these RMTs are non self-replicating –that is, for any configuration, each leaf cell changes its state during the evolution. Hence, there exists no fixed point.

A CA over Cayley tree is considered a convergent CA if it always converges to some fixed points during its evolution. Therefore, we can obtain the following corollary.

Corollary 1. *A finite CA over Cayley tree under null boundary conditions is not convergent if the RMTs 0, 4, 8 and 12 are not self-replicating RMTs.*

In a CA rule σ_2 in Table 1, the RMTs 0, 4, 8 and 12 are non self-replicating RMTs. Therefore, a finite CA over Cayley tree under null boundary conditions that consists of the rule σ_2 is an example of non-convergent CA.

Proposition 5. *A finite CA over Cayley tree contains only one fixed point if it satisfies any of the following conditions –*

- *the RMTs 0, 1, 2, 3, 8, 9, 10 and 11 are not self-replicating and the RMTs 12 and 15 are self-replicating RMTs.*
- *the RMTs 4, 5, 6, 7, 12, 13, 14 and 15 are not self-replicating and the RMT 0 is self-replicating RMT.*

Proof. Let us consider that a finite CA over Cayley tree consists of a local update rule σ.

Let us now first consider that the RMTs 0, 1, 2, 3, 8, 9, 10 and 11 of σ are not self-replicating. Whenever the state of a cell in a configuration is 0 or switches from 1 to 0 during evolution, it changes its state to 1 in the next iteration.

The all-1 configuration can be represented using only two RMTs 12 and 15. If both the RMTs 12 and 15 in σ are self-replicating then any configuration moves to the all-1 configuration and remains there during evolution of the CA. Here, the all-1 configuration is a fixed point.

The second condition is the 0/1 exchange symmetry of the first condition. Therefore, the condition can be proved with a similar argument. In this case, the all-0 configuration is a fixed point.

Hence the proof.

Example 2. Let us consider a CA on Cayley tree of order 2, which considers a local update rule σ_4 (σ_4 is defined in the sixth row of Table 1). The all 1 configuration is the only fixed point of the CA. On the other hand, the all 0 configuration is the only fixed point of another CA with a local update rule σ_5 (σ_5 is defined in the seventh row of Table 1).

Corollary 2. *A finite CA over Cayley tree contains at least one fixed point if it satisfies any of the following conditions:*

- *the RMT 0 is self-replicating.*
- *the RMTs 12 and 15 are self-replicating.*

The RMT 0 of the local update rule of a finite CA over Cayley tree is self-replicating implies that the all-0 configuration is a fixed point. Similarly, the RMTs 12 and 15 are self-replicating implies that the all-1 configuration is a fixed point. Therefore, the Corollary 2 can be proved with a similar argument of the Proposition 5.

4 Fixed Point Graph

Next, we report the possible fixed points of a CA over Cayley tree. Here, we propose a directed graph, named *Fixed Point Graph* (FPG), that identifies the fixed points of the CA.

A Fixed Point Graph (FPG) graph of a CA is a directed graph where the vertices are the self-replicating RMTs of the CA rule. To draw an FPG from a given CA rule, let us first consider the self-replicating RMTs of the rule only. Each self-replicating RMT is treated as an individual vertex of a forest of vertices. Then, we draw the directed edges between the vertices. An edge is drawn from vertex v_1 to vertex v_2 if v_1 and v_2 are related in the above Table 2 and Table 3. There exist three types of edges – (1) if v_1 is in $(i, j)^{th}$ RMT and v_2 is in the corresponding set of $(i + 1, 2j)^{th}$ RMTs (in Table 2) then a solid edge is drawn from v_1 to v_2; (2) if v_1 is in $(i, j)^{th}$ RMT and v_2 is in the corresponding set of $(i + 1, 2j + 1)^{th}$ RMTs (in Table 2) then a dashed edge is drawn from v_1 to v_2; (3) if v_1 is in $(0, 0)^{th}$ RMT and v_2 is in the corresponding set of $(1, 0)^{th}$ RMTs (in Table 3), then a dash-dot edge is drawn from v_1 to v_2 (see Fig. 3). Let us illustrate this concept using the following example.

Example 3. This example illustrates the procedure of drawing the Fixed Point Graph (FPG) from a CA rule σ_3 from Table 1. The RMTs 5(0101), 7(0111), 11(1011), 12(1100), 13(1101) and 14(1110) are self-replicating RMTs of the rule σ_3. When the $(i, j)^{th}$ RMT is 5 then the $(i + 1, 2j)^{th}$ RMT is 11 and the $(i + 1, 2j + 1)^{th}$ RMTs are 12, 13 and 14. Hence, there exists a solid edge from vertex 5 to vertex 11, and dashed edges from vertex 5 to vertices 12, 13 and 14, the FPG of the CA. When $(0, 0)^{th}$ RMT is 5 then the $(1, 0)^{th}$ RMT is 11. Therefore, a dash-dot edge exists from vertex 5 to vertex 11. Other vertices and edges are drawn in a similar fashion in the graph. Figure 3 shows the FPG of the CA.

Let us now identify a configuration for the CA rule σ_3 and the height 2 from the FPG in Fig. 3. Let us first consider that the vertex 11 as the root $(x_{0,0})$ of the configuration. The next possible vertices for the nodes at level 1 are – RMTs 5 and 7. Let us consider vertex 7 as the neighbor of the root –that is, all the three neighbors of the root are vertex 7. Now, the possible children of the vertex

7 are – RMTs 12, 13 and 14. Since it is the leaf level (height 2), we choose vertex 12 (as per Proposition 2) as a child of vertex 7. The configuration is shown in Fig. 4, which is a fixed point for the CA.

This implies, fixed point of a CA over Cayley tree can be identified by utilizing the FPG of the CA. The procedure of getting fixed point configurations from an FPG is discussed below.

4.1 The Strategy

Start from each vertex as root (level 0 vertex), recursively visit the next level (level 1) vertices as the neighbors of the root. For each vertex at level i, recur-

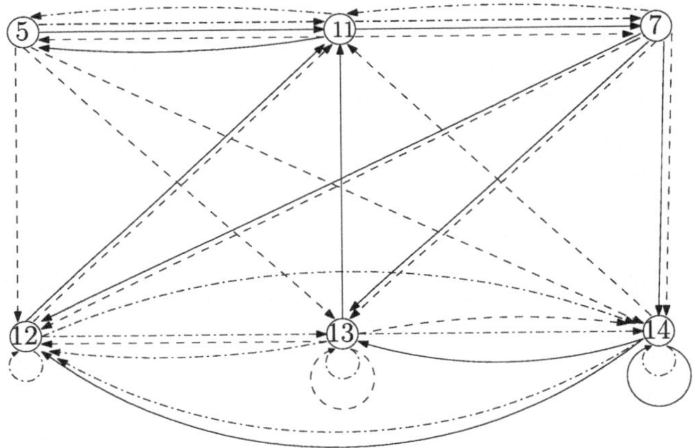

Fig. 3. Fixed Point Graph of the rule σ_3 of Table 1.

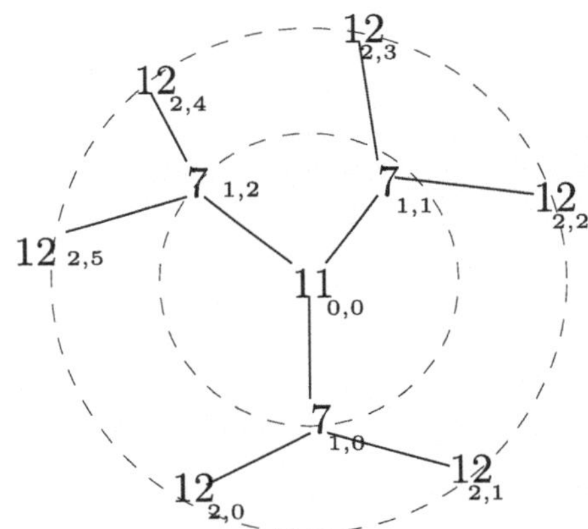

Fig. 4. A configuration of a CA over Cayley tree of order 2 and height 2 from the FPG of Fig. 3.

sively visit its next level $(i + 1)$ vertices up to the given height of the tree. If each leaf vertex, say $x \in \{0, 4, 8, 12\}$, consider that a fixed point is identified.

The procedure `BranchVertices` takes an FPG G and a vertex v and computes the possible vertices of the next level of v. If the vertex v is the root, then the procedure returns three sets of vertices: the list of vertices that can be the parent of v, the list of vertices that can be the $child_1$ of v and the list of vertices that can be the $child_2$ of v (see line numbers 21 of the procedure `BranchVertices`). If the vertex v is not the root, on the other hand, the procedure returns two sets of vertices: the list of vertices that can be the $child_1$ of v and the list of vertices that can be the $child_2$ of v (see line numbers 23 of the procedure `BranchVertices`). This `BranchVertices` function is called from another procedure `NextLevelVertices` during the computation of the proposed Algorithm 1.

```
 1 function BranchVertices(G, v, isRoot)
 2     l = []
 3     r = []
 4     if isRoot then
 5         p = []
 6     end if
 7     for all e in G do
 8         if e[0] = v then
 9             if e[2] = 0 then
10                 Append e[1] in l
11             else if e[2] = 1 then
12                 Append e[1] in r
13             else if e[2] = 2 then
14                 if isRoot then
15                     Append e[1] in p
16                 end if
17             end if
18         end if
19     end for
20     if isRoot then
21         return p, l, r
22     else
23         return l, r
24     end if
25 end function
```

The procedure `NextLevelVertices` takes an FPG (as G for calling the function `BranchVertices`), the height of the Cayley tree (as h) and the set of vertices (as $vset$). This procedure is first considered (calling in Algorithm 1) the root node, say v_r, in its $vset$. When h is 0, the procedure considers the vertices in $vset$ as the leaf nodes. Therefore, it returns 0 (invalid configuration) if there exists a vertex v in $vset$ such that $v \notin \{0, 4, 8, 12\}$ (Proposition 2), it returns 1 (valid configuration) otherwise. For the other value of h, the procedure computes the possible combinations of vertices at its next level using the function `BranchVertices` (see the function call at the line numbers 12 and 24 in the Algorithm 1) and recursively calls `NextLevelVertices` for each combination of vertices. Finally, the procedure returns the counts of fixed points for the root vertex v_r.

The counting algorithm (Algorithm 1) considers each vertex v in the FPG G of a CA as a root and calls the procedure `NextLevelVertices` for the v in its $vset$ (function call at line number 45 in Algorithm 1). The procedure `NextLevelVertices` computes the number of fixed points for the v. The proposed Algorithm 1 computes the total number of fixed points of the CA by summing the fixed point counts over all vertices in the FPG G.

Algorithm 1. Counts the number of fixed point configurations

```
Require: A fixed point graph G
Ensure: Count
function NextLevelVertices(G, h, vset)
    if h = 0 then
        for all v in vset do
            if v not in [0, 4, 8, 12] then return 0
            end if
        end for
        return 1
    end if
    cnt ← 0
    if vset contains the root then
        content = []
        Pbuff, Lbuff, Rbuff = BranchVertices(G, vset[0], True)
        for all p in Pbuff do
            for all l in Lbuff do
                for all r in Rbuff do
                    Append [p, l, r] to the content
                end for
            end for
        end for
    else
        temp = []
        for all v in vset do
            k = []
            Lbuff, Rbuff = BranchVertices(G, v, False)
            for all l in Lbuff do
                for all r in Rbuff do
                    Append [l, r] to k
                end for
            end for
            Append k to the temp
        end for
        content = []
        for combo in product(*temp) do
            com = sum(combo, [])
            Append com to the content
        end for
    end if
    for all vertexset in content do
        cnt+ =NextLevelVertices(G, h − 1, vertexset)
    end for
    return cnt
end function
Count ← 0
for all vertex in G do
    Count+ =NextLevelVertices(G, h − 1, [vertex])
end for
```

For example, let us consider a CA rule σ_3 (see Table 1) with a height of 2. The FPG is shown in Fig. 3. The proposed algorithm (Algorithm 1) returns 1 for the CA. That is, the number of fixed points of the CA is 1, where the fixed point is the configuration of Fig. 4.

5 Conclusion

In this paper, we have studied Cellular Automata over Cayley tree of order 2. Here, we have constructed Fixed Point Graph (FPG) using self-replicating RMTs of the local rule of a CA. The graph has been utilized as a tool to identify the fixed point configurations of a CA. Finally, we have proposed an algorithm to compute the number of fixed points of a CA over Cayley tree.

This is an initial step of exploring the behavior of Cellular Automata over Cayley tree of order 2. Hence, the natural extension of this work is to explore the convergence of such CAs. The other dynamical behaviors of the model, such as reversibility, phase transitions, etc., remain to be explored.

Acknowledgment. The author is grateful to Dr. Sukanta Das for his insightful comments and discussions which have been useful for this work.

References

1. Akın, H., Chang, C.H.: The entropy and reversibility of cellular automata on cayley tree. Int. J. Bifurcat. Chaos **30**(04), 2050061 (2020)
2. Bhattacharjee, K., Das, S.: A search for good pseudo-random number generators: survey and empirical studies. Comput. Sci. Rev. **45**, 100471 (2022)
3. Das, S., Mukherjee, S., Naskar, N., Sikdar, B.K.: Characterization of single cycle ca and its application in pattern classification. Electr. Notes Theor. Comput. Sci. **252**, 181–203 (2009)
4. Fici, G., Fiorenzi, F.: Topological properties of cellular automata on trees. arXiv preprint arXiv:1208.2766 (2012)
5. Games, M.: The fantastic combinations of john conway's new solitaire game "life" by martin gardner. Sci. Am. **223**(120–123), 11 (1970)
6. Ganikhodjaev, N., Temir, S., Akin, H.: Modulated phase of a potts model with competing binary interactions on a cayley tree. J. Stat. Phys. **137**(4), 701–715 (2009)
7. Khan, A.R., Choudhury, P.P., Dihidar, K., Mitra, S., Sarkar, P.: VLSI architecture of a cellular automata machine. Comput. & Math. Appl. **33**(5), 79–94 (1997)
8. Paul, S.: Cellular Automata: Temporal Stochasticity and Computability (2022). https://arxiv.org/abs/2210.13971
9. Paul, S., Bhattacharjee, K.: Nonuniform temporally stochastic cellular automata as model of contagious disease spread. Complex Syst. **34**(1) (2025)
10. Paul, S., Das, S., Sikdar, B.K.: In-Memory Sorting-Searching with Cayley Tree (2025). https://arxiv.org/abs/2506.19379
11. Paul, S., Roy, S., Das, S.: Pattern classification with temporally stochastic cellular automata. In: Manzoni, L., Mariot, L., Roy Chowdhury, D. (eds.) International Workshop on Cellular Automata and Discrete Complex Systems. pp. 137–152. Springer, Cham (2023). https://doi.org/10.1007/978-3-031-42250-8_10
12. Raghavan, R.: Cellular automata in pattern recognition. Inf. Sci. **70**(1–2), 145–177 (1993)
13. Sethi, B., Roy, S., Das, S.: Asynchronous cellular automata and pattern classification. Complexity **21**(S1), 370–386 (2016)
14. Von Neumann, J., Burks, A.W., et al.: Theory of Self-reproducing Automata (1966)

Design of Light-Weight Combined Pseudo Random Number Generator Using Cellular Automaton

Bhuvaneswari Arumugam[1] and Kamalika Bhattacharjee[2(✉)]

[1] Department of Computer Science and Engineering, National Institute of Technology, Tiruchirappalli 620015, Tamil Nadu, India
[2] Department of Information Technology, Indian Institute of Engineering Science and Technology, Shibpur 711103, West Bengal, India
kamalika.it@gmail.com

Abstract. Most of the current day computer applications require Pseudo Random Number Generators (PRNGs) that generate sequence of pseudo-random numbers efficiently inside CPUs or GPUs and with good theoretical quality merits. Uniformity is the most crucial quality criterion of PRNG; in the case of linear PRNG, it is measured using a theoretical quality measure called *equidistribution*. This work targets to design light-weight PRNGs with cellular automata (CAs) having maximal equidistribution and close to maximal period. For that, first the theoretical quality merit of linear cellular automaton (CA)-based PRNGs whose sizes are close to the computer word size (32 and 64) are studied and it is observed that they lack in uniformity. To address this problem, this work introduces light-weight two-component combined CAs-based PRNGs with time spacing that achieve maximal equidistribution. We show that, the proposed combined CA-based PRNGs also pass almost all tests in the standard statistical testbeds, and the results are comparable to the existing state-of-the-art linear PRNGs.

Keywords: Maximal length CA · Uniformity · Maximal Equidistribution · Period · Combined CA-based PRNG · bitwise XOR

1 Introduction

A Pseudo Random Number Generator (PRNG), a deterministic algorithm, generates a sequence of pseudo-random numbers that seem to be random, but are replicatable as per requirement. Any number generated by PRNGs is not *truly* random since it is based on an initial value called *seed* that can be easily reproduced if the *seed* is known. The important quality criteria of PRNGs are period

---

This work is done as a project in the Indian Summer School on Cellular Automata 2025.
The original version of the chapter has been revised. The author's name in chapter 5 is updated in correct order. A correction to this chapter can be found at https://doi.org/10.1007/978-3-032-18612-6_21

H. Raju et al. (Eds.): ASCAT 2026, CCIS 2801, pp. 62–76, 2026.
https://doi.org/10.1007/978-3-032-18612-6_5

length, uniformity, and efficiency. Long periods are required for PRNGs; otherwise, the generated sequence can be predicted easily. Then, the numbers should be uniformly distributed, meaning that each number should occur with an equal probability. And the implementation used by the PRNG should be efficient such that very less resource is used for generating a number. In the case of a linear PRNG, this uniformity is assessed by a theoretical *figure of merit* called an *equidistribution*. This equidistribution examines whether the numbers generated by the PRNG are evenly distributed or not. If the numbers are evenly distributed in all possible dimensions, it is called *maximally equidistributed*. In addition to these theoretical criteria, a PRNG also needs to pass the standard statistical tests, such as Dieharder and TestU01 suite, that empirically find evidence against non-randomness.

There exist three major classes of PRNGs, such as Linear Congruential Generators (LCGs), Linear Feedback Shift Register (LFSRs) and Cellular Automaton (CA) based generators [5]. Due to the ease of bitwise implementation and availability of theories from linear algebra for characterization, the most popular PRNGs have been linear PRNGs based on LFSRs, such as Mersenne Twister [2], WELL [9], XORShift PRNGs [10], etc. LFSR produces the maximal length period $2^k - 1$ when the corresponding polynomial is primitive. However, there are two problems on using these as PRNGs. First, a single-component Linear PRNG often fails to achieve the *maximal equidistribution* [1]. Secondly, there is a scarcity of primitive polynomials of large degree to be utilized as PRNGs. To address these issues, combined PRNGs were proposed in Ref. [1]. In these kind of PRNGs, more than one Tausworthe generator, a variant of LFSR having primitive polynomials of smaller degree using the bitwise XOR operation, are combined. This combined Tausworthe generator achieves maximal equidistribution and can also increase the period length.

However, recent developments on big data and high-performance computing demand the most efficient use of GPUs and multicore CPUs. In such cases, the applications require PRNGs which are light-weight, meaning that they can fit inside the word size of a computer and it is easily implementable in both software as well as hardware. In this aspect, Cellular Automaton (CA) provides a good alternative solution. A CA is a mathematical model that utilizes a grid of cells. Every cell uses a local rule, which is a transition function, to change its state based on its neighbors. All the cells are updated in parallel, and the group of states of all cells at each time step is called a configuration. CAs are utilized to design PRNGs since they provide a source of randomness and use simple local rules to generate sequences of pseudo-random numbers in a parallel manner.

Elementary Cellular Automaton (ECA) is a one-dimensional, 2-state, and 3-neighborhood system. ECAs have been very popular due to their ease of bitwise and hardware implementation. In fact, Stephen Wolfram initiated the research on cellular automata as origin of randomness in physical systems where he systematically studied and analyzed ECA rule 30 as source of randomness [16,17]. It is shown that ECA rule 30 generates highly chaotic pseudo-random numbers, which uses an initial configuration consisting of a single 1 in the middle cell and all other cells set to 0. It also demonstrates high performance and can be efficiently implemented in parallel [5,18]. Since ECA rule 30 is uniformly applied to each cell in the CA, such a CA is referred to as a *uniform* CA, meaning that all cells follow the same rule. However, maximal period length is not achievable by such a CA.

To enhance the period length, non-uniform CA were introduced in designing PRNGs [5]. In a *non-uniform* CA, instead of using a single rule $\mathcal{R}$, a rule vector $\mathcal{R} = \langle R_0, R_1, \ldots, R_{n-1} \rangle$ is used. Here, rule R_i is assigned to the ith cell in CA. Some non-uniform CAs with linear ECA rules 90 and 150 can generate maximal period length [6]. Such CAs are named as linear *maximal length* cellular automata for which the characteristic polynomial is primitive, and similar to LFSRs, they have been used in designing PRNGs based on CAs [13,19]. These CA-based PRNGs can generate numbers in an efficient way; specifically, linear CA-based PRNGs can generate pseudo-random numbers at a speed comparable to the Mersenne Twister [8], and they also pass most of the empirical tests, such as the Dieharder and BigCrush test results, demonstrating their suitability for high-performance random number generation applications. Such CAs have also been utilized as light-weight multiple-stream parallel PRNGs [11,12].

Nevertheless, although they are efficient to implement, they fail many statistical tests for randomness if used without tempering [12]. It indicates that these CA-based PRNGs are probably not *maximally equidistributed*, indicating they lack in uniformity. However, equidistribution of CA-based PRNGs have never been studied before. In this situation, this work evaluates the equidistribution property of linear CA-based PRNGs with CA sizes close to the current computer's word sizes (32 and 64). Our primary objective is to design a lightweight CA-based PRNG with maximal equidistribution (without using any tempering) and a good period length in an efficient manner. To achieve the maximal equidistribution, we introduce combined linear maximal length CAs-based PRNGs. To be lightweight, we restrict to only 2-component PRNGs with linear maximal length CAs of sizes close to 32 or 64 combined using the XOR operation.

This paper is structured as follows. The general framework of linear PRNG, equidistribution, and maximally equidistribution is described in Sect. 2. In Sect. 3, the equidistribution of linear CA-based PRNG is examined. Then, the proposed combined CAs-based PRNGs with period length and equidistribution properties are presented in Sect. 4. Finally, the proposed PRNGs are tested with statistical tests like Dieharder, BigCrush, and SmallCrush, and the results are displayed in Sect. 5.

2 Background

2.1 General Framework of Linear PRNG

The general framework of the linear PRNG is represented by the following equations. The k-bit state vector at step n is termed x_n, and the w-bit output vector at step n is called y_n. Both k and w are positive integers. The outcome at step n is the real number $u_n \in [0, 1)$. Here, A is the $k \times k$ characteristic matrix, while B is the $w \times k$ matrix used for output transformation (tempering). This framework is used by several linear PRNGs by selecting the proper matrices A and B [3].

$$x_n = Ax_{n-1} \tag{1}$$
$$y_n = Bx_n \tag{2}$$
$$u_n = \sum_{l=1}^{w} y_{n,l-1}2^{-l} = .\, y_{n,0}y_{n,1}y_{n,2}\cdots \tag{3}$$

In the case of a combined PRNG, multiple components are combined in the form of (1) and (2). Assume that the J components are $J = 1, 2, \ldots, j$. Let A_J be the $k_J \times k_J$ transition matrix (characteristic matrix) and B_J be the $w \times k_J$ output transformation matrix for the generator J. The following equations [3] define the output of step n of this combined generator: $\oplus$ denotes the bitwise exclusive-or (XOR) operation in this case.

$$y_n = B_1x_{1,n} \oplus B_2x_{2,n} \oplus \cdots \oplus B_jx_{j,n} \tag{4}$$
$$u_n = \sum_{\ell=1}^{w} y_{n,\ell}\, 2^{-\ell} \tag{5}$$

The following proposition [4] states that the combined PRNG period becomes the Least Common Multiple (LCM) of the individual component periods when we choose components of the combined PRNG with periods of $2^k - 1$ each and component periods are relatively prime.

Proposition 1 ([4]). *A combined pseudo-random sequence is defined as follows: for* $n = 1, 2, \ldots,$

$$U_n = u_n^{(1)} \; XOR \; \cdots \; XOR \; u_n^{(J)},$$

where for $j = 1, \ldots, J$ *each sequence* $u_n^{(j)}, n = 1, 2, \ldots,$ *is a pseudo-random sequence represented in Eq. 3 with characteristic polynomial* $M^{(j)}$*, and the periods of the sequences,* $2^{\deg(M^{(j)})} - 1$, $j = 1, \ldots, J$*, are pairwise co-prime. Then, the combined sequence has a period of* $\prod_{j=1}^{J} \left(2^{\deg(M^{(j)})} - 1\right)$.

2.2 Equidistribution and Maximally Equidistribution

Consider that $\mathbf{q} = (q_1, \ldots, q_t)$ is a vector of non-negative integers. The unit hypercube $[0, 1)^t$ is partitioned into $2^{q_1+\cdots+q_t}$ equal-sized rectangular boxes in 2^{q_j} equal-length intervals along each axis $j = 1, \ldots, t$. When a point set Ψ_I has exactly 2^q points in each box, where $\Psi_I = \{(u_{i_1}, \ldots, u_t) : x_0 \in \mathbb{F}_2\}$ and q satisfies $k - q = q_1 + \cdots + q_t$, it is said to be **q**-equidistributed. It is only possible to meet this condition when $q_1 + \cdots + q_t \leq k$. The uniformity of a linear PRNG can be measured via the **q**-equidistribution. This equidistribution is checked by building a binary matrix and checking the rank. The following steps are taken for building the binary matrix ($\mathcal{B}$) [3]:

- For $j = 1, \ldots, k$ initialize the generator with $x_0 = e_j$, where e_j is the unit vector with a 1 at position j and a 0 at all other positions.
- The generator is run for t steps.

- Extract the q_i most significant bits from each step i.
- These bits form the jth column of the binary matrix.

When k is large, checking all possible $\mathbf{q}$ vectors is too difficult. For this problem, use vectors of the form $\mathbf{q} = (l, \ldots, l)$, for some constant $l \geq 1$, which partitions the space into 2^{tl} cubic boxes. Ψ_I is $\mathbf{q}$-equidistributed for this case, it is said to be *t-distributed with l-bits of accuracy* [3]. This (t, l)-equidistribution can be verified using Proposition 2 for single component generators [1].

Proposition 2 ([1]). *The sequence is (t, ℓ)-equidistributed if and only if the matrix $\mathcal{B}_{t,\ell,s}$ has (full) rank $t\ell$. If $\ell_t = \lfloor k/t \rfloor < k/t \leq L$, then the sequence is also $CF(t)$ if and only if the matrix $\mathcal{B}_{t,\ell_t+1,s}$ has rank k.*

For J components, the binary matrix $\mathcal{B}$ is the juxtaposition of $\mathcal{B}_1, \mathcal{B}_2, \ldots, \mathcal{B}_j$. The first k_1 columns are provided by $\mathcal{B}_1$, the following k_2 columns by $\mathcal{B}_2$, and so on $\mathcal{B}_j$ provides the final k_j columns. Proposition 3 [1] states about the equidistribution of this combined generator, where $\widetilde{\mathcal{B}}_{t,l,s}$ is the combined matrix.

Proposition 3 ([1]). *The sequence is (t, l)-equidistributed if and only if the matrix $\widetilde{\mathcal{B}}_{t,l,s}$ has full rank tl. If $lt = \lfloor k/t \rfloor < k/t \leq L$, then the sequence is also CF(t) if and only if the matrix $\widetilde{\mathcal{B}}_{t,lt+1,s}$ has rank k.*

Let us examine a unit hypercube of size t that has been divided into equal-sized cubic cells of size $2^{t\ell}$. A series of vectors, or t-dimensional points, are produced by the generator and then transferred into the cells based on their values. The resolution in dimension t is represented by the parameter ℓ_t, which is the maximum ℓ for which each cell contains an equal number of points. The generator is referred to as *maximally equidistributed (ME)* [1] when it reaches the maximum value of ℓ_t in all dimensions. Here, l_t denotes the greatest value of $l \leq L$ for which the sequence is (t, l) equidistributed. The following proposition gives the set of values of (t, l) to confirm this (t, l) equidistribution [1].

Proposition 4 ([1]). *A maximal period sequence is ME if and only if $\Lambda_t = 0$ for all $t \in \Phi_1 \cup \Phi_2$. It is also ME if and only if $\Delta_l = 0$ for all $l \in \Psi_1 \cup \Psi_2$.*

$$\Phi_1 = \left\{ \max\left(2, \left\lfloor \frac{k}{L} \right\rfloor\right), \ldots, \left\lfloor \sqrt{k} \right\rfloor \right\}, \tag{6}$$

$$\Phi_2 = \left\{ t = \left\lfloor \frac{k}{\ell} \right\rfloor \,\middle|\, \ell = 1, \ldots, \left\lfloor \sqrt{k} \right\rfloor \right\} \tag{7}$$

$$\Psi_1 = \{1, \ldots, \lfloor \sqrt{k} \rfloor\} \tag{8}$$

$$\Psi_2 = \{\ell = \lfloor k/t \rfloor \mid t = \max(2, \lfloor k/L \rfloor), \ldots, \lfloor \sqrt{k-1} \rfloor\}. \tag{9}$$

2.3 Maximal Length CA

Maximal length CA is a type of non-uniform CA; it has the period length of $2^k - 1$, where k is the CA size. A linear CA is said to be maximal length CA if and only if its rule vector uses the ECA rules 90 and 150 and the corresponding polynomial is primitive over GF(2) [6]. The characteristic matrix for maximal length CA is represented in the following way: the ith row illustrates that the ith cell is dependent on its neighbor. For example, the 4-cell maximal length CA with rule vector $\mathcal{R} = \langle 150, 90, 150, 90 \rangle$ can be represented by:

$$A = \begin{bmatrix} 1 & 1 & 0 & 0 \\ 1 & 0 & 1 & 0 \\ 0 & 1 & 1 & 1 \\ 0 & 0 & 1 & 0 \end{bmatrix}$$

Its characteristic polynomial is $\lambda^4 - \lambda^3 - 3\lambda^2 + 2\lambda + 1$, which is primitive, indicating that the CA has period $2^4 - 1 = 15$. There are several linear maximal length CAs from degree 1 to 500 mentioned in Ref. [7]. For each of these CAs, only a maximum of two cells use rule 150, and the remaining cells use rule 90. In this paper, we utilize some CAs from this list for developing our light-weight combined PRNGs. Before that, we need to know the status of uniformity of these CAs if used directly as PRNGs.

3 Linear CA-Based PRNG

Today's computers use 32-bit and 64-bit word sizes. In this study, we first examine the equidistribution of linear CA-based PRNG for sizes of 32 and 64. Table 1 shows the corresponding rule vectors for these CAs. To assess the equidistribution of these linear CA-based PRNGs, first fit each PRNG into the basic framework described in Subsect. 2.1. For this CA-based PRNG, the matrices A and B are as follows: A is the $k \times k$ characteristic matrix for the corresponding CA rule vector as described in Subsect. 2.3, and B is a $w \times k$ identity matrix, where k is the CA's size. The maximal equidistribution is then tested using Proposition 3, which is discussed in Subsect. 2.2.

Table 1. Linear CA Rule Vectors with the size of 32 and 64

Linear CA Size	Rule Vector
32	150, 90, 90, 90, 90, 90, 90, 90, 90, 90, 90, 90, 90, 90, 150, 90, 90, 90, 90, 90, 90, 90, 90, 90, 90, 90, 90, 90, 90, 90, 90
64	90, 90, 150, 90, 150, 90, 90, 90, 90, 90, 90, 90, 90, 90, 90, 90, 90, 90,90,90, 90, 90,90,90, 90, 90, 90, 90, 90, 90, 90, 90, 90, 90, 90, 90, 90, 90,90,90, 90, 90, 90, 90, 90, 90,90, 90, 90, 90, 90, 90, 90, 90, 90, 90, 90, 90,90, 90, 90, 90, 90, 90

For instance, to determine the maximal equidistribution we first compute the t values using Proposition 3 for a CA of size 64 ($k = 64$) shown in Table 1. The possible values are $t \in \{2, 3, 4, 5, 6, 7, 8, 9, 10, 12, 16, 21, 32, 64\}$. The equidistribution is then examined for every (t, l) value. In this case, only one t value meets the (t, l) equidistribution; the others do not satisfy it, as displayed in Table 2. Therefore, the linear CA with a size of 64 is not maximally equidistributed. Similarly, CA with size 32 also satisfies the (t, l) equidistribution for only one t value. As a result, both linear CAs with the size of 32 and 64 fail to achieve the maximal equidistribution property.

Table 2. Equidistribution of linear CA based PRNG with size of 64

t	$l_t^* = \min(L, \lfloor k/t \rfloor)$	l_t ($l_t \leq l_t^*$)	Rank	Equidistribution
2	32	32	34	not (t, l) equidistributed
3	21	21	24	not (t, l) equidistributed
4	16	16	20	not (t, l) equidistributed
5	12	12	17	not (t, l) equidistributed
6	10	10	16	not (t, l) equidistributed
7	9	9	16	not (t, l) equidistributed
8	8	8	16	not (t, l) equidistributed
9	7	7	16	not (t, l) equidistributed
10	6	6	16	not (t, l) equidistributed
12	5	5	17	not (t, l) equidistributed
16	4	4	20	not (t, l) equidistributed
21	3	3	24	not (t, l) equidistributed
32	2	2	33	not (t, l) equidistributed
64	1	1	64	(t, l) equidistributed

In the next section, we design our combined PRNGs, which achieve maximal equidistribution.

4 Combined Maximal Length CAs-Based PRNG

Since the linear CA-based PRNGs fail to achieve the maximal equidistribution, we opt for combining two linear maximal length CAs from Ref. [7], each close to the size of a computer word (32 and 64), to improve theoretical characteristics (equidistribution) and the period length. The list of possible candidate CAs are given in Table 3. The cell positions are defined in Table 3 using rule 150 with positions starting from 1 to n, and the remaining positions use rule 90. For example, if $k = 32$, the cell positions 1 and 15 use rule 150; the remaining cells use rule 90 giving the rule vector $\mathcal{R} = \langle 150, 90, 90, 90, 90, 90, 90, 90, 90, 90, 90, 90, 90, 90, 150, 90, 90, 90, 90, 90, 90, 90, 90, 90, 90, 90, 90, 90, 90, 90, 90, 90 \rangle$.

Table 3. Linear maximal length CAs with sizes close to 32 and 64

k	Position of Rule 150	k	Position of Rule 150
29	1	60	2,38
30	1	61	1,10
31	11	62	5
32	1,15	63	31
59	4,15	64	3,5

4.1 Combined CAs-Based PRNG Without Time-Spacing

In the combined CAs-based PRNG, each linear maximal length CA is considered as a component. Initially, we select two linear maximal-length CAs as components whose period lengths are relatively prime. So, according to Proposition 1, the combined PRNG has a very large degree. Both CAs are then evolved. At each time step, each component produces a configuration based on the corresponding rule vector. These configurations are considered output for each time step. These two component outputs are combined using the exclusive OR (XOR) operation at every timestep. It produces a new output sequence, which is called the combined pseudo-random sequence. This process is represented in Algorithm 1. In this combined CAs-based PRNG, two components have the sizes of k_1 and k_2. Each has a period of $2^{k_1} - 1$ and $2^{k_2} - 1$, and also the periods of the two components are relatively prime. So, it achieves the period close to $2^{k_1+k_2}$ as discussed in Subsect. 2.1.

Algorithm 1. Combined CAs-based PRNG using linear maximal length CA

Require: Two linear maximal length CAs CA_1 and CA_2 with relatively prime periods
Require: Number of time steps t and Rule vector sizes k_1, k_2
Require: Seeds(configuration) $S_1^{(0)} \in \{0,1\}^{k_1}$, $S_2^{(0)} \in \{0,1\}^{k_2}$
Ensure: Combined pseudorandom sequence U

$U \leftarrow [\,]$
Initialize CA_1 with $S_1^{(0)}$ and CA_2 with $S_2^{(0)}$
for $n = 1$ to t **do**
  $u_1^{(n)} \leftarrow$ NextState(CA_1)
  $u_2^{(n)} \leftarrow$ NextState(CA_2)
  $u_n \leftarrow u_1^{(n)} \oplus u_2^{(n)}$ {combine using Bitwise XOR}
  Append u_n to U
end for
return U

From Table 3, the two-component (k_1, k_2) relatively prime combinations close to 32 are $(29, 30), (29, 31), (29, 32), (30, 31)$ and $(31, 32)$, and close to 64 are $(59, 60), (59, 61), (59, 62), (59, 63), (59, 64), (60, 61), (61, 62), (61, 63), (61, 64)$, $(62, 63)$ and $(63, 64)$. Then, these combinations are tested for equidistribution by using the procedure in Subsect. 2.2, but none of the two-component combinations satisfy the maximal equidistribution.

Example 1. Take the combined PRNG with the size of CA $k_1 = 59$ and $k_2 = 64$. The corresponding rule vectors are $\mathcal{R}_1 = \langle 90, 90, 90, 150, 90, 90, 90, 90, 90, 90, 90, 90, 90, 90, 150, 90, 90, 90, 90, 90, 90, 90, 90, 90, 90, 90, 90, 90, 90, 90, 90, 90, 90, 90, 90, 90, 90, 90, 90, 90, 90, 90, 90, 90, 90, 90, 90, 90, 90, 90, 90, 90, 90, 90, 90, 90, 90, 90, 90 \rangle$ and $\mathcal{R}_2 = \langle 90, 90, 150, 90, 150, 90, 90, 90, 90, 90, 90, 90, 90, 90, 90, 90, 90, 90, 90, 90, 90, 90, 90, 90, 90, 90, 90, 90, 90, 90, 90, 90, 90, 90, 90, 90, 90, 90, 90, 90, 90, 90, 90, 90, 90, 90, 90, 90, 90, 90, 90, 90, 90, 90, 90, 90, 90, 90, 90, 90, 90, 90, 90, 90 \rangle$. The periods of each component are $2^{59} - 1$ and $2^{64} - 1$, and $\gcd(2^{59} - 1,\ 2^{64} - 1) = 1$. So it gives a period length that is close to 2^{123}. To test the equidistribution, the t values are identified using Proposition 4; these values are $t \in \{2, 3, 4, 5, 6, 7, 8, 9, 10, 11, 12, 13, 15, 17, 20, 24, 30, 41, 61, 123\}$. The equidistribution is satisfied for only $t = 123$ and $l = 1$. The remaining (t,l) values do not meet the criteria (full rank of matrix) of equidistribution. Thus, it attains a period close to 2^{123} but does not reach the maximal equidistribution.

4.2 Combined CAs-Based PRNG with Time Spacing

In [13], the concept of time spacing of CAs was introduced, where instead of extracting numbers from the configuration at each timestep, some time steps are skipped. In [1], it has been shown that LFSR-based combined PRNGs with skip can still achieve a period close to maximal and have an improved equidistribution property. So, in this section, we utilize this technique to achieve a period close to maximal as well as maximal equidistribution for a two-component PRNG over maximal length CAs.

To improve the theoretical properties (equidistribution) of the combined CAs-based PRNG discussed in Subsect. 4.1, the following steps are taken: We take two linear maximal length CAs with relatively prime period lengths and evolve for t time steps. Instead of combining at every step, skip each $(s-1)$ steps and combine to generate a pseudo-random sequence. For example, if $s = 3$, the configurations at every third step are combined. This is shown in Algorithm 2.

Algorithm 2. Combined CAs-based PRNG using maximal length CA with time spacing

Require: Two linear maximal length CA CA_1 and CA_2 with relatively prime periods
Require: Number of time steps t and Rule vector sizes k_1, k_2
Require: Skipping steps $2 \leq s \leq 10$
Require: Seeds (configurations) $S_1^{(0)} \in \{0,1\}^{k_1}$, $S_2^{(0)} \in \{0,1\}^{k_2}$
Ensure: Combined pseudorandom sequence U

```
U ← [ ]
Initialize CA_1 with S_1^(0) and CA_2 with S_2^(0)
for n = 1 to t do
    u_1^(n) ← NextState(CA_1)
    u_2^(n) ← NextState(CA_2)
    if n mod s = 0 then
        u_n ← u_1^(n) ⊕ u_2^(n)
        Append u_n to U
    end if
end for
return U
```

In this study, we take all the two-component relatively prime combinations shown in Subsect. 4.1 with $2 \leq s \leq 10$. Initially, we analyze the period length of these combined CAs-based PRNG components. These PRNGs with time spacing follows the general framework for linear PRNG if we consider the matrix A as $(T)^s$ where T is the characteristic matrix of a component CA and B is an identity matrix, as discussed in Sect. 3. The choice of skip step s is important since increasing s definitely improves randomness but at the cost of more computation. Since we are targeting light-weight PRNGs, we set s as $2 \leq s \leq 10$.

If the s values are relatively prime to the product of the period of the individual component, it gives a period length close to $\approx 2^{k_1} + 2^{k_2}$. That is, it achieves the period close to maximal $\approx 2^{k_1} + 2^{k_2}$ only if $\gcd(\rho, s) = 1$, where $\rho = \rho_1.\rho_2$, ρ_1 is the period of the first component, and ρ_2 is the period of the second component. However, if these s values are not relatively prime to ρ, it gives the period of $\frac{\rho}{\gcd(s,\rho)}$. This result, which can be proved using linear algebra, are stated in the following Proposition 5.

Proposition 5. *Let k_1 and k_2 be the size of the linear maximal length CAs, ρ_1 and ρ_2 be the periods of two individual CA components with $\gcd(\rho_1, \rho_2) = 1$, $\rho = \text{lcm}(\rho_1, \rho_2) = \rho_1 \cdot \rho_2$ and $s > 1$ is the skip size. If $\gcd(s, \rho) = 1$, then the period of the combined generator is close to maximal, that is,*

$$\rho = \text{lcm}(\rho_1, \rho_2) \approx 2^{k_1 + k_2}$$

If $\gcd(s, \rho) \neq 1$, then the period is

$$\rho = \frac{\rho}{\gcd(s, \rho)}$$

Then, each of the combinations are tested for equidistribution, we see that some combinations achieve maximal equidistribution, and some achieve almost maximal equidistribution for specific s values. Table 4 shows these values. For example, in Table 4, the following combinations (k_1, k_2, s) satisfy both period $\approx 2^{k_1} + 2^{k_2}$ and maximally equidistribution: $(29, 31, 5)$, $(29, 31, 6)$, $(29, 31, 8)$, $(29, 31, 9)$, $(29, 31, 10)$, $(29, 32, 10)$, $(30, 31, 5)$, $(30, 31, 8)$, $(30, 31, 10)$, $(31, 32, 7)$, $(31, 32, 8)$, $(51, 61, 10)$, $(59, 62, 10)$, $(59, 63, 9)$, $(59, 63, 10)$, $(61, 63, 10)$, $(62, 63, 10)$. Whereas, the combinations which satisfy close to maximal period and are almost maximally equidistributed are: $(29, 30, 10)$, $(29, 31, 7)$, $(59, 60, 8)$, $(59, 61, 7)$, $(59, 61, 8)$, $(58, 61, 9)$, $(59, 62, 7)$, $(59, 62, 8)$, $(59, 63, 8)$, $(59, 64, 7)$, $(59, 64, 8)$, $(60, 61, 8)$, $(61, 62, 7)$, $(61, 62, 8)$, $(61, 62, 10)$, $(61, 63, 8)$, $(61, 63, 9)$, $(61, 64, 7)$, $(61, 64, 8)$, $(62, 63, 8)$, $(63, 64, 8)$. It is observed that the combined PRNG $(59, 64)$ with $s = 10$ achieves maximal equidistribution but does not reach a period close to maximal.

Table 4. Equidistribution Combined CAs-based PRNG with time spacing

k_1	k_2	s period ($2 \leq s \leq 10$)	s ME ($2 \leq s \leq 10$)	ρ and ME	ρ and almost ME
29	30	2,4,5,8,10	-	-	10
29	31	2,3,4,5,6,7,8,9,10	5,6,8,9,10	5,6,8,9,10	7
29	32	2,4,7,8	8	8	-
30	31	2,4,5,8,10	5,8,10	5,8,10	-
31	32	2,4,7,8	5,6,7,8,9,10	7,8	-
59	60	2,4,8	9,10	-	8
59	61	2,3,4,5,6,7,8,9,10	10	10	7,8,9
59	62	2,4,5,7,8,10	10	10	7,8
59	63	2,3,4,5,6,8,9,10	9,10	9,10	8
59	64	2,4,7,8	10	-	7,8
60	61	2,4,8	9,10	-	8
61	62	2,4,5,7,8,10	-	-	7,8,10
61	63	2,3,4,5,6,8,9,10	10	10	8,9
61	64	2,4,7,8	-	-	7,8
62	63	2,4,5,8,10	10	10	8
63	64	2,4,8	10	-	8

Example 2. We take the same combined PRNG $(k1, k2) = (59,\ 64)$ shown in Example 1 with $2 \leq s \leq 10$. It gives a period length that is close to 2^{123} when $s \in \{2, 4, 7, 8\}$ because these s values are relatively prime to $\rho = (2^{59} - 1) \cdot (2^{64} - 1))$. It also achieves the maximal equidistribution for $s = 10$ and almost maximal equidistribution for $s = 7$ and 8. Here, we notice that while $s = 10$ attains the maximal equidistribution, it fails to meet the period close to the maximal one because $\gcd(\rho, 10) \neq 1$. It has a period of $\frac{\rho}{\gcd(10,\rho)} \approx 2^{121}$. However, for $s = 7$ and 8, it attains almost maximal equidistribution and a period close to 2^{123}.

In the next section, we verify their randomness qualities over the standard empirical testbeds.

5 Experiment Results and Analysis

5.1 Empirical Test

Statistical tests such as *BigCrush* [15], *SmallCrush* [15], and *Dieharder* [14] are used to test the randomness quality. This section evaluates the randomness quality of the proposed combined CAs-based PRNGs using these statistical tests. First, we test the combined PRNGs, which satisfy the maximally equidistribution and almost maximally equidistribution with close to maximal period in Dieharder. For this, we generate a sequence of pseudo-random numbers with a 1.5 GB binary file, and give it as input to the Dieharder testbed. Only the following combined PRNGs $(k_1, k_2, s) \in \{(29, 32, 8), (31, 32, 7), (31, 32, 8), (59, 64, 7), (59, 64, 8), (61, 64, 7), (61, 64, 8), (63, 64, 8)\}$ have passed more than 100 s of Dieharder's tests. These PRNGs were again tested with SmallCrush and BigCrush tests, these results are displayed in Table 5.

Table 5. Period Length, Equidistribution, and Statistical test results of combined CAs-based PRNG

PRNGs	Period ($\approx \rho$)	Equidistribution	Statistical Results (failed)		
(k_1, k_2, s)			Dieharder	SmallCrush	BigCrush
Combined CAs-based Generators (ρ not close to maximal)					
CA-PRNG (29, 32, 8)	2^{61}	ME	6	5	24
CA-PRNG (31, 32, 7)	2^{63}	ME	1	1	11
CA-PRNG (31, 32, 8)	2^{63}	ME	1	1	7
CA-PRNG (59, 64, 7)	2^{123}	almost ME	1	3	11
CA-PRNG (59, 64, 8)	2^{123}	almost ME	3	2	11
CA-PRNG (61, 64, 7)	2^{125}	almost ME	1	3	13
CA-PRNG (61, 64, 8)	2^{125}	almost ME	3	2	11
CA-PRNG (63, 64, 8)	2^{127}	almost ME	3	2	12
Combined CAs-based Generators (ρ not close to maximal)					
			Dieharder	SmallCrush	BigCrush
CA-PRNG (31, 32, 5)	2^{61}	ME	3	3	20
CA-PRNG (31, 32, 6)	2^{61}	ME	2	1	15
CA-PRNG (31, 32, 9)	2^{61}	ME	All passed	1	6
CA-PRNG (31, 32, 10)	2^{61}	ME	All passed	2	6
CA-PRNG (59, 64, 10)	2^{121}	ME	All passed	1	11
CA-PRNG (63, 64, 10)	2^{125}	ME	1	1	13
Existing Linear Generators					
Tausworthe (combined)	2^{88}	ME	All passed	All passed	7
Mersenne Twister	$2^{19937} - 1$	not ME	All passed	All passed	3
WELL512a	2^{512}	ME	All passed	All passed	6
WELL1024a	2^{1024}	ME	All passed	All passed	7

Next, we test some combined PRNGs that do not have a period close to maximal but achieve maximal equidistribution for randomness. In particular, the combined PRNG (59,64) with $s = 10$ has the period close to 2^{121} (not close to the maximal period $\approx 2^{123}$), but it achieves the maximal equidistribution. Similarly, the following combined PRNGs (k_1, k_2, s): (31,32,5), (31,32,6), (31,32,9), (31,32,10), (59,60,9), (59,60,10), (60,61,9), (60,61,10), (63,64,10) have periods as $\frac{\rho}{\gcd(s,\rho)}$ (not close to the maximal) where $\rho = 2^{k_1+k_2}$ and attain maximal equidistribution. The results of the statistical tests performed on these combined PRNGs are also displayed in Table 5.

Finally, the proposed combined CAs-based PRNGs period length, theoretical properties (maximally equidistribution), and statistical test results are compared with existing linear PRNGs such as combined Tausworthe generator [1], Mersenne Twister [2] and WELL [9]. Table 5 show that even though Mersenne Twister has a larger period ($2^{19937} - 1$) with no failures in SmallCrush and Dieharder and minimal failures in BigCrush, but it does not satisfy the maximal equidistribution property. The WELL generator also has a good period length, no failures in Dieharder, SmallCrush and minimal failures in BigCrush. But it achieves the maximally equidistribution using the tempering for some k values [9]. In comparison to them, our proposed a light-weight combined CAs-based PRNGs that do not use any tempering, achieve periods close to the maximal one, are maximally equidistributed or almost maximally equidistributed and have comparable performance with majority of tests passed in all the testbeds.

5.2 Speed Test

The proposed combined CAs-based PRNGs are implemented in 32-bit and 64-bit computers. Then the performance is assessed by computing the execution time required to generate the 10^9 random numbers. This performance is compared with the existing linear generators. It is represented in Table 6. We see that, the speed is good but not the best as time spacing has been used in generating the numbers.

Table 6. Comparison of PRNGs based on CPU time to generate 10^9 random numbers

PRNG	CPU Time (s)
Combined CA-PRNG $(29, 32, 8)$	57
Combined CA-PRNG $(31, 32, 7)$	54
Combined CA-PRNG $(31, 32, 8)$	60
Combined CA-PRNG $(59, 64, 7)$	116
Combined CA-PRNG $(59, 64, 8)$	108
Combined CA-PRNG $(61, 64, 7)$	109
Combined CA-PRNG $(61, 64, 8)$	109
Combined CA-PRNG $(63, 64, 8)$	110
Mersenne Twister	116
Tausworthe (combined)	76
GFSR4	70
WELL512a	35
WELL1024a	42

6 Conclusion and Future Work

In this work, we first examine the theoretical characteristics of linear CAs-based PRNGs with sizes of 32 and 64 and observe that they lack in equidistribution. Then, we have designed light-weight, two-component combined CAs-based PRNGs with time spacing using linear maximal length CAs of sizes close to the computer word size (32 and 64). The proposed combined CAs-based PRNGs provide the period length $\approx 2^{k_1+k_2}$ and satisfy the theoretical quality measure (maximally equidistribution or almost maximally equidistribution). They also pass most of the tests in Dieharder and has minimal failures in TestU01 (SmallCrush and BigCrush), comparable to the existing state-of-the-art PRNGs. However, these PRNGs are still not at their fullest potential due to speed. Future work involves further efficient implementation of the PRNGs as well as finding alternative maximally equidistributed PRNGs that do not need time skipping.

Acknowledgments. The authors are grateful to Prof. Sukanta Das and all the mentors of the Indian Summer School on Cellular Automata 2025 for their continuous support, guidance and valuable feedback that shaped this work. This work is partially supported by Visvesvaraya PhD Scheme, Department of Electronics and Information Technology, Ministry of Communication and IT, Govt. of India.

References

1. L'ecuyer, P.: Maximally equidistributed combined Tausworthe generators. Math. Comput. **65**(213), 203–213 (1996)
2. Matsumoto, M., Nishimura, T.: Mersenne Twister: a 623-dimensionally equidistributed uniform pseudo-random number generator. ACM Trans. Model. Comput. Simul. (TOMACS) **8**(1), 3–30 (1998)
3. L'Ecuyer, P., Panneton, F.: F2-linear random number generators. In: Advancing the Frontiers of Simulation: A Festschrift in Honour of George Samuel Fishman, pp. 169–193. Springer, Boston (2009)
4. Tezuka, S., L'Ecuyer, P.: Efficient and portable combined Tausworthe random number generators. ACM Trans. Model. Comput. Simul. (TOMACS) **1**(2), 99–112 (1991)
5. Bhattacharjee, K., Das, S.: A search for good pseudo-random number generators: survey and empirical studies. Comput. Sci. Rev. **45**, 100047 (2022)
6. Adak, S., Das, S.: (Imperfect) strategies to generate primitive polynomials over GF (2). Theoret. Comput. Sci. **872**, 79–96 (2021)
7. Cattell, K., Zhang, S.: Minimal cost one-dimensional linear hybrid cellular automata of degree through 500. J. Electron. Test. **6**(2), 255–258 (1995)
8. Bhattacharjee, K., More, N., Singh, S.K., Verma, N.: Cellular automaton-based emulation of the Mersenne twister. Complex Syst. **32**(2) (2023)
9. Panneton, F., L'ecuyer, P., Matsumoto, M.: Improved long-period generators based on linear recurrences modulo 2. ACM Trans. Math. Softw. (TOMS) **32**(1), 1–16 (2006)
10. Marsaglia, G.: Xorshift rngs. J. Stat. Softw. **8**, 1–6 (2003)
11. Jaleel, H.A., Kaarthik, S., Sathish, S., Bhattacharjee, K.: Multiple-stream parallel pseudo-random number generation with cellular automata. In: International Workshop on Cellular Automata and Discrete Complex Systems, pp. 90–104. Springer, Cham (2023)

12. Bhattacharjee, K., Kumar, S.: Cellular automata based multiple stream parallel random number generator for 64-bit computing. In: Asian Symposium on Cellular Automata Technology, pp. 109–122. Springer, Cham (2024)
13. Hortensius, P.D., Mcleod, R.D., Pries, W., Miller, D.M., Card, H.C.: Cellular automata-based pseudorandom number generators for built-in self-test. IEEE Trans. Comput. Aided Des. Integr. Circuits Syst. **8**(8), 842–859 (2002)
14. Brown, R.G., Eddelbuettel, D., Bauer, D.: Dieharder Duke University Physics Department Durham (2018)
15. L'Ecuyer, P., Simard, R.: A software library in ANSI C for empirical testing of random number generators. Technical report, Département d'Informatique et de Recherche Opérationnelle Université de Montréal (2002)
16. Wolfram, S.: Origins of randomness in physical systems. Phys. Rev. Lett. **55**, 449–452 (1985)
17. Wolfram, S.: Random sequence generation by cellular automata. Adv. Appl. Math. **7**(2), 123–169 (1986)
18. Wolfram, S.: Cryptography with cellular automata. In: Williams, H.C. (ed.) CRYPTO 1985. LNCS, vol. 218, pp. 429–432. Springer, Heidelberg (1986). https://doi.org/10.1007/3-540-39799-X_32
19. Tomassini, M., Sipper, M., Zolla, M., Perrenoud, M.: Generating high-quality random numbers in parallel by cellular automata. Futur. Gener. Comput. Syst. **16**(2–3), 291–305 (1999)

Implementing the Morita Gate in an Asynchronous Cellular Automaton

Binmao Liu and Jia Lee(✉)

College of Computer Science, ChongQing University, Chongqing, China
lijia@cqu.edu.cn

Abstract. The Morita Gate is the simplest reversible logic element that can be used to realize all reversible sequential machines. In this paper, we show that this universal element can be efficiently implemented in an asynchronous cellular automaton, resulting in computational universality of the cellular automaton model.

1 Introduction

Reversible sequential machines (RSMs) offer a formal abstract model for reversible computing, which can be built out of reversible logic gates [2]. Unlike these logic gates, reversible logic elements (RLMs) that were exclusively proposed by Morita [6] enable each element to take a memory inside to record a certain number of states, whereby each element can be formulated in itself as an RSM. Especially, Morita [5] presented an RLM called Rotary Element which takes four input lines and four output lines, along with two (binary) states, and showed explicitly that this element is capable of constructing reversible Turing machines, and hence, it is computationally universal. After that, Ogiro et al. [8] showed that all non-degenerate elements with three input lines, three output lines and binary states are universal.

Further reduction of the inputs and outputs gives rise to RLMs with only two input lines, two output lines and binary states. After a thorough and meticulous screening, Mukai et al. [7] found that most of such elements are trivial, and only four of them have available logical functions. In particular, two elements are mutually inverse to each other and are able to realize any RSM according to a general design scheme [10]. Recently, Cook and Palmiere [1] demonstrated that among all of the four elements, the most active element in the sense that every input to the element will always switch its state, is able to construct all RSMs, thereby proving the element's universality. They call this element Morita Gate in memory of Morita's pioneering work on RLMs and reversible computing.

In general, circuits composed by reversible elements only allow a single signal to run around in a circuit at any time [6]. Such extremely serialized operation tends to remove the need of a central clock from the circuit for the sake of driving all elements to operate in parallel. Thus, the circuit can possibly operate in asynchronous mode, whereby at any time only the element receiving an input

H. Raju et al. (Eds.): ASCAT 2026, CCIS 2801, pp. 77–82, 2026.
https://doi.org/10.1007/978-3-032-18612-6_6

signal need to be activated, while all other elements may keep idle. This eminent characteristic can even facilitate the implementation of RLMs as well as RSMs composed by them into asynchronous cellular automata (ACAs) [3,4]. In this paper, we newly implement the Morita Gate into a self-timed cellular automaton (STCA [9]), a special type of ACAs in which each cell may be chosen at random and update the state independently from other cells. As a result, the new STCA model is capable of constructing reversible Turing machines into the cell space, whereby it holds universality in computations.

2 Morita Gate

A *reversible sequential machine* is a finite-state machine with output, which can be defined as follows:

$$M = (Q, \Sigma, \Gamma, \delta)$$

where Q is a non-empty finite set of states. Σ and Γ ($\Sigma \cap \Gamma = \emptyset$) are non-empty finite sets of input symbols and output symbols, respectively. Also, $\delta : \Sigma \times Q \rightarrow \Gamma \times Q$ is a bijective function called *transition function*. Because δ is bijective, it is reasonable to assume that $|\Sigma| = |\Gamma|$, i.e., an RSM takes an equal number of input and output symbols.

A *reversible logic element* is a module which consists of a finite number of input and output lines. Communication between an element and other elements is done via exchanging tokens through interconnection lines among them. Formally, an RLE can be defined as

$$(\mathcal{M}, I, O, \rho, \varrho, \Psi)$$

in which $\mathcal{M} = (Q, \Sigma, \Gamma, \delta)$ is an RSM. I and O are finite sets of input and output lines ($I \cap O = \emptyset$), respectively, and both $\rho : I \rightarrow \Sigma$ and $\varrho : O \rightarrow \Gamma$ are injective functions. In addition, $\Psi \subseteq I \times Q \times O \times Q$ is a set of operations. For any $a \in I$, $b \in O$, and $p, q \in Q$,

$$a, p \rightarrow b, q \in \Psi \implies \delta(\rho(a), p) = (\varrho(b), q).$$

The above operation $a, p \rightarrow b, q$ works as follows: When the RLE (RSM $\mathcal{M}$) is in state p and a token appears on its input line a, the element will assimilate the token from line a, then generates a token on output line b and changes the state to q.

A *Morita Gate* is one of the simplest RLEs defined as

$$(\mathcal{M}, \{T, T'\}, \{T_0, T_1\}, \rho_M, \varrho_M, \Psi_M)$$

which takes two input lines: T and T', two output lines: T_0 and T_1, and two states: 0 and 1. In particular, Ψ_M contains four operations as illustrated in Fig. 1(a). Assume a Morita Gate is in state $s \in \{0, 1\}$. A token arriving at input line T (resp. T') will be processed and give rise to a token on output line T_s (resp. T_{1-s}), followed by switching the element's state to $1 - s$ as demonstrated in Fig. 1(b).

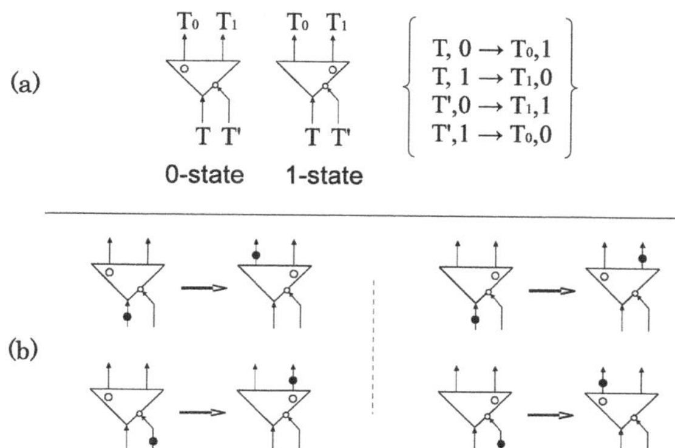

Fig. 1. (a) Morita Gate and its set of operations. (b) Morita Gate operating on a token arriving at one of the input lines. A token appearing on a line is denoted by a black blob on that line.

3 Embedding Morita Gate into STCA

A *self-timed cellular automaton* is a two-dimensional array of identical cells with von Neumann neighborhood. Each cell is partitioned into four sub-cells in one-to-one correspondence with its four nearest neighbors, and each sub-cell takes state either 0 or 1 at any time. In addition, at any time, only one cell that may be selected randomly from the cell space will undergo state transitions via a transition function $f:\ \{0,1\}^8 \rightarrow \{0,1\}^8$, which operate on the cell itself along with the nearest sub-cells of each of its four neighbors. Thus, a transition rule $f(n,w,s,e,u,l,d,r) = (n',w',s',e',u',l',d',e')$ can be depicted in the form of

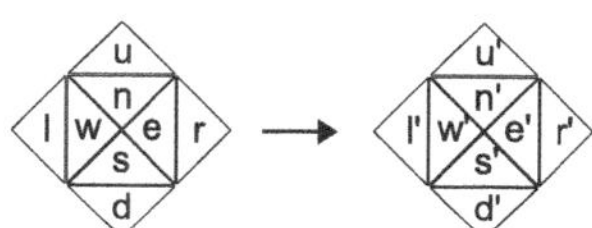

where $n,w,s,e,u,l,d,r,n',w',s',e',u',l',d',r' \in \{0,1\}$. Because the update of cells are timed randomly and independently of each other, a STCA is an ACA.

Figure 2 provides some basic patterns specifically designed for the STCA, including two local configurations representing the Morita Gate. In particular, the pattern in Fig. 2(a) represents a token that will be transferred to the right indicated by the arrow, driven by the rule $r1$ in Fig. 3. Also, the pattern in Fig. 2(b) can be used to change the direction of a token to the left or right, which is essential for laying out circuits and transmitting tokens in the two-dimensional cell space. Moreover, the local configurations in Fig. 2(c) and (c')

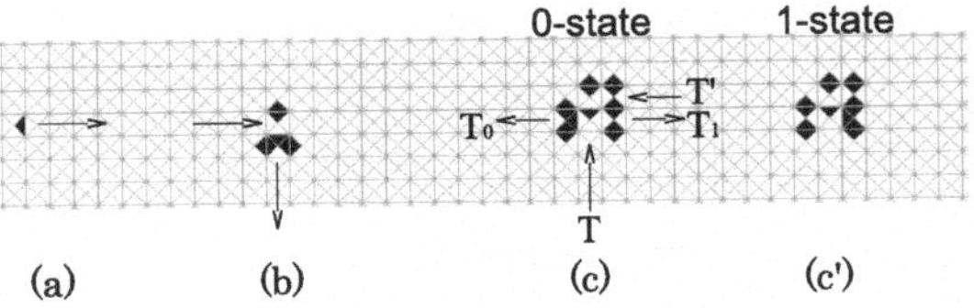

Fig. 2. (a) A token (signal). For simplicity, a sub-cell in state 1 is denoted by a filled triangle, while a sub-cell in state 0 is denoted by a blank. (b) A left or right turn element. Configurations representing a Morita Gate in (c) state 0 and (c') state 1, respectively.

represent a Morita Gate in state 0 and state 1, respectively. The update of all the patterns in Fig. 2 are controlled by the six transition rules given in Fig. 3, as demonstrated in Fig. 4.

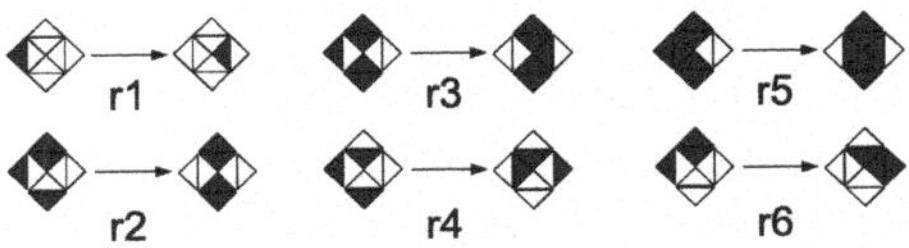

Fig. 3. Transition rules of STCA. The STCA satisfies rotational and reflective symmetry, such that rotating or flipping the lefthand and righthand sides of any rule simultaneously will generate an equivalent rule of the STCA. For simplicity, the rotational and reflective equivalents of each rule are omitted.

The so-called Subroutine Gadget in [1] is a useful module as illustrated in Fig. 5(a), along with its construction using the Morita Gate. This module can be easily characterized as an RLM with input lines: I_0, I_1 and D, output lines: O_0, O_1 and A, along with three states: λ, 0 and 1. In addition, its operation set include operations $I_s, \lambda \rightarrow s, A$ and $D, s \rightarrow O_s, \lambda$ where $s \in \{0, 1\}$. Thus, a token arriving on input line I_s will change the module's state from λ to s, outputting a token on line A to the outside. and after that a token receiving from input line D will generate a token on output line O_s, and revert the module's state to λ. This module, therefore, may possibly used as a memory to record temporarily where an input token comes from. Based on the circuit scheme in Fig. 5(a), Fig. 5(b) shows the construction of the memory module in the STCA using those local patterns in Fig. 2.

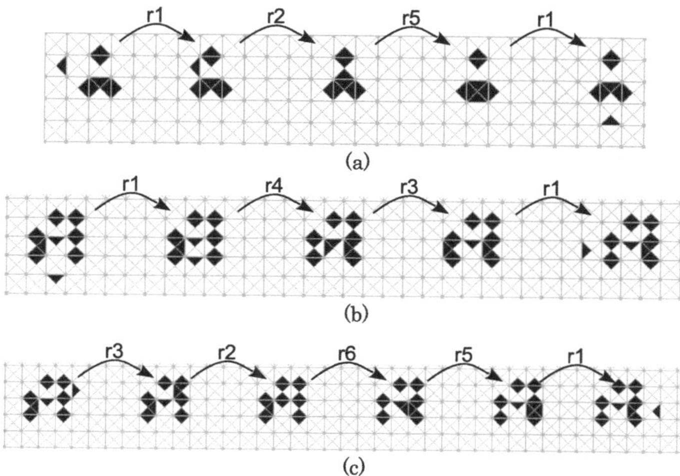

Fig. 4. (a) The turn element in Fig. 2 operating on a token coming from a horizontal line, which will transfer the token to the vertical line. Each arc indicates one transition step of cells, whereby its label indicates the transition rule in Fig. 3 used for the transition. Processes of the Morita Gate in state 0 operating on an input token receiving from the line (b) T and (c) T'. Similar processes when the element in state 1 can be verified.

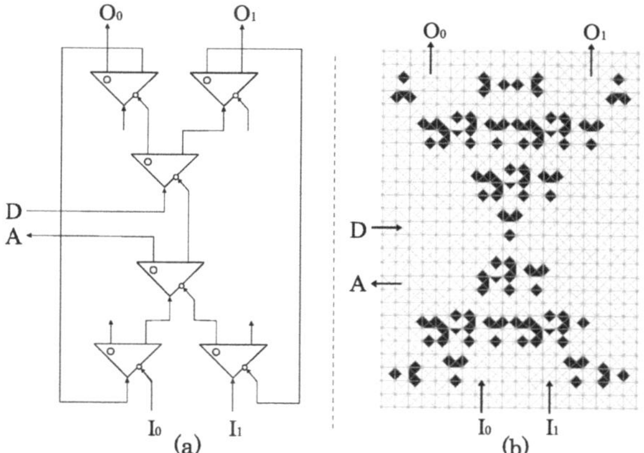

Fig. 5. (a) Memory module and its realization by Morita Gate [1]. (b) Configuration representing the memory in accordance with the circuit scheme on the left.

4 Conclusions

This paper presented a self-timed cellular automaton with six transition rules that can implement the reversible logic element: Morita Gate. The successful implementation of a universal RLM demonstrates our new STCA can accomplish universal computing via constructing reversible sequential machines, or even reversible Turing machines. On the other hand, unlike the STCA in [3] for implementing the Rotary Element [5], the STCA here is irreversible at the local scale, because the rules $r2$ and $r5$ in Fig. 3 have symmetrical righthand sides. Moreover, implementing a Rotary Element in the STCA needs only four rules, though it has twice the number of input and output lines as compared to the Morita Gate. Thus, how to reduce the number of transition rules, and especially how to design locally reversible transition functions for implementing reversible elements, would be an interesting problem for ACAs.

Acknowledgements. We are grateful to Prof. Katsunobu Imai at Fukuyama University, Japan, for the helpful suggestions.

References

1. Cook, M., Palmiere, E.: The Morita Gate is universal. In: Proceedings of the 22th International Conference on Unconventional Computation and Natural Computation, UCNC2005, Nice, France (2025)
2. Fredkin, E., Toffoli, T.: Conservative logic. Int. J. Theor. Phys. **21**(3–4), 219–253 (1982)
3. Lee, J., Peper, F., Adachi, S., Morita, K., Mashiko, S.: Reversible computation in asynchronous cellular automata. In: Calude, C.S., Dinneen, M.J., Peper, F. (eds.) UMC 2002. LNCS, vol. 2509, pp. 220–229. Springer, Heidelberg (2002). https://doi.org/10.1007/3-540-45833-6_19
4. Lee, J., Peper, F., Adachi, S., Morita, K.: An asynchronous cellular automaton implementing 2-state 2-input 2-output reversed-twin reversible elements. In: Umeo, H., Morishita, S., Nishinari, K., Komatsuzaki, T., Bandini, S. (eds.) ACRI 2008. LNCS, vol. 5191, pp. 67–76. Springer, Heidelberg (2008). https://doi.org/10.1007/978-3-540-79992-4_9
5. Morita, K.: A simple universal logic element and cellular automata for reversible computing. In: Margenstern, M., Rogozhin, Y. (eds.) MCU 2001. LNCS, vol. 2055, pp. 102–113. Springer, Heidelberg (2001). https://doi.org/10.1007/3-540-45132-3_6
6. Morita, K.: Reversible computing and cellular automata – a survey. Theor. Comput. Sci. **395**(1), 101–131 (2008)
7. Mukai, Y., Ogiro, T., Morita, K.: Universality problems on reversible logic elements with 1-bit memory. Int. J. Unconventional Comput. **10**(5), 353–373 (2014)
8. Ogiro, T., Kanno, A., Tanaka, K., Kato, H., Morita, K.: Nondegenerate 2-state 3-symbol reversible logic elements are all universal. Int. J. Unconventional Comput. **1**, 47–67 (2005)
9. Peper, F., Lee, J., Adachi, S., Mashiko, S.: Laying out circuits on asynchronous cellular arrays: a step towards feasible nanocomputers? Nanotechnology **14**, 469–485 (2003)
10. Tang, M.X., Lee, J., Morita, K.: General design of reversible sequential machines based on reversible logic elements. Theor. Comput. Sci. **568**, 19–27 (2015)

Developing NSRTD for 5-Neighborhood Periodic Boundary Cellular Automata

Som Banerjee(✉) and Mamata Dalui

Department of Computer Science and Engineering, National Institute of Technology, Durgapur, Durgapur 713209, West Bengal, India
sb.21cs1105@phd.nitdgp.ac.in, mdalui.cse@nitdgp.ac.in

Abstract. The 5-neighborhood cellular automaton (CA) has demonstrated significant effectiveness in modeling applications such as cryptography and random number generation, owing to its enhanced randomness and diffusion characteristics derived from the larger neighborhood size. However, most widely used CA characterization tools—such as the Rule Vector Graph (RVG), Explicit Rule Vector Graph (ERVG), Reachability Tree (RT), and Next State RMT Transition Diagram (NSRTD)—have primarily been developed for 3-neighborhood cellular automata (3N CAs). This creates a clear need for tools that can effectively support 5-neighborhood CAs.

To address this gap, this paper presents the development of NSRTD framework, specifically designed for 2-state, 5-neighborhood periodic boundary CA (5N PBCA). The proposed framework serves as a potential analytical tool for studying the spatio-temporal behavior of Rule Min Terms (RMTs) in 5N PBCA. Furthermore, algorithms detailing the step-by-step working procedure of the framework have been developed and analyzed. The proposed NSRTD framework effectively accommodates both uniform and nonuniform CAs.

Keywords: Cellular automata · Periodic boundary CA · NSRTD · Attractor · 5-neighborhood CA

1 Introduction

Cellular automaton (CA) was first introduced in the literature by John von Neumann [1]. Later, Stephen Wolfram made significant contributions to the theoretical foundation and development of Cellular Automata (CAs) [2]. One of his most influential works involved the formulation of Elementary Cellular Automata (ECA) and their classification framework, widely known as Wolfram's Classification [3]. This pioneering study inspired researchers across diverse fields to explore the structural and dynamic properties of CAs. Since then, numerous research studies have utilized ECA rules due to their ability to form basins of attractor. These attractors have been effectively employed in designing various CA-based applications, including cryptography [4], fault detection [5], pattern recognition [6], memory testing [7], and cache coherence verification [8], among others.

H. Raju et al. (Eds.): ASCAT 2026, CCIS 2801, pp. 83–100, 2026.
https://doi.org/10.1007/978-3-032-18612-6_7

Further, some works have been reported in the literature for modeling CA-based applications, like test pattern generator [9] and random number generator [10, 11] etc. by using 2-dimensional 5N CA. The authors in [9] demonstrated that the patterns generated by two-dimensional CA exhibit superior randomness compared to those produced by one-dimensional CA or Linear Feedback Shift Registers (LFSRs). However, it is also to be noted that for 2-dimensional 5 or higher-neighborhood CA, the hardware cost and the computational complexity is more than 1-dimensional CA.

The next state RMT transition diagram (NSRTD) is a graph based CA characterization tool which has been first reported in the literature in [12]. As NSRTD is capable of identifying both the self-loop(s) as well as multi-length cycle(s), it has emerged to be one of the potential CA characterization tool having the capability of characterizing diverse classes of CAs. Further, some works on CA characterization have been done using different CA characterization tools, like reachability tree [13, 14], rule vector graph [15, 16], explicit rule vector graph [7], and NSRTD [12, 17], etc. However, all such works are mainly focused on 3N CA. Although, only a few works are there in the literature that uses matrix algebra as a CA characterization tool for 5N CA [18, 19], most of the other CA characterization tools described above are mainly developed for 3N CA and for 5 or extended-neighborhood CA, "run and watch" is the only option to investigate the behavior and dynamics of a CA [20].

From the literature review, it is evident that there is an urgent need for a tool that can support 5N CAs. This motivates us to pursue research for developing the NSRTD framework for 2-state 5N PBCA, that can be used as a specialized tool to study the spatio-temporal behavior of RMTs of a 5N PBCA. To achieve this, we have designed six algorithms which demonstrate the step-wise procedure for the construction of the NSRTD framework for 2-state 5N PBCA. The proposed NSRTD framework has been designed to work for both uniform as well as nonuniform PBCA.

2 Cellular Automata

CA is an arrangement of cells in the form of lattice structure. A CA evolves in discrete time and space. During evolution, the present state (PS) of each CA cell i is updated to a next state (NS) based upon a next state function expressed as $f_i(S^t_{i-2}, S^t_{i-1}, S^t_i, S^t_{i+1}, S^t_{i+2})$ where S^t_i, S^t_{i+1}, S^t_{i+2}, S^t_{i-1} and S^t_{i-2} are the present states at time t of the ith CA cell, its first right, second right, first left and second left neighbors, respectively. Truth table (Table 1) depicts next state function f_i. The decimal equivalent of the 32 outputs of f_i is referred to as rule $\mathcal{R}$. Here, four rules – 3275539260, 90960, 1010580540, 15 – have been illustrated in Table 1. Table 1 is split into two sub-tables for ease of visibility, where the sub-table 1 illustrates the NS of RMTs 16–31, and the sub-table 2 illustrates the NS of RMTs 0–15, for each of the four rules $\mathcal{R}$.

In a PBCA, the rightmost and leftmost cells are adjacent neighbors to each other as shown in Fig. 1. The combination of the present states (i.e. S^t_{i-2}, S^t_{i-1},

Table 1. RMTs and their corresponding NS based on rule $\mathcal{R}$. (sub-table 1)

PS	11111	11110	11101	11100	11011	11010	11001	11000	10111	10110	10101	10100	10011	10010	10001	10000
RMT	$T(31)$	$T(30)$	$T(29)$	$T(28)$	$T(27)$	$T(26)$	$T(25)$	$T(24)$	$T(23)$	$T(22)$	$T(21)$	$T(20)$	$T(19)$	$T(18)$	$T(17)$	$T(16)$
NS	1	1	0	0	0	0	1	1	0	0	1	1	1	1	0	0
NS	0	0	0	0	0	0	0	0	0	0	0	0	0	0	0	1
NS	0	0	1	1	1	1	0	0	0	0	1	1	1	1	0	0
NS	0	0	0	0	0	0	0	0	0	0	0	0	0	0	0	0

Table 1. RMTs and their corresponding NS based on rule $\mathcal{R}$. (sub-table 2)

PS	01111	01110	01101	01100	01011	01010	01001	01000	00111	00110	00101	00100	00011	00010	00001	00000	Rule $\mathcal{R}$
RMT	$T(15)$	$T(14)$	$T(13)$	$T(12)$	$T(11)$	$T(10)$	$T(9)$	$T(8)$	$T(7)$	$T(6)$	$T(5)$	$T(4)$	$T(3)$	$T(2)$	$T(1)$	$T(0)$	
NS	1	1	0	0	0	0	1	1	0	0	1	1	1	1	0	0	3275539260
NS	0	1	1	0	0	0	1	1	0	1	0	1	0	0	0	0	90960
NS	0	0	1	1	1	1	0	0	0	0	1	1	1	1	0	0	1010580540
NS	0	0	0	0	0	0	0	0	0	0	0	0	1	1	1	1	15

S_i^t, S_{i+1}^t, S_{i+2}^t for a 5N CA) create rule min term (RMT) as illustrated in first row of Table 1. The RMTs are expressed as $T(x)$, where $x = \{0, 1, ..., 30, 31\}$. For space-time representation of a CA during evolution, a RMT is expressed by $T_i^t(x)$, where i is the CA cell index and t is the time (i.e., $T_4^t(31)$ represents cell-4 RMT $T(31)$ at time t). However, for illustration within figures, RMTs are expressed as decimal number x (i.e., $T_i^t(31)$ is expressed as 31). The sequence of RMTs appearing in a CA state is called RMT string RS, represented as T_1, ..., T_{i-2}, T_{i-1}, T_i, T_{i+1}, T_{i+2}, ..., T_n for an n-cell CA, where $T_i \in \{T(0), T(1), ..., T(31)\}$. For 5N PBCA, a 6-cell CA configuration 110011 is represented by RS (T_1, T_2, T_3, T_4, T_5, T_6) = ($T(30)$, $T(28)$, $T(25)$, $T(19)$, $T(7)$, $T(15)$).

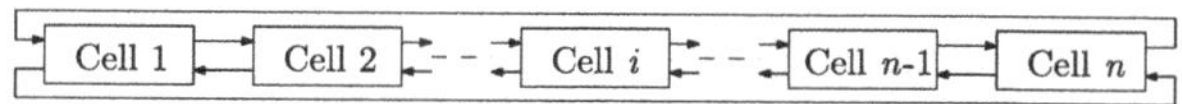

Fig. 1. Periodic Boundary CA

A set of states that forms cycles during a CA state transition is called an attractor. An attractor formed with self-loop (i.e., 0→0, or 10→10 of Fig. 2) is called single-length cycle attractor (also called fixed point attractor). On the other hand, an attractor formed with multiple states (i.e., 1 → 11 → 1, 4 → 14 → 4, 2 → 3 → 8 → 12 → 2 of Fig. 2) is known as multi-length cycle attractor.

3 Proposed Work

This section reports the development of NSRTD framework for 5N CA, motivated by both modeling capability and theoretical necessity. Unlike 3N CA, which can only capture two immediate nearest neighbors, this 5N CA can incorporate a broader spatial context, enabling the modeling of longer-range dependencies, richer local correlations, and more complex dynamical behaviors that

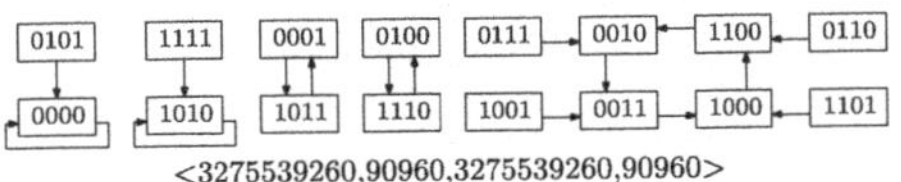

Fig. 2. State transition of PBCA

cannot be modeled in 3N CA. The number of RMTs increases exponentially from 8 in 3N CA to 32 in 5N CA, leading to the increased rule space, thereby allowing the emergence of a large set of NSRs, non-trivial NSRSs (cycles), complex NSRS graphs, and complex compatibility structures of NSRSs, which are otherwise absent or limited in 3N CA. The 3N NSRTD framework may provide a foundational insight, but it cannot adequately evaluate the framework's scalability with increased neighborhood complexity. In this context, a 5N CA serves as a critical intermediate step for understanding how the NSRTD framework can behave under increased neighborhood complexity, which leads to increased combinatorial and structural complexities, without becoming computationally intractable. Many real-world systems, e.g., biological networks, distributed computation models, and error-resilient systems, exhibit interactions extending beyond immediate neighbors. Hence, a 5N CA can offer a more realistic abstraction, thereby making theoretical developments more applicable to practical domains.

3.1 NSRTD Terminologies for 5N PBCA

The NSRTD is a graph based tool that has an inherent capability to identify the self-loops(s) and multi-length cycle(s) formed during the dynamic evolution of a CA. NSRTD can provide a detailed perspective of the spatio-temporal relationship of RMTs of a CA by using the NCR relationship and NSR relationship, respectively. This ensures an in-depth analysis about the properties and structure of a CA, that are essential for modeling different real-life CA based applications. This work is the first to address the issue for developing a NSRTD framework for 5N PBCA. Some of the important NSRTD terminologies that has been developed for 5N PBCA are discussed below:

If we consider T_i^t to be the RMT of a CA cell i at time t, which changes its state based on rule $\mathcal{R}_i$, then T_{i+1}^t will be the next cell RMT (NCR) and is denoted by $T_{i+1}^t = (2\times T_i^t \bmod 32)$, $((2\times T_i^t{+}1) \bmod 32)$. For example, if T_i^t is $T(1)$, then the corresponding NCRs of $T(1)$ will be $T(2)$ (i.e., $2 \times 1 \bmod 32 = 2$) and $T(3)$ (i.e., $2 \times 1 + 1 \bmod 32 = 3$). Similarly, if T_i^t is $T(17)$, then the corresponding NCRs of $T(17)$ will be $T(2)$ (i.e., $2 \times 17 \bmod 32 = 2$) and $T(3)$ (i.e., $2 \times 17 + 1 \bmod 32 = 3$). In Table 2, columns 1, 3, 5, and 7 represent the 32 RMTs, and their corresponding NCRs are represented in columns 2, 4, 6 and 8, respectively.

If we denote T_i^t as the RMT of a CA cell i at time t, whose state evolves according to rule $\mathcal{R}_i$, then T_i^{t+1} is represented as the next state RMT (NSR) of T_i^t. For 2-state 5N PBCA, we consider *abcde* as RMT corresponding to rule $\mathcal{R}_i$

Table 2. RMTs and their corresponding Next Cell RMTs (NCRs)

RMTs (1)	Corresponding NCRs (2)	RMTs (3)	Corresponding NCRs (4)	RMTs (5)	Corresponding NCRs (6)	RMTs (7)	Corresponding NCRs (8)
T(0)	T(0), T(1)	T(8)	T(16), T(17)	T(16)	T(0), T(1)	T(24)	T(16), T(17)
T(1)	T(2), T(3)	T(9)	T(18), T(19)	T(17)	T(2), T(3)	T(25)	T(18), T(19)
T(2)	T(4), T(5)	T(10)	T(20), T(21)	T(18)	T(4), T(5)	T(26)	T(20), T(21)
T(3)	T(6), T(7)	T(11)	T(22), T(23)	T(19)	T(6), T(7)	T(27)	T(22), T(23)
T(4)	T(8), T(9)	T(12)	T(24), T(25)	T(20)	T(8), T(9)	T(28)	T(24), T(25)
T(5)	T(10), T(11)	T(13)	T(26), T(27)	T(21)	T(10), T(11)	T(29)	T(26), T(27)
T(6)	T(12), T(13)	T(14)	T(28), T(29)	T(22)	T(12), T(13)	T(30)	T(28), T(29)
T(7)	T(14), T(15)	T(15)	T(30), T(31)	T(23)	T(14), T(15)	T(31)	T(30), T(31)

depending on which the state of CA cell i changes, and l_1, l_2, r_1, and r_2 are the present states of the neighboring cells (i-3), (i-4), (i+3), and (i+4), respectively. Then the corresponding next state is represented by $a'b'c'd'e'$, where c' = next state of RMT $abcde$ of cell i using rule $\mathcal{R}_i$, a' = next state of RMT l_2l_1abc of cell (i-2) using rule $\mathcal{R}_{i-2}$, b' = next state of RMT l_1abcd of cell (i-1) using rule $\mathcal{R}_{i-1}$, d' = next state of RMT $bcder_1$ of cell (i+1) using rule $\mathcal{R}_{i+1}$, and e' = next state of RMT $cder_1r_2$ of cell (i+2) using rule $\mathcal{R}_{i+2}$, respectively. It is to be noted that for 5N PBCA, l_1, l_2, r_1, and r_2 can take any value, either 0 or 1.

The next state RMT sequence (expressed by NSRS) refers to the NSR's sequences, based on which the state change of a CA cell occurs during its state transition. For a CA cell i, NSRS can be represented by T_i^0, ..., T_i^{t-1}, T_i^t, T_i^{t+1}, ..., T_i^n (here, T_i^t is the NSR of T_i^{t-1}).

The next state RMT sequence graph (NSRS-G) is a directed graph formed by merging all possible NSRSs of a CA cell, where each of the vertex denotes a RMT and the edge denotes a NSR.

NSRSs of two adjacent CA cells (cell i and cell (i+1), respectively) are said to form compatible class of NSRSs, iff all the RMTs involved in the cycle (i.e., self-loop(s) and/or multi-length cycle(s)) of cell (i+1) satisfies NCR relationship with the RMTs involved in the cycle (i.e., self-loop(s) and/or multi-length cycle(s)) of cell i.

Further, the next state RMT transition diagram (NSRTD) is expressed by using a 2-dimensional directed graph, where one dimension depicts the NCR relation (i.e., spatial relationships) between two consecutive CA cells and the other dimension depicts the NSR relation (i.e., temporal relationships) of the RMTs of CA cell. The NSRTD is always created by utilizing the compatibility class of NSRSs.

3.2 Proposed NSRTD Framework for 5N PBCA

This subsection presents the NSRTD framework for 5N PBCA. We have designed six algorithms (Algorithms 1–6) that formalize the step-wise procedures for constructing the framework. The proposed NSRTD framework can work for both uniform and nonuniform CA.

The inputs of Algorithm 1 are rule vector RV, NCRs for RMTs, length of CA (n), and a uniformity flag Z, where $Z = 0$ indicates uniform CA, and $Z = 1$ indicates nonuniform CA. The outputs of Algorithm 1 are the set of NSRTD graph(s) with self-loop NSRSs only ($NSRTD_{SL}$), which corresponds to fixed point attractor(s), and the set of NSRTD graph(s) with self-loop and/or multi-length cycle NSRSs ($NSRTD_{MLC}$), which corresponds to multi-length cycle attractor(s).

In step 1, initialization of CA cell index set I is done, where, for uniform CA (Z=0), I consist of element 1, and for nonuniform CA (Z=1), I consist of elements 1 to n. In step 2, for each element i (CA cell index) of set I, steps 3–5 are executed. In step 3, the set of NSRs of a CA cell i (NSR_SET_i) is identified by calling Algorithm 2 and passing the parameters RV, and i.

Algorithm 2 takes as input RV, and i and generates NSR_SET_i as output. To identify NSR in 5N PBCA, we represent a RMT configuration as a 9-bit tuple $(l_2, l_1, abcde, r_1, andr_2)$ denoting a 5-cell RMT along with its two neighbors in each side. We denote $abcde$ set as A, l_1 set as L_1, l_2 set as L_2, r_1 set as R_1, and r_2 set as R_2. In step 1, initialization of sets A, L_2, L_1, R_1, and R_2 are done. Step 2 computes the Cartesian product of the sets A, L_2, L_1, R_1, and R_2 represented as Ω, where $\Omega = L_2 \times L_1 \times A \times R_1 \times R_2 = \{0, 1\}^9$ and $|\Omega| = 512$. In step 3, NSR_SET_i is initialized as empty set. In step 4, for each element $(l_2, l_1, abcde, r_1, r_2)$ of set Ω, steps 5–6 are executed. In step 5, five local update rules $\mathcal{R}_{i-2}$, $\mathcal{R}_{i-1}$, $\mathcal{R}_i$, $\mathcal{R}_{i+1}$, $\mathcal{R}_{i+2}$ are applied on five RMTs l_2l_1abc, l_1abcd, $abcde$, $bcder_1$, $cder_1r_2$ to create the respective next states a', b', c', d', e' (refer Table 1). In step 6, a NSR is created by a transition ($\rightarrow$) from decimal equivalent of RMT $abcde$ to RMT $a'b'c'd'e'$ (represented as dec($abcde$) $\rightarrow$ dec($a'b'c'd'e'$), where dec(x) represents the decimal equivalent of the binary x), and it is added to NSR_SET_i by union operation. Finally, NSR_SET_i is returned in step 7.

In step 4 of Algorithm 1, the NSRS graph of a CA cell i ($NSRS\text{-}G_i$) is created by calling Algorithm 3 and passing the parameter NSR_SET_i.

Algorithm 3 takes as input set of NSRs (NSR_SET) and generates NSRS graph ($NSRS\text{-}G$) as output. Step 1 initializes an empty directed graph $NSRS\text{-}G$ with vertex set V and edge set E. In step 2, for each NSR of NSR_SET, steps 3–4 are executed. Here, a NSR in NSR_SET is represented as $u \rightarrow v$, where u represents the present state RMT, v is the next state RMT, and '$\rightarrow$' represents the NSR transition. In step 3, $u \rightarrow v$ are examined, and the vertices are added to the vertex set V. Similarly, in step 4, $u \rightarrow v$ are examined, and the edge is added to the edge set E. Finally, after examining all the $u \rightarrow v$ (NSR), the $NSRS\text{-}G$ is completely formed and in step 5, it is returned as output of Algorithm 3.

In step 5 of Algorithm 1, the set of cycle(s) of a CA cell i (CYC_i) is identified by calling Algorithm 4 and passing the parameter $NSRS\text{-}G_i$.

Algorithm 4 takes as input NSRS graph ($NSRS\text{-}G$) and generates the cycle set CYC as output. In step 1, the cycle set CYC is initialized as an empty set. In step 2, all the cycle(s) (i.e., self-loop(s) and multi-length cycle(s)) are identified from the $NSRS\text{-}G$. In steps 3–4, each cycle CY identified in step 2 is added to CYC. Finally, in step 5, CYC is returned as the output of Algorithm 4.

In step 6 of Algorithm 1, the set of compatibility class ($\{C\}$) is initialized as an empty set. In step 7, the uniformity flag (Z) is checked, and if the CA is uniform, steps 8–9 are executed. Otherwise, the CA is nonuniform, and steps 10–15 are executed. In step 8, $\{C\}$ (i.e., set of compatibility class $C_{i(i+1)}$, $\forall_{i=1}^{n-1}$

Algorithm 1: NSRTD_Framework_5N_PBCA

Input : Rule vector *RV*, NCRs for RMTs, length of CA (n), uniformity flag Z=0/1 (0 for uniform CA, 1 for nonuniform CA)
Output: Set of NSRTD graph(s) with self-loop NSRSs only ($NSRTD_{SL}$) and set of NSRTD graph(s) with self-loop and/or multi-length cycle NSRSs ($NSRTD_{MLC}$)

Index set $I = \{1\}$ if $Z = 0$, else $I = \{1, 2, \ldots, n\}$
foreach $i \in I$ **do**
  $NSR\_SET_i \leftarrow$ Calculate_NSR_SET(RV, i);
  $NSRS\text{-}G_i \leftarrow$ Construct_NSRS-G(NSR_SET_i);
  $CYC_i \leftarrow$ Extract_Cycles($NSRS\text{-}G_i$);
end
$\{C\} \leftarrow \emptyset$;
if $Z = 0$ **then**
  $\{C\} \leftarrow$ Find_Compatibility(CYC_1, CYC_1);
  $\{S_u\} \leftarrow$ Generate_RMT_Sequences($\{C\}, n$);
else
  for $i = 1$ *to* $n - 1$ **do**
    $C_{i(i+1)} \leftarrow$ Find_Compatibility(CYC_i, CYC_{i+1});
    $\{C\} \leftarrow \{C\} \cup C_{i(i+1)}$;
  end
  $C_{n1} \leftarrow$ Find_Compatibility(CYC_n, CYC_1);
  $\{C\} \leftarrow \{C\} \cup C_{n1}$;
  $\{S_u\} \leftarrow$ Generate_RMT_Sequences($\{C\}, n$);
end
if $\{S_u\} = \emptyset$ **then**
  Exit
else
  foreach $S_u \in \{S_u\}$ **do**
    if *all transitions in* S_u *are self-loops* **then**
      $NSRTD_{SL} \leftarrow S_u$
    else
      $NSRTD_{MLC} \leftarrow S_u$
    end
  end
end

Algorithm 2: Calculate_NSR_SET

Input : Rule vector *RV*, CA cell index i
Output: Set of NSRs NSR_SET_i

Initialize $A = \{0,1\}^5$, $L_2 = L_1 = R_1 = R_2 = \{0,1\}$;
$\Omega \leftarrow L_2 \times L_1 \times A \times R_1 \times R_2$;
$NSR\_SET_i \leftarrow \emptyset$;
foreach $(l_2, l_1, abcde, r_1, r_2) \in \Omega$ **do**
  $(a', b', c', d', e') \leftarrow$
  $(\mathcal{R}_{i-2}(l_2 l_1 abc), \mathcal{R}_{i-1}(l_1 abcd), \mathcal{R}_i(abcde), \mathcal{R}_{i+1}(bcder_1), \mathcal{R}_{i+2}(cder_1 r_2))$;
  $NSR\_SET_i \leftarrow NSR\_SET_i \cup \{dec(abcde) \rightarrow dec(a'b'c'd'e')\}$;
end
return NSR_SET_i

and compatibility class C_{n1}) is identified by calling Algorithm 5 and passing same CYC_1 as the first and second parameters. Here, for uniform CA, the compatibility class between the first and second CA cells, i.e., C_{12}, is identified using the same CYC_1 and used to represent any $C_{i(i+1)}$ $\forall_{i=1}^{n-1}$ or C_{n1}.

Algorithm 5 takes as input cycle set CYC_i and CYC_{i+1} and generates the compatibility class $C_{i(i+1)}$ for CA cell index i. In step 1, set $C_{i(i+1)}$ is initialized as empty set. In step 2, for each cycle CYC_{ia} of cycle set CYC_i, steps 3–5 are

Algorithm 3: Construct_NSRS-G

Input : Set of next state RMTs NSR_SET
Output: NSRS graph ($NSRS\text{-}G$)

Initialize an empty directed graph $NSRS\text{-}G = (V, E)$;
foreach $(u \rightarrow v) \in NSR\_SET$ **do**
    $V \leftarrow V \cup \{u, v\}$;
    $E \leftarrow E \cup \{(u, v)\}$;
end
return $NSRS\text{-}G$

Algorithm 4: Extract_Cycles

Input : NSRS graph ($NSRG\text{-}G$)
Output: Set of cycles CYC

$CYC \leftarrow \emptyset$;
Identify all cycle(s) (self-loop(s) and multi-length cycle(s)) from the $NSRG\text{-}G$;
foreach *cycle* CY *identified* **do**
    $CYC \leftarrow CYC \cup CY$;
end
return CYC

Algorithm 5: Find_Compatibility

Input : Cycle sets CYC_i and CYC_{i+1}
Output: Compatibility class $C_{i(i+1)}$

1 $C_{i(i+1)} \leftarrow \emptyset$;
2 **foreach** *cycle* $CYC_{ia} \in CYC_i$ **do**
3   **foreach** *cycle* $CYC_{(i+1)b} \in CYC_{i+1}$ **do**
4   **if** CYC_{ia} *is compatible with* $CYC_{(i+1)b}$ **then**
5             $C_{i(i+1)} \leftarrow C_{i(i+1)} \cup \{(CYC_{ia}, CYC_{(i+1)b})\}$;
        end
    end
end
6 **return** $C_{i(i+1)}$

Algorithm 6: Generate_RMT_Sequences

Input : Set of compatibility class $\{C\}$, length of CA n.
Output: Set of S_us $\{S_u\}$.

1 $\{S_u\} \leftarrow \emptyset$;
2 Generate each unique RMT sequence $S_u = S_{1u} \rightarrow S_{2u} \rightarrow \cdots \rightarrow S_{(n-1)u} \rightarrow S_{nu}$ following step 3,
3 **if** $(S_{iu}, S_{(i+1)u}) \in C_{i(i+1)}, \forall_{i=1}^{n-1}$ *and* $(S_{nu}, S_{1u}) \in C_{n1}$; **then**
4     $\{S_u\} \leftarrow \{S_u\} \cup S_u$
end
5 **return** $\{S_u\}$

executed. In step 3, for each cycle $CYC_{(i+1)b}$ of cycle set CYC_{i+1}, steps 4–5 are executed. In step 4, if cycle CYC_{ia} is compatible with cycle $CYC_{(i+1)b}$, then step 5 adds the compatible pair $(CYC_{ia}, CYC_{(i+1)b})$ in $C_{i(i+1)}$. So, in steps 2–4, all the cycle(s) of two consecutive CA cells (i and $i+1$) are examined with each other for a possible compatible pair. In step 6, $C_{i(i+1)}$ is returned as output of Algorithm 5.

In step 9 of Algorithm 1, the RMT sequence set ($\{S_u\}$) is generated by calling Algorithm 6 and passing the parameters, the set of compatibility class ($\{C\}$), and the length of CA (n).

Algorithm 6 takes as input $\{C\}$, and n and generates the set of unique RMT sequences $\{S_u\}$ as output. As reported earlier, $\{C\}$ is $(\{C_{i(i+1)}, \forall_{i=1}^{n-1}\} \cup C_{n1})$. In step 1, $\{S_u\}$ is initialized as an empty set. Step 2 generates each unique RMT sequence S_u following the two conditions in step 3. Step 3 states that each (S_{iu} and $S_{(i+1)u}$) in a S_u must be formed by using a compatible pair from $C_{i(i+1)}$, where i runs from 1 to $n-1$ for an n-cell CA, and also S_{nu} and S_{1u} in a S_u must be formed by using a compatible pair from C_{n1}. Further, if conditions of step 3 are satisfied, the n-length S_u generated are added to $\{S_u\}$ in step 4. Step 5 returns the $\{S_u\}$.

Next, in step 10 (in case of nonuniform CA), for each CA cell i from 1 to $n-1$, steps 11–12 are executed. In step 11, the compatibility class $C_{i(i+1)}$ between two consecutive CA cells i and $i+1$ are identified by calling Algorithm 5 and passing CYC_i, and CYC_{i+1} as parameters. The working procedures of Algorithm 5 have already been described in step 8. In step 12, each $C_{i(i+1)}$ identified in step 11 are added to $\{C\}$. In step 13, the compatibility class C_{n1} between the last and the first CA cells is identified by calling Algorithm 5 and passing CYC_n, and CYC_1 as parameters. Step 14 adds C_{n1} identified in step 13 to $\{C\}$. Step 15 is the same as step 9 as discussed before. Further, in step 16, if $\{S_u\}$ is the empty set, the algorithm terminates in step 17. Otherwise, in step 18, for each S_u in $\{S_u\}$, steps 19–21 are executed. Step 19 examines whether all S_{iu} in S_u are made up of self-loop NSRSs only. If so, then the S_u is added to $NSRTD_{SL}$ in step 20. Otherwise, in step 21, the S_u is added to $NSRTD_{MLC}$.

Analysis: Step 1 of Algorithm 1 initializes the index set I with 1 for uniform CA, and it takes constant time. For nonuniform CA, I is initialized with 1 to n values, and it takes O(n) time. Each individual operation of steps 3–5 is executed once for uniform CA. Otherwise, for a nonuniform CA, the operation of steps 3–5 is executed n times for an n-cell CA.

In step 3, Algorithm 2 is called to identify NSR_SET_i. In Algorithm 2, step 1 initializes the sets A, L_2, L_1, R_1, and R_2 in constant time. Step 2 performs a Cartesian product (Ω) of sets A, L_2, L_1, R_1, and R_2, and as the cardinality of Ω is 512, it takes constant time (independent of CA length n). Step 3 initializes NSR_SET_i with an empty set in constant time. Step 4 runs for 512 times to perform operations in steps 5–6. In step 5, five local update rules are applied over five RMTs to create their next states, and this takes constant time. In step 6, an NSR is identified and added to NSR_SET_i in constant time. Step 7 returns NSR_SET_i. So, the overall time to run Algorithm 2 becomes constant (independent of n). So, step 3 of Algorithm 1 takes constant time for uniform CA and O(n) time for an n-cell nonuniform CA.

Step 4 identifies $NSRS\text{-}G_i$ by calling Algorithm 3. In Algorithm 3, step 1 initializes an empty set $NSRS\text{-}G$ in constant time. Step 2 runs for each NSR in NSR_SET (512 NSRs) to perform operations in steps 3–4. In step 3, the vertices of an NSR are added to V in constant time. Similarly, in step 4, an edge of an NSR is added to E in constant time. Step 5 returns *NSRS-G*. So, the overall time to run Algorithm 3 becomes constant (independent of n). Hence, step 4 of Algorithm 1 takes constant time for uniform CA and $O(n)$ time for an n-cell nonuniform CA.

Step 5 identifies the cycle set CYC_i by calling Algorithm 4. In Algorithm 4, step 1 initializes CYC with an empty set in constant time. Step 2 identifies all the cycle(s) from the *NSRS-G*. There can be a maximum of 32 self-loops corresponding to the 32 RMTs of a CA cell, and the count of multi-length cycles is also finite and independent of CA length n, thereby taking constant time. Since the total count of cycles is finite and independent of n, in steps 3–4, each identified cycle is added to CYC in constant time. Step 5 returns CYC. So, the overall time to run Algorithm 4 becomes constant (independent of n). So, step 5 of Algorithm 1 takes constant time for uniform CA and $O(n)$ time for an n-cell nonuniform CA.

In step 6, $\{C\}$ is initialized as an empty set, and it takes constant time. Step 7 checks the uniformity flag Z to verify whether the CA is uniform or nonuniform, in constant time.

In step 8, $\{C\}$ is identified by calling Algorithm 5. Step 1 of Algorithm 5 initializes $C_{i(i+1)}$ with an empty set in constant time. Step 2 runs for each cycle in CYC_i, and step 3 runs for each cycle in CYC_{i+1}. So, in steps 2–3, each cycle of CYC_i is checked with each cycle of CYC_{i+1} to perform the operation specified in steps 4–5. Since the total cycles in CYC_i and CYC_{i+1} are finite, steps 2–3 also take constant time (independent of n). In step 4, the two cycles CYC_{ia} and $CYC_{(i+1)b}$ are checked for compatibility (i.e., verifying that all the RMTs of cycle 2 satisfy the NCR relationship with all the RMTs of cycle 1). There can be a maximum of 32 RMTs involved in a cycle, and for each individual RMT, there can be a maximum of two out-degree edges (corresponding to its NCRs). So, in step 4, the time to find a compatible pair is independent of n, and takes a constant time. Step 5 adds each compatible pair $(CYC_{ia}, CYC_{(i+1)b})$ in $C_{i(i+1)}$ in constant time. Step 6 returns $C_{i(i+1)}$. So, the overall time to run Algorithm 5 also becomes constant (independent of n). So, step 8 of Algorithm 1 takes constant time.

In step 9, $\{S_u\}$ is identified by calling Algorithm 6. Step 1 of Algorithm 6 initializes $\{S_u\}$ with an empty set in constant time. Step 2 generates each S_u by following the conditions of step 3. Step 3 states that each (S_{iu} and $S_{(i+1)u}$) in a S_u must be formed by using a compatible pair from $C_{i(i+1)}$, where i runs from 1 to $n-1$ for an n-cell CA, and also S_{nu} and S_{1u} must be formed by using a compatible pair from C_{n1}. Further if conditions of step 3 are satisfied, the n-length S_u generated are added to $\{S_u\}$ in step 4. For an n-cell CA, generation of each S_u comprising of self-loop NSRSs only, takes $O(n)$ time. The total number of S_u is σ ($\sigma = \sum S_u$), which depends on CA configured with rule vector *RV*. The upper bound of σ is 2^n (e.g., a uniform CA configured with rule 4042322160), and the lower bound of σ is 1 (e.g., a uniform CA configured with rule 40). So, in steps 2–4, the overall complexity to generate $\{S_u\}$ comprising of self-loop NSRSs only, takes $O(n) * O(\sigma) = O(n\sigma)$ time. On the other hand, generation of a S_u comprising of multi-length cycle(s) NSRSs for n-cell CA takes $O(n) * O(\eta) = O(n\eta)$ time, where η represents the total count of RMT strings involved in the creation of the RMT sequence S_u. The lower bound of η is 2, i.e., a S_u (involving multi-length cycle NSRSs) with two RMT strings, which complies with a state transition diagram having a multi-length cycle involving two states. The upper bound of η is 2^n, i.e., a S_u (involving multi-length cycle NSRSs) with 2^n number of RMT strings, which complies with a state transition diagram having a multi-length cycle involving all the 2^n states. Step 5 returns the $\{S_u\}$.

Step 10 runs for $n - 1$ times to perform the operations of steps 11–12. In step 11, $C_{i(i+1)}$ is determined by calling Algorithm 5. The overall time to run Algorithm 5 is constant (independent of n), which has been discussed in step 8. Since step 11 executes $n - 1$ times, it takes $O(n - 1)$ time. Step 12 adds each $C_{i(i+1)}$ in $\{C\}$ in constant time. Since step 12 also executes $n - 1$ times, it also takes $O(n - 1)$ time. In step 13, C_{n1} is determined by calling Algorithm 5. Since Algorithm 5 takes constant time to run, step 13 also takes constant time. Step 14 adds C_{n1} in $\{C\}$ in constant time. The time complexity of step 15 is exactly the same as step 9, which has already been discussed earlier. Step 16 verifies if $\{S_u\}$ is an empty set in constant time. Finally, in steps 18–21, each S_u in $\{S_u\}$ are examined and grouped appropriately into $NSRTD_{SL}$ or $NSRTD_{MLC}$ set in $O(n\sigma)$ or $O(n\eta)$ time, respectively.

Example 1. To demonstrate the working of Algorithms 1–6, we have taken an example scenario where Algorithm 1 takes as input a rule vector <3275539260, 90960, 3275539260, 90960>, CA length $n = 4$, and $Z = 1$. In step 1, as Z is 1 (nonuniform CA), I is initialized as {1, 2, 3, 4}. Step 2 runs for each i from 1 to 4 (i.e., first cell ($i = 1$), second cell ($i = 2$), third cell ($i = 3$), and last cell ($i = 4$)). In step 3, the NSR_SET_i for each of the first, second, third, and last CA cell is identified by making successive calls to Algorithm 2. In this example scenario, we consider identifying the NSR_SET_1 (NSRs set of first cell).

Step 1 of Algorithm 2 initializes the sets A, L_2, L_1, R_1, and R_2. Step 2 computes the Cartesian product of sets L_2, L_1, A, R_1, and R_2 represented as Ω. Each element of Ω is represented by the tuple $(l_2, l_1, abcde, r_1, r_2)$. The total number of such tuple $(l_2, l_1, abcde, r_1, r_2)$ are 512 and are represented as {(0, 0, 00000, 0, 0), (0, 0, 00000, 0, 1), (0, 0, 00000, 1, 0), ..., (1, 1, 11111, 1, 1)}. In Fig. 3, we have shown how each tuple $(l_2, l_1, abcde, r_1, r_2)$ is grouped as RMT and the corresponding neighborhood combinations. There are 32 RMTs, and for each RMT, there are 16 possible neighborhood combinations. In step 3, NSR_SET_1 is initialized as empty set. Steps 5 and 6 are performed for each tuples $(l_2, l_1, abcde, r_1, r_2)$ in Ω. In step 5, five local update rules 3275539260, 90960, 3275539260, 90960, 3275539260 are applied on five RMTs l_2l_1abc, l_1abcd, $abcde$, $bcder_1$, and $cder_1r_2$ to create the five next states a', b', c', d', e' as shown in Fig. 3. For RMT(0) and neighborhood combination (1) (i.e., $l_2, l_1, abcde, r_1, r_2$ = (0, 0, 00000, 0, 0)), the identified a', b', c', d', e' is 0, 0, 0, 0, 0. Similarly, for RMT(1) and neighborhood combination (16) (i.e., $l_2, l_1, abcde, r_1, r_2$ = (1, 1, 00001, 1, 1)), the identified a', b', c', d', e' is 1, 1, 0, 0, 0. In step 6, a NSR is created by a transition ($\rightarrow$) from the decimal equivalent of RMT *abcde* to

RMT *a'b'c'd'e'*. For RMT(0) and neighborhood combination (1), the NSR is 0 → 0. Similarly, for RMT(1) and neighborhood combination (16), the NSR is 1 → 24. Now all the NSRs identified are added to NSR_SET_1 and are returned in step 7. The RMTs (as depicted in columns 1 and 4 of Table 3) and their corresponding identified NSRs (as depicted in columns 2, 3, and 5, 6, respectively, of Table 3) for each CA cell (i.e., first cell, second cell, third cell, and last cell) are reported in Table 3. In a 4-cell non-uniform PBCA <3275539260, 90960, 3275539260, 90960>, the NSRs of the first cell will be the same as the third cell. Similarly, the NSRs of the second cell will be the same as that of the last cell.

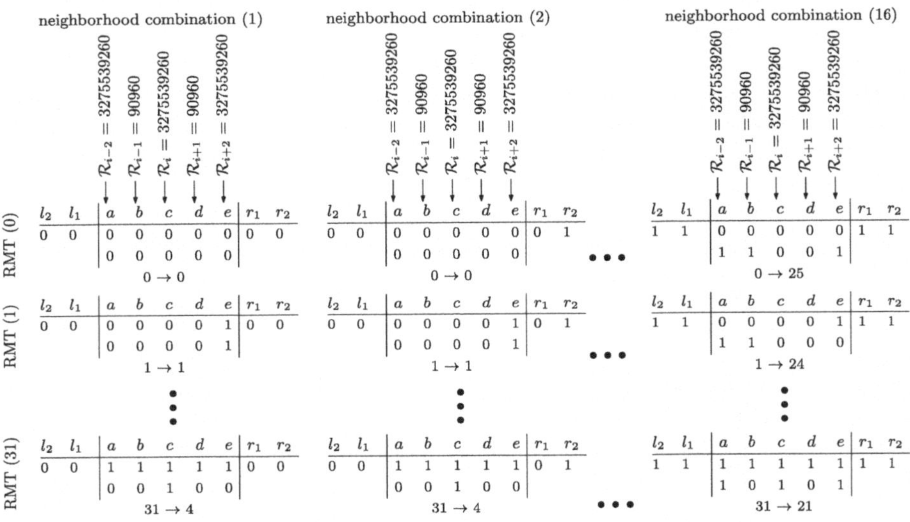

Fig. 3. Identification of NSR for first cell CA configured with RV $<3275539260, 90960, 3275539260, 90960>$

Following step 4 of Algorithm 1, the *NSRS-G*$_i$ for each of the first, second, third, and last CA cell is identified by making successive calls to Algorithm 3. In this example scenario, we consider identifying the *NSRS-G*$_1$ (NSRS-G of the first cell). Step 1 of Algorithm 3 initializes an empty directed graph *NSRS-G*. Further, for each NSR in NSR_SET (step 2), represented as RMT(present state) → RMT(next state), the RMTs are added to V, and the edge between the two RMTs is added to E in steps 3 and 4, respectively. For example, the NSR RMT(0) → RMT(8) of the first cell (NSR_SET_1) will create two vertices 0 and 8, and there will be a directed edge from 0 to 8. Similarly, the NSR RMT(3) → RMT(7) of the first cell (NSR_SET_1) will create two vertices 3 and 7, and there will be a directed edge from 3 to 7. So, there will be 32 vertices corresponding to 32 RMTs. In this way, the *NSRS-G* of all CA cells are formed. In a 4-cell PBCA <3275539260, 90960, 3275539260, 90960>, the *NSRS-G*$_1$, and *NSRS-G*$_3$ are the

Table 3. RMTs and their corresponding identified Next State RMTs (NSRs) for CA $< 3275539260, 90960, 3275539260, 90960 >$

RMTs (1)	Corresponding NSRs		RMTs (4)	Corresponding NSRs	
	First Cell/Third Cell (2)	Second Cell/Last Cell (3)		First Cell/Third Cell (5)	Second Cell/Last Cell (6)
$T(0)$	$T(0)$, $T(1)$, $T(8)$, $T(9)$, $T(16)$, $T(17)$, $T(24)$, $T(25)$	$T(0)$, $T(16)$	$T(16)$	$T(0)$, $T(1)$, $T(8)$, $T(9)$, $T(16)$, $T(17)$, $T(24)$, $T(25)$	$T(12)$, $T(28)$
$T(1)$	$T(0)$, $T(1)$, $T(8)$, $T(9)$, $T(16)$, $T(17)$, $T(24)$, $T(25)$	$T(2)$, $T(3)$, $T(18)$, $T(19)$	$T(17)$	$T(0)$, $T(1)$, $T(8)$, $T(9)$, $T(16)$, $T(17)$, $T(24)$, $T(25)$	$T(10)$, $T(11)$, $T(26)$, $T(27)$
$T(2)$	$T(4)$, $T(5)$, $T(6)$, $T(7)$, $T(20)$, $T(21)$, $T(22)$, $T(23)$	$T(2)$, $T(3)$, $T(18)$, $T(19)$	$T(18)$	$T(4)$, $T(5)$, $T(6)$, $T(7)$, $T(12)$, $T(13)$, $T(14)$, $T(15)$, $T(20)$, $T(21)$, $T(22)$, $T(23)$, $T(28)$, $T(29)$, $T(30)$, $T(31)$	$T(10)$, $T(11)$, $T(26)$, $T(27)$
$T(3)$	$T(4)$, $T(5)$, $T(6)$, $T(7)$, $T(20)$, $T(21)$, $T(22)$, $T(23)$	$T(0)$, $T(1)$, $T(16)$, $T(17)$	$T(19)$	$T(4)$, $T(5)$, $T(6)$, $T(7)$, $T(12)$, $T(13)$, $T(14)$, $T(15)$, $T(20)$, $T(21)$, $T(22)$, $T(23)$, $T(28)$, $T(29)$, $T(30)$, $T(31)$	$T(8)$, $T(9)$, $T(24)$, $T(25)$
$T(4)$	$T(6)$, $T(7)$, $T(22)$, $T(23)$	$T(14)$, $T(15)$, $T(30)$, $T(31)$	$T(20)$	$T(6)$, $T(7)$, $T(22)$, $T(23)$	$T(2)$, $T(3)$, $T(18)$, $T(19)$
$T(5)$	$T(4)$, $T(5)$, $T(20)$, $T(21)$	$T(8)$, $T(24)$	$T(21)$	$T(4)$, $T(5)$, $T(20)$, $T(21)$	$T(0)$, $T(16)$
$T(6)$	$T(0)$, $T(1)$, $T(2)$, $T(3)$, $T(16)$, $T(17)$, $T(18)$, $T(19)$	$T(12)$, $T(28)$	$T(22)$	$T(0)$, $T(1)$, $T(2)$, $T(3)$, $T(16)$, $T(17)$, $T(18)$, $T(19)$	$T(0)$, $T(16)$
$T(7)$	$T(0)$, $T(1)$, $T(2)$, $T(3)$, $T(16)$, $T(17)$, $T(18)$, $T(19)$	$T(10)$, $T(26)$	$T(23)$	$T(0)$, $T(1)$, $T(2)$, $T(3)$, $T(16)$, $T(17)$, $T(18)$, $T(19)$	$T(2)$, $T(18)$
$T(8)$	$T(4)$, $T(5)$, $T(6)$, $T(7)$, $T(12)$, $T(13)$, $T(14)$, $T(15)$, $T(20)$, $T(21)$, $T(22)$, $T(23)$, $T(28)$, $T(29)$, $T(30)$, $T(31)$	$T(12)$	$T(24)$	$T(4)$, $T(5)$, $T(6)$, $T(7)$, $T(20)$, $T(21)$, $T(22)$, $T(23)$	$T(16)$
$T(9)$	$T(4)$, $T(5)$, $T(12)$, $T(13)$, $T(20)$, $T(21)$, $T(28)$, $T(29)$	$T(14)$, $T(15)$	$T(25)$	$T(4)$, $T(5)$, $T(20)$, $T(21)$	$T(18)$, $T(19)$
$T(10)$	$T(0)$, $T(1)$, $T(16)$, $T(17)$	$T(10)$, $T(11)$	$T(26)$	$T(0)$, $T(1)$, $T(8)$, $T(9)$, $T(16)$, $T(17)$, $T(24)$, $T(25)$	$T(18)$, $T(19)$
$T(11)$	$T(0)$, $T(1)$, $T(16)$, $T(17)$	$T(8)$, $T(9)$	$T(27)$	$T(0)$, $T(1)$, $T(8)$, $T(9)$, $T(16)$, $T(17)$, $T(24)$, $T(25)$	$T(16)$, $T(17)$
$T(12)$	$T(0)$, $T(1)$, $T(8)$, $T(9)$, $T(16)$, $T(17)$, $T(24)$, $T(25)$	$T(2)$, $T(3)$	$T(28)$	$T(0)$, $T(1)$, $T(8)$, $T(9)$, $T(16)$, $T(17)$, $T(24)$, $T(25)$	$T(10)$, $T(11)$
$T(13)$	$T(0)$, $T(1)$, $T(8)$, $T(9)$, $T(16)$, $T(17)$, $T(24)$, $T(25)$	$T(4)$	$T(29)$	$T(0)$, $T(1)$, $T(8)$, $T(9)$, $T(16)$, $T(17)$, $T(24)$, $T(25)$	$T(8)$
$T(14)$	$T(4)$, $T(5)$, $T(20)$, $T(21)$	$T(4)$	$T(30)$	$T(4)$, $T(5)$, $T(20)$, $T(21)$	$T(8)$
$T(15)$	$T(4)$, $T(5)$, $T(20)$, $T(21)$	$T(2)$	$T(31)$	$T(4)$, $T(5)$, $T(20)$, $T(21)$	$T(10)$

same and are shown in Fig. 4(a). Similarly, the *NSRS*-G_2, and *NSRS*-G_4 being same are shown in Fig. 5(a).

Following step 5 of Algorithm 1, the CYC_i for each of the first, second, third, and last CA cell is identified by making successive calls to Algorithm 4. Step 1 of Algorithm 4 initializes CYC as an empty set. Step 2 identifies all the cycles (self-loop(s) and multi-length cycle(s)) from the *NSRS-G* (Fig. 4(a) and 5(a)), and they are added to CYC in steps 3–4. The self-loops for the first and third CA cells are the same and are represented as $\{(T(0), T(0)), (T(1), T(1)), (T(5), T(5)), (T(16), T(16)), (T(17), T(17)), (T(21), T(21))\}$. The self-loops for the second and last CA cells are the same and are represented as $\{(T(0), T(0)), (T(2), T(2)), (T(10), T(10))\}$. Similarly, the multi-length cycles for the first and third CA cells are the same and are represented as $\{(T(0), T(1), T(0)), (T(2), T(23), T(2)), (T(8), T(29), T(8)), (T(4), T(6), T(17), T(25), T(4)), ...\}$. The multi-length cycles for the second and last CA cells are the same and are represented as $\{(T(4), T(14), T(4)), (T(17), T(27), T(17)), (T(0), T(16), T(12), T(3), T(0)), (T(2), T(19), T(8), T(12), T(2)), ...\}$. Some of the self-loops and multi-length cycles identified for the first or third CA cell and second or last CA cell considering this example scenario above, are shown within the NSRS graphs in Fig. 4(b) and 5(b), respectively.

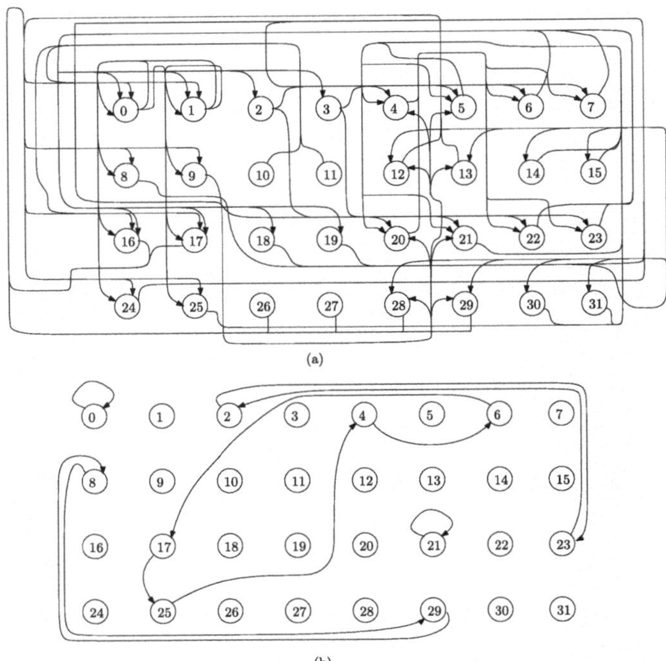

Fig. 4. First Cell NSRS-G (*NSRS-G*$_1$)/Third Cell NSRS-G (*NSRS-G*$_3$) for CA $< 3275539260, 90960, 3275539260, 90960 >$ (a) consisting of all the cycles (b) filtered cycles used to create NSRTD graphs.

Following step 6 of Algorithm 1, $\{C\}$ is initialized as an empty set. As the CA is nonuniform, the condition of step 7 is not satisfied, and it goes to step 10. In step 11, each of the compatibility class ($C_{i(i+1)}$) between the first-second cell (C_{12}), second-third cell (C_{23}), and third-last cell (C_{34}) is identified by calling Algorithm 5. Step 1 of Algorithm 5 initializes $C_{i(i+1)}$ as an empty set. In steps 2–4, each of the cycles between the two CA cells i and $i+1$ are examined for a possible compatible pair. For example, the self-loop ($T(21)$, $T(21)$) of the first cell forms a compatible pair with the self-loop ($T(10)$, $T(10)$) of the second cell. Similarly, the multi-length cycle ($T(8)$, $T(29)$, $T(8)$) of the first cell forms a compatible pair with the multi-length cycle ($T(17)$, $T(27)$, $T(17)$) of the second cell. In step 5, any compatible pair (CYC_{ia} and $CYC_{(i+1)b}$) identified are added to $C_{i(i+1)}$. The identified compatibility classes of NSRSs are:
$C_{12} = \{((T(0), T(0)), (T(0), T(0))); ((T(21), T(21)), (T(10), T(10))); ((T(8), T(29), T(8)), (T(17), T(27), T(17))); ((T(2), T(23), T(2)), (T(4), T(14), T(4))); ((T(17), T(25), T(4), T(6), T(17)), (T(2), T(19), T(8), T(12), T(2))); ...\}$
$C_{23} = \{((T(0), T(0)), (T(0), T(0))); ((T(10), T(10)), (T(21), T(21))); ((T(17), T(27), T(17)), (T(2), T(23), T(2))); ((T(4), T(14), T(4)), (T(8),$

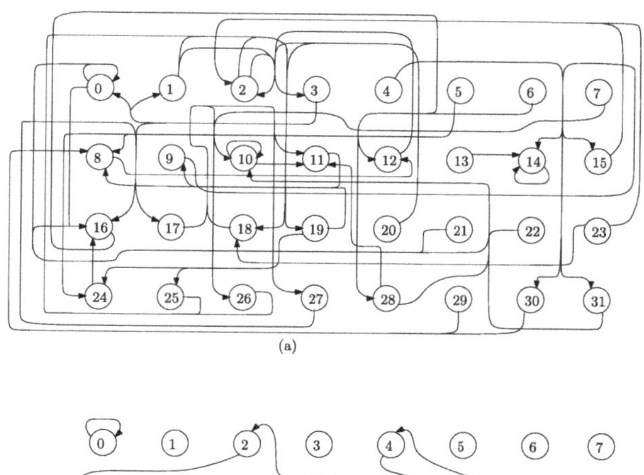

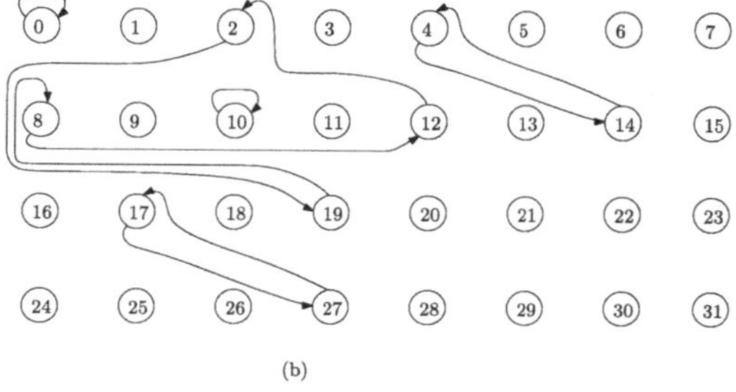

Fig. 5. Second Cell NSRS-G ($NSRS\text{-}G_2$)/Last Cell NSRS-G ($NSRS\text{-}G_4$) for CA $< 3275539260, 90960, 3275539260, 90960 >$ (a) consisting of all the cycles (b) filtered cycles used to create NSRTD graphs.

$T(29)$, $T(8)$)); (($T(2)$, $T(19)$, $T(8)$, $T(12)$, $T(2)$), ($T(4)$, $T(6)$, $T(17)$, $T(25)$, $T(4)$); ...}
C_{34} = {(($T(0)$, $T(0)$), ($T(0)$, $T(0)$)); (($T(21)$, $T(21)$), ($T(10)$, $T(10)$)); (($T(2)$, $T(23)$, $T(2)$), ($T(4)$, $T(14)$, $T(4)$)); (($T(8)$, $T(29)$, $T(8)$), ($T(17)$, $T(27)$, $T(17)$)); (($T(4)$, $T(6)$, $T(17)$, $T(25)$, $T(4)$), ($T(8)$, $T(12)$, $T(2)$, $T(19)$, $T(8)$)); ...}

In step 12, the compatibility classes C_{12}, C_{23}, C_{34} are added to $\{C\}$. In step 13, the compatibility class between the last-first cell (C_{41}) is identified by calling Algorithm 5. The identified compatibility class of NSRSs is:
C_{41} = {(($T(0)$, $T(0)$), ($T(0)$, $T(0)$)); (($T(10)$, $T(10)$), ($T(21)$, $T(21)$)); (($T(4)$, $T(14)$, $T(4)$), ($T(8)$, $T(29)$, $T(8)$)); (($T(17)$, $T(27)$, $T(17)$), ($T(2)$, $T(23)$, $T(2)$)); (($T(8)$, $T(12)$, $T(2)$, $T(19)$, $T(8)$), ($T(17)$, $T(25)$, $T(4)$, $T(6)$, $T(17)$)); ...}.

Following step 14, the compatibility class C_{41}, is added to $\{C\}$. In step 15, $\{S_u\}$ is determined by calling Algorithm 6. Step 1 of Algorithm 6 initialize S_u as an empty set. In step 2, a S_u is generated by picking one compatible pair from each compatibility class C_{12}, C_{23}, C_{34}, and C_{41} of {C} following step 3. In this scenario, two S_us are formed with self-loop NSRSs only i.e. S_1 = (($T(0)$, $T(0)$) → ($T(0)$, $T(0)$) → ($T(0)$, $T(0)$) → ($T(0)$, $T(0)$)), and S_2 = (($T(21)$, $T(21)$)

$\rightarrow (T(10),\ T(10)) \rightarrow (T(21),\ T(21)) \rightarrow (T(10),\ T(10)))$. Similarly, three S_us are formed with multi-length cycle NSRSs i.e. $S_3 = ((T(8),\ T(29),\ T(8)) \rightarrow (T(17),\ T(27),\ T(17)) \rightarrow (T(2),\ T(23),\ T(2)) \rightarrow (T(4),\ T(14),\ T(4)))$, $S_4 = ((T(2),\ T(23),\ T(2)) \rightarrow (T(4),\ T(14),\ T(4)) \rightarrow (T(8),\ T(29),\ T(8)) \rightarrow (T(17),\ T(27),\ T(17)))$, and $S_5 = ((T(17),\ T(25),\ T(4),\ T(6),\ T(17)) \rightarrow (T(2),\ T(19),\ T(8),\ T(12),\ T(2)) \rightarrow (T(4),\ T(6),\ T(17),\ T(25),\ T(4)) \rightarrow (T(8),\ T(12),\ T(2),\ T(19),\ T(8)))$. In step 4, each S_us (S_1 to S_5) are added to $\{S_u\}$ and they are represented as graphs G_1, G_2, G_3, G_4, and G_5 in Fig. 6.

In step 16 of Algorithm 1, as $\{S_u\}$ is not an empty set, it goes to step 18. In steps 18–19, each of the S_u in $\{S_u\}$ is examined, and if all the transitions are made of self-loops (such as S_1 and S_2, in this scenario), they are stored in $NSRTD_{SL}$ in step 20. Otherwise, in step 21, the S_us (such as S_3, S_4, and S_5, in this scenario) are stored in $NSRTD_{MLC}$.

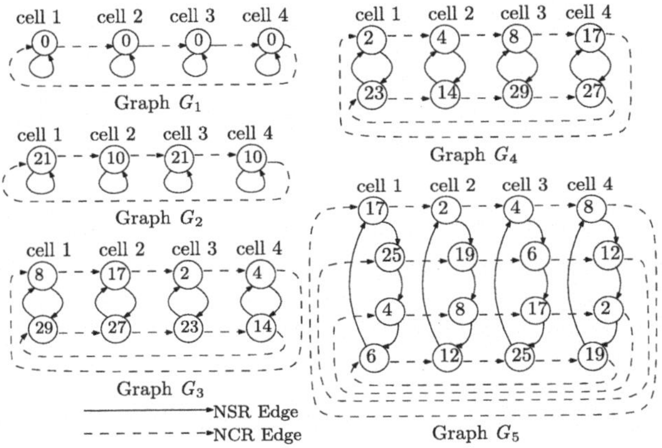

Fig. 6. NSRTD for CA $< 3275539260, 90960, 3275539260, 90960 >$

4 Conclusion

As reported in the literature, till date there is no formal CA characterization tool for 5N CA. In light of the above, this work is the first to report the development of NSRTD framework for 2-state 5N PBCA. Algorithms have been developed that formalize the steps of the 2-state 5N NSRTD framework. The proposed NSRTD framework is effective for both, uniform and nonuniform CAs. Furthermore, this developed framework can effectively be used as a tool to study the spatio-temporal behavior of the RMTs of a 5N CA, which is essential for characterizing this class of CAs.

References

1. Neumann, J., Burks, A.W.: Theory of Self-reproducing Automata. University of Illinois Press, Urbana and London (1966)
2. Wolfram, S.: Cellular Automata and Complexity: Collected Papers. CRC Press, Boca Raton (2018)
3. Martinez, G.J.: A note on elementary cellular automata classification. J. Cell. Autom. **8**(3–4), 233–259 (2013)
4. Chaudhuri, P.P., Chowdhury, D.R., Nandi, S., Chattopadhyay, S.: Additive Cellular Automata: Theory and Applications. IEEE Computer Society Press, Washington (1997)
5. Das, S., Naskar, N.N., Mukherjee, S., Dalui, M., Sikdar, B.K.: Characterization of CA rules for SACA targeting detection of faulty nodes in WSN. In: Bandini, S., Manzoni, S., Umeo, H., Vizzari, G. (eds.) ACRI 2010. LNCS, vol. 6350, pp. 300–311. Springer, Heidelberg (2010). https://doi.org/10.1007/978-3-642-15979-4_32
6. Maji, P.: Cellular automata evolution for pattern recognition. Ph.D. thesis, Jadavpur University, Kolkata, India (2005)
7. Saha, M., Dalui, M., Sikdar, B.K.: A cellular automata based highly accurate memory test hardware realizing march C-. Microelectron. J. **52**, 91–103 (2016)
8. Dalui, M., Sikdar, B.K.: Design of directory based cache coherence protocol verification logic in CMPs around TACA. In: International Conference on High Performance Computing & Simulation 2013, pp. 318–325. IEEE, Helsinki (2013). https://doi.org/10.1109/HPCSim.2013.6641433
9. Chowdhury, D.R., Sengupta, I., Chaudhuri, P.P.: A class of two-dimensional cellular automata and their applications in random pattern testing. J. Electron. Test. **5**(1), 67–82 (1994)
10. Maiti, S., Roy Chowdhury, D.: Study of five-neighborhood linear hybrid cellular automata and their synthesis. In: Giri, D., Mohapatra, R.N., Begehr, H., Obaidat, M.S. (eds.) ICMC 2017. CCIS, vol. 655, pp. 68–83. Springer, Singapore (2017). https://doi.org/10.1007/978-981-10-4642-1_7
11. Tomassini, M., Sipper, M., Perrenoud, M.: On the generation of high-quality random numbers by two-dimensional cellular automata. IEEE Trans. Comput. **49**(10), 1146–1151 (2000)
12. Dalui, M., Chakraborty, B., Das, N., Sikdar, B.K.: NSRT diagram for identification of SACA and TACA rules in null-boundary. Int. J. Mod. Phys. C **33**(6), 2250071 (2022)
13. Das, S., Mukherjee, S., Naskar, N., Sikdar, B.K.: Characterization of single cycle CA and its application in pattern classification. Electron. Notes Theor. Comput. Sci. **252**, 181–203 (2009)
14. Adak, S., Naskar, N., Maji, P., Das, S.: On synthesis of non-uniform cellular automata having only point attractors. J. Cell. Autom. **12**(1–2), 81–100 (2016)
15. Maitii, N.S., Ghosh, S., Sikdar, B.K., Chaudhuri, P.P.: Rule vector graph (RVG) to design linear time algorithm for identifying the invertibility of periodic-boundary three neighborhood cellular automata. J. Cell. Autom. **7**(4), 335–362 (2012)
16. Maiti, N.S., Ghosh, S., Munshi, S., Chaudhuri, P.P.: Linear time algorithm for identifying the invertibility of null-boundary three neighborhood cellular automata. Complex Syst. **19**(1), 89–113 (2010)
17. Banerjee, S., Dalui, M., Hazra, S.: A generic NSRTD framework for characterizing cellular automaton rules in constant (0/1) boundary conditions. Complex Syst. **33**(1), 61–86 (2024)

18. Martı, A., Rodrı, G.: Reversibility of linear cellular automata. Appl. Math. Comput. **217**(21), 8360–8366 (2011)
19. Cinkir, Z., Akin, H., Siap, I.: Reversibility of 1D cellular automata with periodic boundary over finite fields. J. Stat. Phys. **143**(4), 807–823 (2011)
20. Bhattacharjee, K., Naskar, N., Roy, S., Das, S.: A survey of cellular automata: types, dynamics, non-uniformity and applications. Nat. Comput. **19**(2), 433–461 (2020)

Equivalence Classes of Complement-Invariant CA on Finite 2D Triangular Lattice

A. Bhavadharani[1], D. K. Sheena Christy[1](✉), and Sukanya Mukherjee[2]

[1] Department of Mathematics, SRM Institute of Science and Technology, Faculty of Engineering and Technology, Kattankulathur 603203, Tamil Nadu, India
{ba7497,sheenac}@srmist.edu.in

[2] Department of Computer Science and Engineering, Institute of Engineering and Management Kolkata, University of Engineering and Management, Kolkata, India
sukanyaiem@gmail.com

Abstract. Stephen Wolfram's classification of elementary cellular automata into 88 equivalence classes marked a significant milestone in reducing and systematizing the study of one-dimensional cellular automaton rules. Motivated by this, we present a classification framework for two-dimensional complement-invariant binary cellular automata defined on a finite triangular lattice with a von Neumann neighborhood of radius one. The complete rule space consists of 65,536 rules. By introducing the concept of complement invariance, we consider 256 rules. Within the class of complement-invariant rules, each rule and its bitwise complement produce mutually complementary evolution patterns, which reduces the 256 rules to 128 equivalence classes under periodic boundary. Applying vertical symmetry refines the classification to a final set of 80 equivalence classes. Vertical symmetry works under both null and periodic boundaries. Our results demonstrate that complement symmetry is unique to complement-invariant rules and provide a systematic organization of complement-invariant CA on the triangular lattice. This work proves the isomorphism of vertically symmetric rules and identifies complement-invariant rules that generate self-similar fractal patterns.

Keywords: 2D Cellular Automata · Finite Triangular Lattice · Periodic Boundary · Complement Invariance

1 Introduction

Cellular automata (CA) are discrete, parallel computational systems composed of a regular grid of cells. The state of each cell changes over time based on rules that depend on the states of the cells around it. Von Neumann and Burks [1] were the first to formally define the idea of CA in their groundbreaking paper on

This work is done as a project in the Indian Summer School on Cellular Automata 2025.

H. Raju et al. (Eds.): ASCAT 2026, CCIS 2801, pp. 101–115, 2026.
https://doi.org/10.1007/978-3-032-18612-6_8

self-reproducing automata. Early research, like Moore's machine models of self-reproduction [2], helped us understand how CA could be used for computation. Packard and Wolfram's [3] work on 2D CA made them more popular. They studied the temporal evolution of two-dimensional CA systems and demonstrated the diversity and complexity of their emergent behaviors. Researchers have utilized these systems to replicate a wide range of phenomena, such as biological development and physical systems. Durand [4] looked at the global properties of 2D CAs in more depth, focusing on the computational complexity and structural properties that come from local interactions.

Bhattacharjee et al. [5] conducted thorough surveys that provide an overview of CA types, dynamics, non-uniformity, and applications. After the initial research on CA, scientists have done a lot of work on 2D CA on triangular lattices because they have unique geometric and topological properties. The triangular tessellation enables distinctive neighborhood configurations, which can profoundly affect the evolutionary dynamics and complexity of the CA system. Bays [6] examined the dynamics of CA on triangular tessellations, yielding preliminary insights into pattern formation and localized interactions. Morita [7] examined 8-state reversible and conservative triangular partitioned CA, illustrating the universality and intricate behavior attainable within these lattice structures. Cousin [8] focused on the 256 elementary rules of 2D triangular automata, creating a structure for organized classification and examination of the behavior of the rules. Saadat [9] stressed how important lattice geometry is for changing CA dynamics.

Further applications and structural analyses of triangular lattice CA have been performed to investigate reversibility, periodic behavior, and semitotalistic rules. Zawidzki [10] studied the semitotalistic 2D CA on triangulated 3D surfaces, illustrating their effectiveness in modeling intricate geometries. Uguz et al. [11,12] examined the structure, reversibility, and dynamics of linear 2D von Neumann neighborhood CA in triangular lattices with null and periodic boundary conditions. A fundamental strategy for analyzing and reducing the complexity of CA involves the categorization of rules into equivalence classes predicated on symmetry transformations. Wolfram first put one-dimensional elementary cellular automata (ECA) into 88 equivalence classes by looking at reflection, complementation, and combination symmetries [13]. These kinds of classification systems help find rules that act in the same way or in a similar way when these changes are implemented. This narrows down the rule space. In 2D CA, analogous methodologies have been utilized to categorize rules based on rotational, reflectional, and complementary symmetries [14,15].

Complement invariance and complement symmetry are significant in the examination of 2D CA. Complement symmetry delineates pairs of rules that exhibit reciprocal behavior upon state inversion, facilitating the classification of rules into symmetric equivalence classes [16,17]. Prior studies on 2D linear CA, such as Uguz et al. [18–20], focused on square grids for self-replicating patterns and image processing, whereas our work explores these phenomena in the triangular lattice. The theoretical foundations of 2D CA on triangular lattices remain largely unexplored. Building upon these foundational and structural studies, the present work focuses specifically on the classification of 256 complement-invariant rules of 2D binary CAs on a finite triangular lattice with periodic boundaries. In Sect. 3, using complement invariance, complement symmetry, and vertical symmetry, we reduce the rule space into 80 equivalence classes, providing a systematic framework for future investigations of their dynamics and behavior.

2 Preliminaries

In this section, we present the fundamental concepts, notations, and definitions that form the basis for our study.

Definition 1 (Cellular automaton). *[5] A cellular automaton is defined as a quadruple* $\mathbb{A} = (\mathbb{L}, \mathbb{Q}, \mathbb{N}, \mathbb{R})$, *where*

- $\mathbb{L} \subseteq \mathbb{Z}^d$ *is a d-dimensional lattice (2D triangular grid),*
- $\mathbb{Q}$ *is a finite set of states* ($Q = \{0, 1\}$),
- $\mathbb{N} = \{(n_1, n_2, \cdots, n_q)\}$ *is the set of all neighborhood combinations,*
- $\mathbb{R} : \mathbb{Q}^{|\mathbb{N}|} \rightarrow \mathbb{Q}$ *is the set of local transition rules.*

In this work, we employ a finite 2D triangular lattice, where the first row of the lattice starts with an upright triangle throughout this work. Since the lattice is finite, boundary cells require special treatment. We consider periodic boundary conditions, where cells at the edges interact with cells on the opposite edge, ensuring continuity in evolution. Figure 1a illustrates the von Neumann neighborhood for a cell, considering both upright and inverted triangles. Figures 1b and 1c depict the null and periodic boundary conditions, respectively, in a finite triangular lattice. The neighborhood ordering adopted in this work follows the convention introduced in earlier studies on triangular lattice CA [12], where the central cell is not included. In the present model, the central cell is additionally incorporated while preserving the original relative ordering of the neighboring sites, as shown in Table 1.

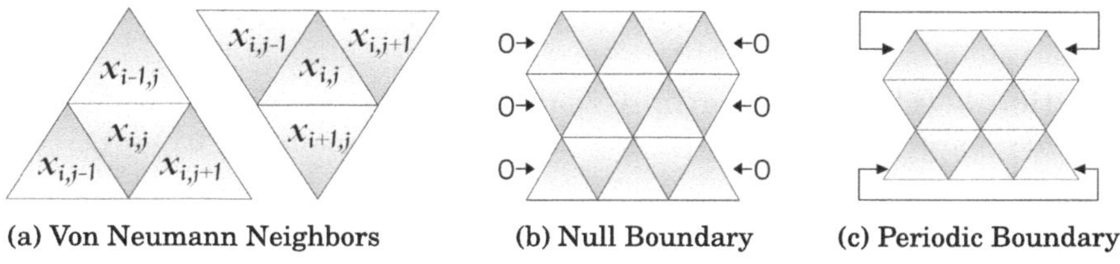

Fig. 1. Neighborhood and boundary conditions in 2D Triangular CA

Since the lattice consists of upright and inverted triangles, the orientation of neighbors differs based on the triangle type. Any 4-bit neighborhood configuration (e.g., `0011`) is interpreted as (n_1, n_2, n_3, n_4), corresponding to the states of the neighbors in the specified order. The transition rule of a 2D cellular automaton specifies the next state of a cell based on the states of its neighbors. In a 2D triangular lattice with a von Neumann neighborhood of radius 1, each cell interacts with four neighbors. Table 2 presents an example of such a rule (Rule 27030), where each 4-bit neighborhood configuration corresponds to a unique next state of the central cell. In ECA, a rule defines the next state of a cell based on its current state and the states of its two immediate neighbors. For every rule among the 256 ECA rules, there exists a unique complement rule, resulting in a total of 128 complementary pairs within the 256-rule space.

Table 1. Neighbor labeling for von Neumann neighborhood in triangular lattice

Triangle Type	n_1	n_2	n_3	n_4
Upright	Central Cell	Right	Bottom	Left
Inverted	Central Cell	Top	Right	Left

Table 2. Rule 27030

1111	1110	1101	1100	1011	1010	1001	1000	0111	0110	0101	0100	0011	0010	0001	0000
0	1	1	0	1	0	0	1	1	0	0	1	0	1	1	0

Definition 2 (Rule complement). *[18] The complement of an ECA rule can be derived from its local rule table by interchanging 0's and 1's.*

Example 1. Consider the 1D ECA Rule 110 with a 3-cell neighborhood. Let the binary expansion of the rule 110 be $f \in \mathbb{R}$.

Neighborhood	111	110	101	100	011	010	001	000
$f(v)$ (Rule 110)	0	1	1	0	1	1	1	0
$\overline{f}(v)$ (Rule 145)	1	0	0	1	0	0	0	1

v denotes the neighborhood RMT, and $\overline{f}(v)$ is the complement of rule f, inverting all outputs of Rule 110.

Definition 3 (Configuration). *A configuration of a d-dimensional cellular automaton on a finite lattice with state set $\mathbb{Q}$ is a mapping $\mathscr{C} : \mathbb{L}^d \to \mathbb{Q}$, where $\mathbb{L}^d$ denotes a finite d-dimensional grid.*

Definition 4 (Vertical flip operator). *Let $\mathscr{C}^{(t)}$ be a configuration of a 2D cellular automaton defined on an $m \times n$ finite lattice at time t. The* vertical flip operator *$\tau : \mathscr{C}^{(t)} \to \mathscr{D}^{(t)}$ acts by reflecting the configuration vertically. If $x_{i,j}^{(t)}$ and $y_{i,j}^{(t)}$ denote the states of cells at position (i, j) in $\mathscr{C}^{(t)}$ and $\mathscr{D}^{(t)}$, respectively, then*

$$y_{i,j}^{(t)} = \tau(x_{i,j}^{(t)}) = x_{m-i+1,j}^{(t)},$$

where m and n denote the number of rows and columns of the lattice.

The state of each cell in the *d-dimensional finite grid is mapped to a state in* $\mathbb{Q}$. The configuration of a 2D CA is represented by a matrix of order $m \times n$. The configuration at time t is denoted by $\mathscr{C}^{(t)}$ where

$$\mathscr{C}^{(t)} = \begin{pmatrix} x_{11}^{(t)} & x_{12}^{(t)} & \cdots & x_{1n}^{(t)} \\ x_{21}^{(t)} & x_{22}^{(t)} & \cdots & x_{2n}^{(t)} \\ \vdots & \vdots & \ddots & \vdots \\ x_{m1}^{(t)} & x_{m2}^{(t)} & \cdots & x_{mn}^{(t)} \end{pmatrix}_{m \times n}$$

3 Complement-Invariant Rules

3.1 Complement-Invariant Rules: Complement Symmetry

In the study of 2D CA on a finite triangular lattice, certain rules exhibit complement invariance, which is a property where each Rule Min Term (RMT) and its complement are mapped to the same output state.

Definition 5 (Complement-Invariant Rule). *A 2D cellular automaton* $\mathbb{A} = (\mathbb{L}, \mathbb{Q}, \mathbb{N}, \mathbb{R})$ *is said to be* complement-invariant *if its local transition rule* $f \in \mathbb{R}$ *is invariant under neighborhood complementation.* $f(v) = f(\overline{v})$ *where* v *denotes a neighborhood RMT and* $\overline{v}$ *denotes its bitwise complement.*

In particular, a 2D von Neumann triangular cellular automaton of radius 1 is complement-invariant if

$$f(n_1, n_2, n_3, n_4) = f(\overline{n_1}, \overline{n_2}, \overline{n_3}, \overline{n_4}),$$

for all $n_i \in \mathbb{Q}$, $1 \leq i \leq 4$, where $\overline{n_i}$ denotes the complement of n_i in $\mathbb{Q}$.

Example. In this work, we restrict ourselves to the case of binary states, i.e., $\mathbb{Q} = \{0, 1\}$. By applying the above definition of complement-invariance, each neighborhood configuration and its bitwise complement must yield the same output under the local transition rule $f \in \mathbb{R}$. Thus, the following equivalences hold:

$$\begin{aligned} f(0000) &= f(1111), & f(0001) &= f(1110), \\ f(0010) &= f(1101), & f(0011) &= f(1100), \\ f(0100) &= f(1011), & f(0101) &= f(1010), \\ f(0110) &= f(1001), & f(0111) &= f(1000). \end{aligned}$$

Complement-invariance forces the two members of each pair to share the same image under $f \in \mathbb{R}$. One independent output bit suffices per pair. Hence the number of distinct complement-invariant local rules is $2^8 = 256$.

Theorem 1. *Let* $r_1 \in \mathbb{R}$ *and its complement* $r_2 \in \mathbb{R}$ *be two complement-invariant local transition rules for a finite triangular 2D cellular automaton. The evolution of* r_1 *and* r_2 *are complementary to each other at every time step t under periodic boundary.*

Proof. Let us consider a finite triangular 2D cellular automaton with synchronous updates and periodic boundary. Let r_1 be a complement-invariant rule, and let r_2 be its bitwise complement. Let $\mathscr{C}^{(0)}$ be any $m \times n$ initial binary configuration. Evolve $\mathscr{C}^{(0)}$ under r_1 and r_2, using the synchronous update scheme and periodic boundary condition. Let $r_1 : \mathscr{C}^{(0)} \to \mathscr{C}^{(t)}$ and $r_2 : \mathscr{C}^{(0)} \to \mathscr{E}^{(t)}$. The orientation of the triangular cell $x_{i,j}$ is determined by the parity of $i+j$. If $x_{i,j} = 1$, then the geometric representation of the cell is an upright triangle when $i \oplus j = 0$ and an inverted triangle when $i \oplus j = 1$. Let $\mathscr{C}^{(0)} = [x_{ij}^{(0)}]_{m\times n}$ be the initial configuration. By applying the local transition rules r_1 and r_2 to the initial configuration $\mathscr{C}^{(0)}$, we get (matrix representation)

$$\mathscr{C}^{(1)} = [x_{ij}^{(1)}]_{m\times n} \quad \text{and} \quad \mathscr{E}^{(1)} = [y_{ij}^{(1)}]_{m\times n}.$$

Since r_1 and r_2 are complements to each other, we have $\overline{x_{ij}^{(1)}} = y_{ij}^{(1)}$. The neighborhood tuple of the cells $x_{ij}^{(1)}$ and $y_{ij}^{(1)}$ is given by

$$v(x_{ij}^{(1)}) = (x_{i,j}^{(1)}, x_{i,j+1}^{(1)}, x_{i+1,j}^{(1)}, x_{i,j-1}^{(1)}) \tag{1}$$

$$\begin{aligned} v(y_{ij}^{(1)}) &= (y_{i,j}^{(1)}, y_{i,j+1}^{(1)}, y_{i+1,j}^{(1)}, y_{i,j-1}^{(1)}) \\ &= (\overline{x_{i,j}^{(1)}}, \overline{x_{i,j+1}^{(1)}}, \overline{x_{i+1,j}^{(1)}}, \overline{x_{i,j-1}^{(1)}}). \end{aligned}$$

By the definition of complement invariance

$$r_1(x_{i,j}^{(1)}, x_{i,j+1}^{(1)}, x_{i+1,j}^{(1)}, x_{i,j-1}^{(1)}) \neq r_2(y_{i,j}^{(1)}, y_{i,j+1}^{(1)}, y_{i+1,j}^{(1)}, y_{i,j-1}^{(1)})$$

$$\Longrightarrow \overline{\mathscr{C}^{(1)}} = \mathscr{E}^{(1)}$$

Assume for some $t \geq 1$,

$$\mathscr{E}^{(t)} = \overline{\mathscr{C}^{(t)}}.$$

Induction on Time Step 't': For each cell x_{ij} at time t, the neighborhood in $\mathscr{E}^{(t)}$ is the bitwise complement of the neighborhood in $\mathscr{C}^{(t)}$. Then

$$y_{ij}^{(t+1)} = r_2(v(y_{ij}^{(t)})) = \overline{r_1(v(x_{ij}^{(t)}))} = \overline{x_{ij}^{(t+1)}}.$$

$$\Longrightarrow \mathscr{E}^{(t+1)} = \overline{\mathscr{C}^{(t+1)}}.$$

□

The periodic boundary condition is essential here, since each cell in $\mathscr{E}^{(t)}$ is exactly the bitwise complement of the corresponding neighborhood in $\mathscr{C}^{(t)}$.

Definition 6 (Complement-variant rule). *A local transition rule $r \in \mathbb{R}$ for a 2D triangular cellular automaton is called* complement-variant *if there exists at least one neighborhood v such that $r(v) \neq r(\overline{v})$, where $\overline{v}$ is the bitwise complement of v.*

Corollary 1. *Let $r_1 \in \mathbb{R}$ and its complement $r_2 \in \mathbb{R}$ be two complement-variant local transition rules ($f(v) \neq f(\overline{v})$) for a finite triangular 2D cellular automaton. The evolutions of r_1 and r_2 are not complementary to each other at each time step t under the periodic boundary.*

Proof. Consider the cell $x_{i,j}^{(t)} \in \mathscr{C}^{(t)}$ at time t. Since r_1 and r_2 are complement-variant,

$$r_1(x_{i,j}^{(t)}, x_{i,j+1}^{(t)}, x_{i+1,j}^{(t)}, x_{i,j-1}^{(t)}) = r_2(y_{i,j}^{(t)}, y_{i,j+1}^{(t)}, y_{i+1,j}^{(t)}, y_{i,j-1}^{(t)})$$

r_1 and r_2 may produce identical next-state values for certain neighborhoods, thereby violating complementary evolution.

$$\mathscr{E}^{(t+1)} \neq \overline{\mathscr{C}^{(t+1)}}$$

□

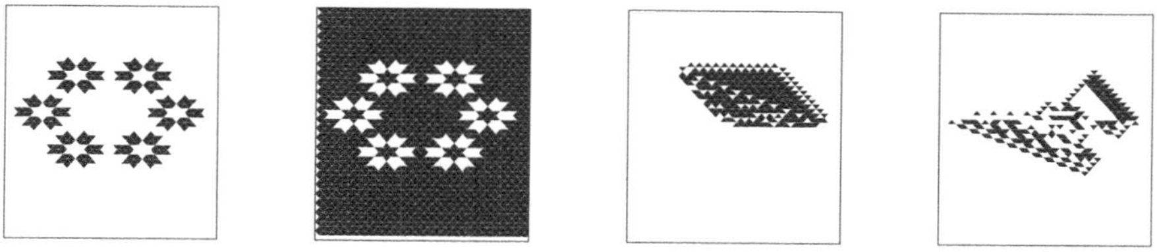

(a) Rule 27030 (b) Rule 38505 (c) Rule 30701 (d) Rule 34384

Fig. 2. Evolution of rules 27030, 38505, 30701, and 34384 at $t = 20$

Figures 2a and 2b represent the evolution of an initial configuration at time step 20 using complement-invariant rules 27030 and 38505. Figures 2c and 2d represent the evolution of an initial configuration at time step 20 using complement variant rules 30701 and 34384. It is observed that the complement-invariant rule and its complement exhibit complementary behavior in their evolution, whereas the complement-variant rule and its complement do not yield complementary evolution at time step 20.

3.2 Complement-Invariant Rules: Vertical Symmetry

Definition 7. *Let $r_1, r_2 \in \mathbb{R}$ be two cellular automata rules, and let $\mathscr{C}$ be an initial configuration. Define $\tau(\mathscr{C})$ as the vertical flip of $\mathscr{C}$. $\mathscr{C}^t$ and $\tau(\mathscr{C}^t)$ are configurations obtained after t time steps of evolution under r_1 and r_2 respectively. The rules r_1 and r_2 are vertically symmetric to each other if, for all $t \geq 0$,*

$$\tau(r_1(\mathscr{C})) = r_2(\tau(\mathscr{C})).$$

Theorem 2. *Let $f, g \in \mathbb{R}$ be local transition rules in a 2D von Neumann triangular lattice cellular automaton. If $f(n_1, n_2, n_3, n_4) = g(n_1, n_3, n_2, n_4)$ for all neighborhood tuples (n_1, n_2, n_3, n_4), then the evolution of any initial configuration $\mathscr{C}^{(0)}$ under f and the evolution of $\tau(\mathscr{C}^{(0)})$ under g will be vertically symmetric to each other at every time step t.*

Proof. Let a cellular automaton be defined on an $m \times n$ 2D triangular lattice. Let the initial configuration be denoted by $\mathscr{C}^{(0)}$, consisting of cells $x_{i,j}^{(0)}$ for $1 \leq i \leq m$ and $1 \leq j \leq n$. The configuration evolves under a transition function $f : \mathscr{C}^{(0)} \rightarrow \mathscr{C}^{(t)}$, where $t \in \mathbb{N}$ and $t \geq 1$. Define a vertical reflection map $\tau : \mathscr{C} \rightarrow \mathscr{D}$ such that

$$\tau(y_{i,j}^{(t)}) = x_{m+1-i,j}^{(t)},$$

Here, $y_{i,j}^{(t)} \in \mathscr{D}^{(t)}$ and $x_{m+1-i,j}^{(t)} \in \mathscr{C}^{(t)}$. Let the vertically flipped initial configuration be $\mathscr{D}^{(0)} = \tau(\mathscr{C}^{(0)})$. Let $\mathscr{D}^{(0)}$ evolve under a transition function

$$g : \mathscr{D}^{(0)} \rightarrow \mathscr{D}^{(t)}.$$

Consider a cell $x_{m+1-i,j} \in \mathscr{C}^{(t)}$. Suppose $x_{m+1-i,j}$ is an upright triangle. Then, by definition,

$$f(v(x_{m+1-i,j}^{(t)})) = f(x_{m+1-i,j}^{(t)}, x_{m+1-i,j+1}^{(t)}, x_{m+2-i,j}^{(t)}, x_{m+1-i,j-1}^{(t)}) \quad (2)$$

Under the map τ, the corresponding cell in the flipped configuration becomes an inverted triangle. Its neighborhood is then

$$g(v(y_{i,j}^{(t)})) = g(x_{m+1-i,j}^{(t)}, x_{m+2-i,j}^{(t)}, x_{m+1-i,j+1}^{(t)}, x_{m+1-i,j-1}^{(t)})2 \quad (3)$$

Applying the vertical symmetry to equations (2) and (3), we obtain:

$$f(x_{m+1-i,j}^{(t)}) = g(y_{i,j}^{(t)})$$

□

3.3 Generation of Equivalence Classes of Complement-Invariant Rules

We generate the equivalence classes of complement-invariant rules for a 2D von Neumann triangular cellular automaton as follows. First, using the definition of complement invariance (Definition 3.1), the original set of 65536 rules is restricted to 256 complement-invariant rules. This ensures that only rules f satisfying $f(v) = f(\overline{v})$ for all neighborhood RMTs v are considered, where $\overline{v}$ denotes the complement of v in the state set. For each of the 256 rules, the procedure is

1. Convert the complement-invariant $\mathbb{C}_{\mathbb{S}}$ to a 16-bit string representation.
2. Compute its complement 16-bit string and convert back to decimal $\overline{\mathbb{C}_{\mathbb{S}}}$.
3. Initialize a new equivalence class with the rule and its complement.
4. Check if a vertically symmetric rule $\mathbb{V}_{\mathbb{S}}(\mathbb{C}_{\mathbb{S}})$ exists for the rule $\mathbb{C}_{\mathbb{S}}$(Definition 3.2). If so, include it in the equivalence class. Similarly, compute the vertically symmetric pair $\mathbb{V}_{\mathbb{S}}(\overline{\mathbb{C}_{\mathbb{S}}})$ for the complement rule $\overline{\mathbb{C}_{\mathbb{S}}}$ if applicable.
5. If the rule is already included in a previous equivalence class, skip further processing.

Algorithm 1 presents the pseudocode for constructing equivalence classes.

Algorithm 1. Equivalence Class Construction of Complement-Invariant Rules

$k \leftarrow 1$, $\mathscr{E} \leftarrow \emptyset$ ▷ Equivalence class index and set of classes
for each rule $\mathbb{C}_S$ **do**
  if $\mathbb{C}_S \notin \bigcup \mathscr{E}$ **then** ▷ If rule not in any existing class
    $\mathscr{E}_k \leftarrow \{\mathbb{C}_S, \overline{\mathbb{C}_S}\}$ ▷ Initialize with rule and its complement
    if $\exists \mathbb{V}_S(\mathbb{C}_S)$ **then**
      $\mathscr{E}_k \leftarrow \mathscr{E}_k \cup \{\mathbb{V}_S(\mathbb{C}_S), \mathbb{V}_S(\overline{\mathbb{C}_S})\}$ ▷ Add vertically symmetric rules
    end if
    $\mathscr{E} \leftarrow \mathscr{E} \cup \{\mathscr{E}_k\}$ ▷ Add the new equivalence class
    $k \leftarrow k + 1$ ▷ Increment class index for next new rule
  end if
end for

This process is repeated until all 256 rules are assigned to equivalence classes. Each equivalence class thus co ntains all rules connected through complementation and vertical symmetry, with no repetition across classes.

Computational Complexity. The classification procedure consists of two stages. First, all 2^{16} possible rules are examined to identify complement-invariant rules. For each rule, conversion to a fixed-length 16-bit binary representation and verification of complement invariance require a constant number of bit-level operations, resulting in an overall time complexity of $O(2^{16})$ for this stage. In the second stage, equivalence classes are constructed from the N complement-invariant rules ($N = 256$ in the present setting). Each unclassified rule is grouped with its complement and vertically symmetric counterparts using fixed permutations on 16-bit strings. Therefore, this stage has time complexity $O(N)$.

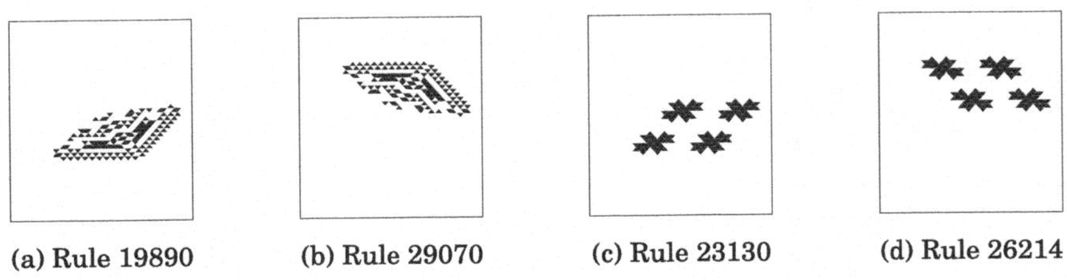

(a) Rule 19890 (b) Rule 29070 (c) Rule 23130 (d) Rule 26214

Fig. 3. Evolution of Vertical symmetric rules at $t = 20$

Figure 3 illustrates the evolution of vertical-symmetric rule pairs from a given initial configuration at time step 20. Subfigures 3a and 3b show the evolution of Rule 19890 and its vertical-symmetric counterpart 29070, while subfigures 3c and 3d depict Rule 23130 and its vertical-symmetric pair 26214. By introducing the concept of vertical symmetry, we observe that 32 pairs of these complement-invariant rules exhibit the property of vertical symmetry. The complement-invariant rules can be regarded as equivalence classes under the relations of complement symmetry and vertical symmetry. Initially, the 256 rules form 128 equivalence classes under complement symmetry. By further introducing vertical symmetry, 96 of these classes merge, resulting in 80 distinct equivalence classes in total.

The equivalence classes of complement-invariant rules for the 2D triangular lattice cellular automaton under periodic boundary conditions are presented in Table 3. In this table, $\mathbb{C}_{\mathbb{S}}$ represents a complement-invariant rule, while $\overline{\mathbb{C}_{\mathbb{S}}}$ denotes its complement. The notation $\mathbb{V}_{\mathbb{S}}(\mathbb{C}_{\mathbb{S}})$ corresponds to the vertically symmetric rule associated with $\mathbb{C}_{\mathbb{S}}$, and $\mathbb{V}_{\mathbb{S}}(\overline{\mathbb{C}_{\mathbb{S}}})$ represents the vertically symmetric rule of $\overline{\mathbb{C}_{\mathbb{S}}}$. While each of the 256 complement-invariant rules has a well-defined complementary rule, vertical symmetry is satisfied only by a subset of rules. Rules that do not admit a vertical reflection within the rule space have no symmetric counterpart, and hence their corresponding table positions remain empty. The evolution of two equivalence classes of complement-invariant rules, as listed in Table 3, is illustrated in Fig. 4. Each equivalence class represents a distinct set of rules exhibiting identical dynamical behavior under complement symmetry and vertical symmetry on a 2D triangular lattice cellular automaton with periodic boundary conditions. The evolution is observed at time step 40, starting from an initial configuration in which a central pattern of live cells (state 1) is embedded within a background of cells in state 0.

Table 3. Equivalence Classes of 256 Complement-Invariant Rules

S. No.	$\mathbb{C}_\mathbb{S}$	$\overline{\mathbb{C}_\mathbb{S}}$	$\mathbb{V}_\mathbb{S}(\mathbb{C}_\mathbb{S})$	$\mathbb{V}_\mathbb{S}(\overline{\mathbb{C}_\mathbb{S}})$	S. No.	$\mathbb{C}_\mathbb{S}$	$\overline{\mathbb{C}_\mathbb{S}}$	$\mathbb{V}_\mathbb{S}(\mathbb{C}_\mathbb{S})$	$\mathbb{V}_\mathbb{S}(\overline{\mathbb{C}_\mathbb{S}})$
1	0	65535			41	16386	49149		
2	384	65151			42	16770	48765		
3	576	64959			43	16962	48573		
4	960	64575			44	17346	48189		
5	1056	64479	4104	61431	45	17442	48093	20490	45045
6	1440	64095	4488	61047	46	17826	47709	20874	44661
7	1632	63903	4680	60855	47	18018	47517	21066	44469
8	2016	63519	5064	60471	48	18402	47133	21450	44085
9	2064	63471	8196	57339	49	18450	47085	24582	40953
10	2448	63087	8580	56955	50	18834	46701	24966	40569
11	2640	62895	8772	56763	51	19026	46509	25158	40377
12	3024	62511	9156	56379	52	19410	46125	25542	39993
13	3120	62415	12300	53235	53	19506	46029	28686	36849
14	3504	62031	12684	52851	54	19890	45645	29070	36465
15	3696	61839	12876	52659	55	20082	45453	29262	36273
16	4080	61455	13260	52275	56	20466	45069	29646	35889
17	5160	60375			57	21546	43989		
18	5544	59991			58	21930	43605		
19	5736	59799			59	22122	43413		
20	6120	59415			60	22506	43029		
21	6168	59367	9252	56283	61	22554	42981	25638	39897
22	6552	58983	9636	55899	62	22938	42597	26022	39513
23	6744	58791	9828	55707	63	23130	42405	26214	39321
24	7128	58407	10212	55323	64	23514	42021	26598	38937
25	7224	58311	13356	52179	65	23610	41925	29742	35793
26	7608	57927	13740	51795	66	23994	41541	30126	35409
27	7800	57735	13932	51603	67	24186	41349	30318	35217
28	8184	57351	14316	51219	68	24570	40965	30702	34833
29	10260	55275			69	26646	38889		
30	10644	54891			70	27030	38505		
31	10836	54699			71	27222	38313		
32	11220	54315			72	27606	37929		
33	11316	54219	14364	51171	73	27702	37833	30750	34785
34	11700	53835	14748	50787	74	28086	37449	31134	34401
35	11892	53643	14940	50595	75	28278	37257	31326	34209
36	12276	53259	15324	50211	76	28662	36873	31710	33825
37	15420	50115			77	31806	33729		
38	15804	49731			78	32190	33345		
39	15996	49539			79	32382	33153		
40	16380	49155			80	32766	32769		

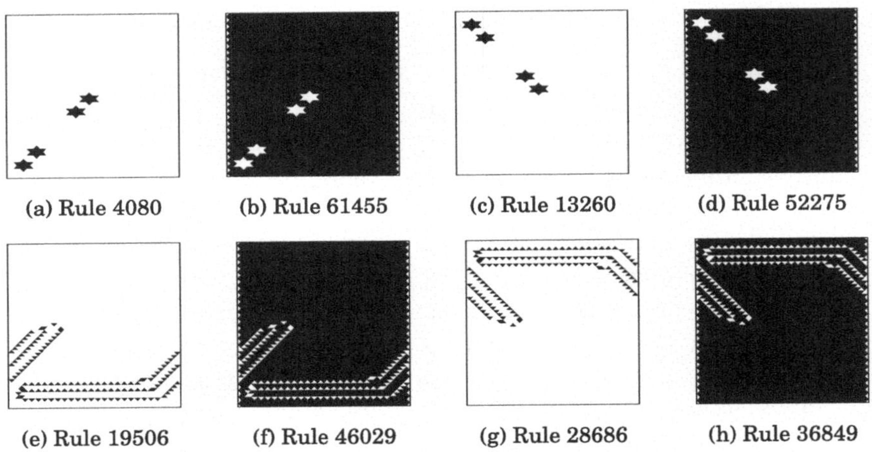

(a) Rule 4080 (b) Rule 61455 (c) Rule 13260 (d) Rule 52275

(e) Rule 19506 (f) Rule 46029 (g) Rule 28686 (h) Rule 36849

Fig. 4. Evolution of rules belonging to equivalence classes 16 and 53 ($t = 20$)

Property 1 (Isomorphism under Vertical Symmetry). Let r_1 and r_2 be any two complement-invariant rules of 2D triangular CA. If r_1 and r_2 are vertically symmetric, their corresponding state transition diagrams g_1 for r_1 and g_2 for r_2 are isomorphic directed graphs.

Proof. Let τ denote the vertical reflection operator acting on configurations, and suppose that the local rules r_1 and r_2 satisfy

$$r_2 = \tau \circ r_1 \circ \tau^{-1}.$$

Let g_1 and g_2 be the state transition graphs induced by r_1 and r_2, respectively, whose vertex sets consist of all $m \times n$ configurations of the cellular automaton. Define a mapping $\Phi : \mathscr{V}(g_1) \to \mathscr{V}(g_2)$ by

$$\Phi(C) = \tau(C),$$

for every configuration $C \in \mathscr{V}(g_1)$.

Bijectivity: Since τ represents vertical reflection, it is reversible with $\tau^{-1} = \tau$. Hence Φ is both injective and surjective, and therefore bijective.

Edge Preservation: Consider an edge $\mathscr{C} \to r_1(\mathscr{C})$ in g_1. We have

$$\begin{aligned}\Phi(r_1(\mathscr{C})) &= \tau(r_1(\mathscr{C}))\\ &= r_2(\tau(\mathscr{C}))\\ &= r_2(\Phi(\mathscr{C})).\end{aligned}$$

Every transition $\mathscr{C} \to r_1(\mathscr{C})$ in g_1 corresponds to a transition $\Phi(\mathscr{C}) \to r_2(\Phi(\mathscr{C}))$ in g_2. □

In matrix form, this corresponds to $A_{r_2} = P_\tau \cdot A_{r_1} \cdot P_\tau^{-1}$, where P_τ is the permutation matrix, where A_{r_1} and A_{r_2} are adjacency matrices corresponding to the graphs g_1 and g_2 respectively.

The above property holds for both null boundary and periodic boundary conditions. Figures 6(a) and 6(b) illustrate this phenomenon for a 2×2 CA under null boundary conditions, while Figs. 6(c) and 6(d) show the same under periodic boundary conditions. In both cases, the two transition diagrams exhibit structural equivalence.

Self-similar Fractals. A *self-similar fractal* is one in which the entire structure is composed of scaled copies of itself. In CA, self-similar fractals emerge when the evolution from a finite initial configuration reproduces its own shape at successive time steps. The resulting configuration shows recursive structures appearing in multiples as shown in Fig. 5 (e.g., doubled, quadrupled, or more).

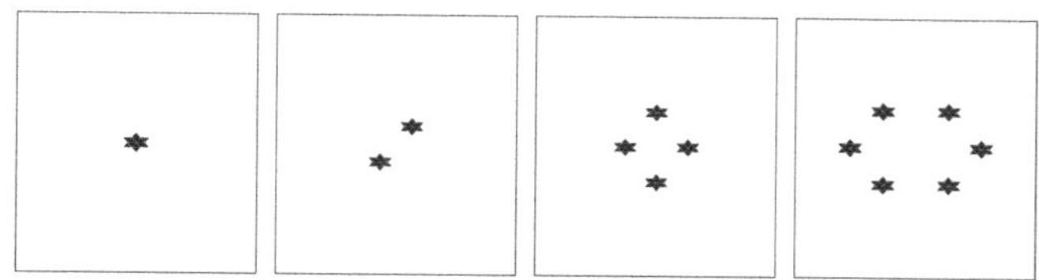

Fig. 5. Self-similar fractals

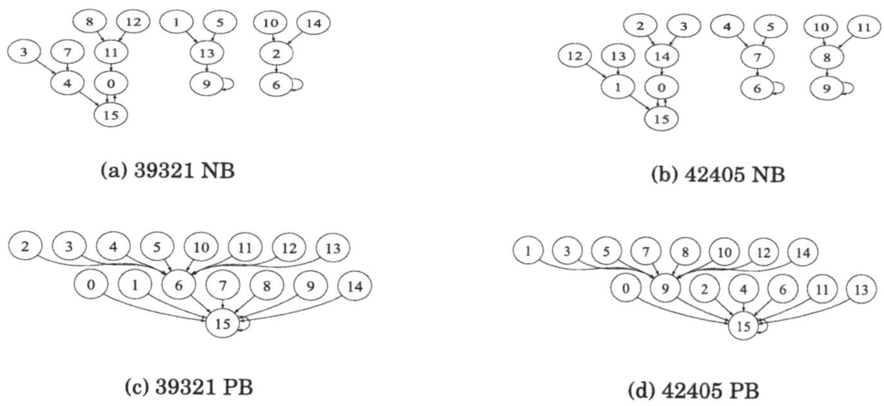

Fig. 6. State transition diagrams for 2×2 CA

After classifying all complement-invariant rules into 80 equivalence classes using vertical symmetry and complement symmetry, we simulated their evolution on finite triangular 2D lattices. Out of all complement-invariant rules, we found 14 rules that exhibit notable fractal patterns. These 7 pairs of rules form a visually and mathematically interesting subset of the complement-invariant rules, demonstrating how local rule properties can dictate global pattern formation. Among these, 7 rules produce exact *self-similar fractals*, i.e., patterns that show recursive and invariant structures. Their bitwise complements produce *complementary evolutions*, in which the self-similar fractal structure is preserved with states 0 and 1 interchanged. Table 4 lists the 7 complement-invariant rules that generate exact self-similar fractals, along with their bitwise complements, which produce complementary fractal patterns.

Table 4. Self-similar fractal-generating rules

Rule ($\mathbb{C}_{\mathbb{S}}$)	4080	13260	15420	21930	23130	26214	27030
Complement ($\overline{\mathbb{C}_{\mathbb{S}}}$)	61455	52275	50115	43605	42405	39321	38505

4 Conclusion

In conclusion, this work provides a systematic classification of 2D complement-invariant CA defined on a finite triangular lattice with a von Neumann neighborhood and periodic boundary conditions. We demonstrate that complement symmetry arises within complement-invariant rules, reducing the 256 complement-invariant rules to a set of 80 equivalence classes. The vertical symmetry property is observed to hold under both null and periodic boundary conditions, whereas the equivalence class formation is valid under periodic boundary conditions alone. For future work, we aim to investigate further equivalence relations based on similarities in global dynamical behavior, with the goal of identifying additional classes and thereby significantly reducing the complexity of the 2D triangular lattice CA rule space. The study proved the isomorphism of vertically symmetric rules and highlighted the self-similar fractal-generating behavior exhibited by complement-invariant rules.

References

1. Von Neumann, J., Burks, A.W.: Theory of self-reproducing automata. IEEE Trans. Neural Networks **5**(1), 3–14 (1966)
2. Moore, E.F.: Machine models of self-reproduction. In: Proceedings of Symposia in Applied Mathematics, vol. 14, no. 5, pp. 17–33 (1962)
3. Packard, N.H., Wolfram, S.: Two-dimensional cellular automata. J. Stat. Phys. **38**(5), 901–946 (1985)
4. Durand, B.: Global properties of 2D cellular automata: some complexity results. In: Borzyszkowski, A.M., Sokołowski, S. (eds.) MFCS 1993. LNCS, vol. 711, pp. 433–441. Springer, Heidelberg (1993). https://doi.org/10.1007/3-540-57182-5_35

5. Bhattacharjee, K., Naskar, N., Roy, S., Das, S.: A survey of cellular automata: types, dynamics, non-uniformity and applications. Nat. Comput. **19**, 433–461 (2020)
6. Bays, C.: Cellular automata in the triangular tessellation. Complex Syst. **8**(2), 127 (1994)
7. Morita, K.: Universality of 8-state reversible and conservative triangular partitioned cellular automata. In: El Yacoubi, S., Wąs, J., Bandini, S. (eds.) ACRI 2016. LNCS, vol. 9863, pp. 45–54. Springer, Cham (2016). https://doi.org/10.1007/978-3-319-44365-2_5
8. Cousin, P.: Triangular automata: the 256 elementary cellular automata of the two-dimensional plane. Complex Syst. **33**(3) (2024)
9. Saadat, M.: Cellular Automata in the Triangular Grid. Master's Thesis, Eastern Mediterranean University (EMU) (2016)
10. Zawidzki, M.: Application of semitotalistic 2D cellular automata on a triangulated 3D surface. Int. J. Des. Nat. Ecodyn. **6**(1), 34–51 (2011)
11. Uguz, S., Acar, E., Redjepov, S.: 2D triangular von Neumann cellular automata with periodic boundary. Int. J. Bifurcat. Chaos **29**(03), 1950029 (2019)
12. Uguz, S., Redjepov, S., Acar, E., Akin, H.: Structure and reversibility of 2D von Neumann cellular automata over triangular lattice. Int. J. Bifurcat. Chaos **27**(06), 1750083 (2017)
13. Wolfram, S.: A New Kind of Science. Wolfram Media, Champaign (2002)
14. Fukś, H.: Second order additive invariants in elementary cellular automata. arXiv:nlin/0502037 (2005)
15. D'Alotto, L.: A classification of two-dimensional cellular automata using infinite computations. The Infinity Computer (2020)
16. Kozlov, V.: Elementary cellular automata as invariant under symmetry transformations. MDPI (2022)
17. Vellarayil Mohandas, N.: Classification of two-dimensional binary cellular automata rules. MDPI (2018)
18. Uguz, S., Sahin, U., Akin, H., Siap, I.: Self-replicating patterns in 2D linear cellular automata. Int. J. Bifurcat. Chaos **24**(01), 1430002 (2014)
19. Uguz, S., Sahin, U., Akin, H., Siap, I.: 2D cellular automata with an image processing application. Acta Phys. Pol., A **125**, 435–438 (2014)
20. Uguz, S., Akin, H., Siap, I., Sahin, U.: On the irreversibility of Moore cellular automata over the ternary field and image application. Appl. Math. Model. **40**, 8017–8032 (2016)

Analysing Couples $(131, X)$ Under Temporally Stochastic Environment

Prem Patel[1], Het Naik[1], Hetvi Shah[1], Twisha Patel[1], Manushree Patel[1], Sumit Adak[2], and Souvik Roy[1(✉)]

[1] School of Engineering and Applied Science, Ahmedabad University, Ahmedabad, Gujarat, India
{prem.p1,het.n,hetvi.s4,twisha.p1,manushree.p,souvik.roy}@ahduni.edu.in
[2] DTU Compute, Technical University of Denmark, Kgs. Lyngby, Denmark
suad@dtu.dk

Abstract. To understand the effect of temporal mixing in cellular automata, this study considers temporally stochastic couples $(131, X)$ where elementary cellular automata 131 acts as a blind rule and X comes from the set of the rest elementary cellular automata rules. Overall, the result shows that periodic rule 131 is associated with a tendency to generate chaotic phenomena under the temporally stochastic environment. To understand the microscopic reason, here, we analyse the following situations (a) couple $(131, 50)$ where two periodic rules together show chaotic dynamics for any mixing rate; (b) couple $(131, 78)$ where it is possible to observe class change (from chaos to simplicity) for changing mixing rate; and (c) couple $(131, 136)$ where this probabilistic system shows important phase transition dynamics. This study also classifies couples $(131, X)$ based on these different interesting behaviours which guides towards overall dynamics of temporally stochastic elementary systems.

Keywords: Elementary Cellular Automata (ECA) · Temporally Stochastic ECA · Chaos · Class transition · Phase transition

1 Introduction

The notion of stochastic[1] cellular automata has gained a lot of recent attention in the cellular automata (CA) research community [2,5,12,14]. Initially, CA researchers have introduced the notion of (spatial) stochastic cellular automata (also known as probabilistic CA) which can be defined as random mixtures of multiple distinct CA rules [2]. Specifically, most of the recent works use random mixtures of two distinct elementary cellular automata (ECA)[2] (also known as diploid CA) to understand the dynamics of these probabilistic systems [3,4,8, 10,23]. In the direction of computing with this noise, Fatès [9] has proposed

[1] Note that, Greek word 'stochos' carries the notion of target or aim [9].
[2] one-dimensional cellular automata after considering two-state and three-neighbourhood dependency [27].

H. Raju et al. (Eds.): ASCAT 2026, CCIS 2801, pp. 116–129, 2026.
https://doi.org/10.1007/978-3-032-18612-6_9

an efficient solution for the density classification problem[3] with diploid cellular automata[4]. Bolt et al. [3,4] have proposed an algorithm for *identification problem* [3] for diploid CA. From the dynamical systems point of view, Fatès [10] has also explained the event of *phase transition* for diploid couples (i.e. two rules) where one of the members of the couples is null, inversion and identify rule[5]. The work of [8] has also identified the event of phase transition for diploid couples with null rule following the theoretical notation of *mean field approximation.* In a recent work [22,23], Roy et al. have identified emergent behaviour of diploid CA after considering the notion of family noise[6] which reports brutal *class transition* and *phase transition* dynamics. Apart from that, in the similar direction, *hybrid*-CA, CA with memory, *asynchronous*-CA [1,6,11,13,16,17] have also gained a lot of attention as a probabilistic CA model.

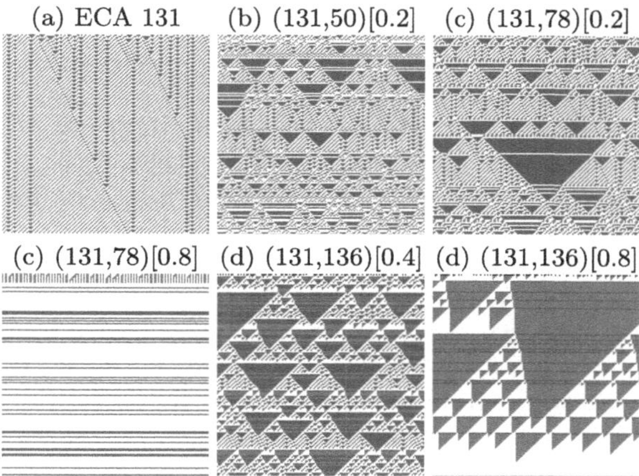

Fig. 1. Space-time pictures of (a) periodic ECA 131; (b) couple (131, 50) [0.2] which shows chaotic dynamics; (c) class transition dynamics of couple (131, 78) which shows chaotic (resp. periodic) dynamics for $\tau = 0.2$ (resp. $\tau = 0.8$); and (d) phase transition dynamics of couple (131, 136) which depicts non-zero (resp. zero) density for $\tau = 0.4$ (resp. $\tau = 0.8$). Here, white box reports state zero. Whereas, blue (resp. red) box captures state one for rule 131 (resp. other rules). (Color figure online)

In this direction, to capture natural complex systems modelling perspective, Roy et al. [21,24] have recently introduced the notion of *temporally stochastic* cellular automata where it allows temporal mixing of two distinct ECA rules. In

[3] In the density classification problem, if the number of 1's in the initial configuration is more, then the system should converge to all-1; otherwise, the destination should be an all-0 configuration [9].

[4] mixing of elementary CA rules 184 and 232 [9].

[5] elementary CA rule 0, 51 and 204 resp. is the null, inversion and identify rule [10].

[6] noise with 0 to 1, left to right, and both transformation [23].

a temporally stochastic system, say $(g, h)[\tau]$, for all the cells, we use (noise) rule h for some time steps. Whereas, we apply (default) rule g for rest of the time steps (for all the cells). Here, the probability of choosing noise rule h for a time step is τ. That is, in a time step, we either choose rule g or select rule h for all the cells (i.e., overall system). Therefore, one can observe the temporal nature of noise in this system. In the context of computational ability, Paul et al. [21] have identified the pattern classification ability of convergent temporal stochastic elementary CA rules where the system shows good efficiency in comparison with benchmark algorithms (like, Bayesian, C4.5 etc.). Paul et al. [20] have also used chaotic temporally non-uniform CA (deterministic version of temporally stochastic CA) for pseudo-random number generation after identifying the theoretical property "surjective" which shows efficient results in comparison with different versions of the Mersenne Twister. Paul [18] has reported the ability of temporally stochastic CA rules in solving *global synchronization problems*[7] with some relaxation. Apart from that, the work of [19] has also provided evidence of the epidemic spread modelling capability of two-dimensional temporally non-uniform cellular automata.

Now, in the first experiment of temporally stochastic elementary CA [24], Roy et al. have reported the dynamics of all possible 3828 couples after considering 88 minimal representative ECA rules. According to the space-time results of Ref. [24], temporally stochastic ECA systems show following exciting dynamics – (a) there are example where two periodic rules together under temporally stochastic update reports chaotic dynamics; (b) there are evidence of *class transition*, i.e., for changing noise rate (τ), the probabilistic system moves from one Wolfram class [27] to another; and (c) It also reports example of phase transition where system changes the status from non-zero density to zero density for changing parameter τ. However, the results are based on qualitative space-time dynamics only which can not be considered as a formal approach.

According to the initial results [24], ECA 131 as a part of the couple rules show all this kind of exciting evidence. Moreover, Roy et al. have also reported in the initial results [24] that ECA 131 has a tendency to generate *chaotic* phenomena under temporally stochastic update. With this background, this paper explores ECA 131 under the temporally stochastic environment to understand the microscopic details. Note that, according to Wolfram's classification, ECA 131 comes under the (simple) periodic class [27]. For example, Fig. 1(a) displays the space-time picture of ECA 131.

In this direction, we first report the dynamics of the temporally stochastic rules $(131, X)[\tau]$ following formal *density* and *entropy* parameters, where X comes from the set of rest ECA rules and $\tau \in [0, 1]$. After that, we provide the microscopic justification behind the following exciting behaviour of $(131, X)$ couples: (i) There is much evidence that ECA 131, along with other periodic ECA rules, shows chaotic behaviour. Here, we provide the reason for the couple $(131, 50)$, see Fig. 1(b) for evidence; (ii) We also discuss the reason behind the

[7] In the global synchronization problem, every initial configuration should converge to the "blinking state" (i.e., all-0 and all-1) [19].

class transition dynamics for the couple (131, 78), see Fig. 1(c); and (iii) Lastly, we report the justification behind the phase transition for the couple (131, 136), see Fig. 1(d). These microscopic results also provide an overall insight into the dynamics of temporally stochastic ECA systems after considering the notion of rule mean term (RMT).

2 ECA 131 and Temporally Stochastic ECA

Here, we consider two-state three-neighbourhood elementary CA (ECA) where the cells (denoted as $\mathcal{L} = \mathbb{Z}/n\mathbb{Z}$) are organized as a one-dimensional ring [27]. A cell is associated with a state ({0,1}) at every time step $t \in \mathbb{N}$. A configuration is the collection of states at a given time. Here, the system follows the local transition function $f : \{0,1\}^3 \rightarrow \{0,1\}$. Indeed, ECA 131 can be written as follows:

(a,b,c) (RMT)	111 (7)	110 (6)	101 (5)	100 (4)	011 (3)	010 (2)	001 (1)	000 (0)	Rule
$f(a,b,c)$	1	0	0	0	0	0	1	1	131

Here, we use the *rule mean term* (RMT) for the naming purpose of the neighbourhood triplet (a, b, c); and we use classical *decimal code* for the naming purpose of the 'rule'. If $f(a, b, c)= b$, then the RMT is *passive*; otherwise, we call it as *active*. Note that, there are 256 rules in this elementary cellular systems, however, we only consider 88 because the rest are equivalent [27].

Recall that, in a temporally stochastic system, say $(g, h)[\tau]$, for all the cells, we use *noise* rule h for some time steps. Whereas, we apply *default* rule g for rest of the time steps (for all the cells). Here, the probability of choosing noise rule h for a time step is τ. That is, in a time step, we either choose rule g with $1 - \tau$ probability or select rule h with τ probability for all the cells. Therefore, one can observe the temporal nature of noise in this system. Formally, the construction of next configuration follows

$$\begin{cases} G_g & \text{with probability } 1 - \tau \\ G_h & \text{with probability } \tau \end{cases}$$

where, G_g and G_h are the corresponding *global* transition functions associated with the local transition functions g and h respectively, where $G_g, G_h : \{0,1\}^{\mathcal{L}} \rightarrow \{0,1\}^{\mathcal{L}}$ and $\tau \in [0, 1]$. In this study, we consider ECA 131 as a blind rule, i.e., $(131, X)$ where, X comes from the set of rest 87 ECA rules.

To capture the space-time pictures, we run the system for 1000 time steps starting with CA size $n = 1000$ with an initial configuration of density $d_0 = 0.5$. Next, to understand the dynamics of temporally stochastic ECA, we use formal parameters density portfolio and entropy [22]. Here, density of a configuration at time t is denoted as $d_t = 1_d/n$ where 1_d notes the number of 1 in the configuration. To capture the density portfolio, we run the system for 2000 time steps after considering CA size $n = 50$. Hereafter, we compute the average density of the system for 2000 to 2100 time steps. In this computation, the τ varies from 0 to 1 with an interval of 0.1 and initial density d_0 varies from 0.1 to 0.9 with an interval of 0.1. Therefore, average density d_a (τ, d_0) gives a two-dimensional sampling surface.

Next, to compute the (Kolmogorov-Sinai) entropy, we store the sequence of configurations (say, Γ) during $t = 10,00,000$ time steps for $n = 50$. Let us consider that, Γ_c is the number of occurrence of configuration c in the sequence Γ. Hence, $\rho(c) = \frac{\Gamma_c}{t}$ depicts the probability of occurrence for configuration c. Now, $H = -\sum_{\forall c} \rho(c) log(\rho(c))$ reports the entropy following the definition of Shannon's entropy. Therefore, $H(\tau, d_0)$ provides the two-dimensional entropy surface. Note that the uncertainty of the system and entropy are directly correlated. In this study, we (mainly) use density (resp, entropy) surface to capture phase (resp, class) transition dynamics. The reader can follow Ref [22] to understand the details of the experimental protocol and corresponding drawbacks.

3 Dynamics of (131,X) Couples

In this section, we report the dynamics of temporally stochastic couples $(131, X)$ following the space-time pictures, density and entropy portfolio. To report the dynamics of the system, we consider Wolfram's classification [27] with some relaxation. We mainly try to classify the temporally stochastic CA dynamics into two classes – *simple behaviour* and *chaotic behaviour*. Here, the notion of simple behaviour includes *null*, *fixed point*, *periodic*, and *locally chaotic dynamics* [26]. On the other hand, the notion of chaotic behaviour includes *chaotic* and *complex* dynamics after considering the formal literature of cellular automata classification problems [15,26]. Apart from that, for changing parameter τ, if the temporally stochastic CA changes its behaviour from simple to chaotic (or, vice versa), then we call the behaviour as *class transition*. Additionally, for the changing parameter τ, if the temporally stochastic CA moves from non all-zero non all-one density to all zero (or all one) density, then we call the behaviour as *phase transition* dynamics.

According to the results, Table 1 reports the behaviour of couples $(131, X)$ where 36 couples (resp, 22 couples) show simple (resp. chaotic) dynamics, where progressive change of τ does not affect the system's dynamics, i.e., τ independent dynamics. Moreover, according to Table 1, 28 couples show class transition dynamics and 1 couple reports phase transition dynamics under temporally stochastic update. Note that Table 1 reports the X rules only (of the couple $(131, X)$) after considering 131 as a blind rule. Next, we discuss these different kinds of interesting behaviour with examples.

First, we discuss the temporally stochastic couples which show *simple* dynamics. Under this umbrella, most of the $(131, X)$ couples show periodic dynamics with little amount of noise. For evidence, Fig. 2 reports the space-time pictures

Table 1. Classification of temporally stochastic CAs (131,X) after considering qualitative (space-time pictures) and quantitative (density and entropy portfolio) parameters.

Simple	0	1	2	3	4	8	10	11	12	13	24	25	32	33	34
	35	36	37	40	42	43	44	56	57	72	74	76	106	108	130
	138	140	162	170	172	204									
Chaotic	18	22	26	28	30	41	50	54	58	60	90	94	110	122	126
	128	146	150	156	160	168	178								
Class transition	5	6	7	9	14	15	19	23	27	29	38	45	46	51	73
	77	78	104	105	132	134	142	152	154	164	184	200	232		
Phase transition	136														

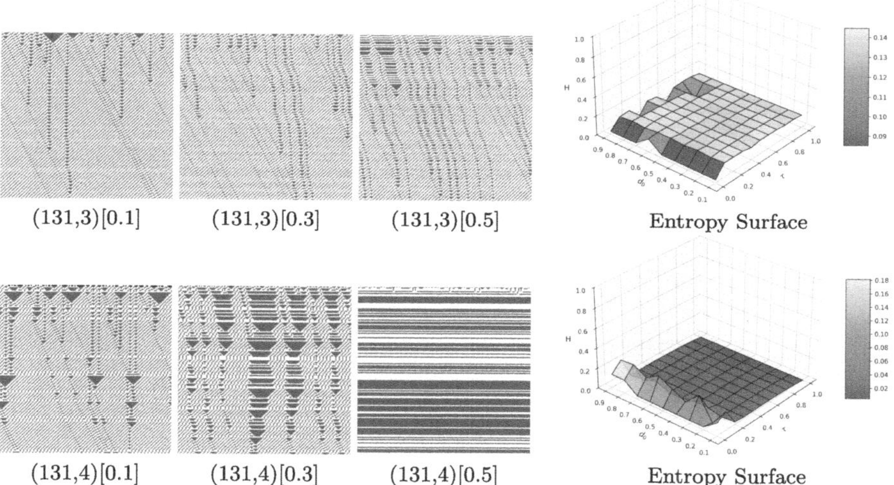

Fig. 2. Evidence space-time pictures and entropy surface for temporally stochastic couples (131, 3) and (131, 4) which show simple class dynamics following Table 1.

of couple (131, 3) for $\tau = \{0.1, 0.3, 0.5\}$. Here, the two-dimensional entropy computation also shows a *flat surface*. On the other hand, some of the couples from this class also show periodic dynamics which solves the *global synchronization problem* with some relaxation (known as *self-synchronization problem*[8] [18]). For evidence, Fig. 2 reports the space-time pictures of couple (131, 4) for $\tau = \{0.1, 0.3, 0.5\}$. Here, ECA 0, 1, 8, 32, 33, 36, 37, 40 and 72 show similar self-synchronization problem solving ability. Moreover, note that all the rules of this simple class come from Wolfram's class I and II [27]. However, it is not possible to claim the system is *robust*. That is, let us consider that $131 = X =$

[8] In the self-synchronization problem, every initial configuration should converge to the blinking states all-0 and all-1 where the blinking states appear with some probability [18].

Periodic (according to Wolfram's class) and $(131, X)$ also shows periodic dynamics according to space-time pictures; however, the periodicity of $(131, X)$ is not the same as the periodicity of synchronous update. Here, the periodic dynamics under temporally stochastic update also reflects the effect of noise.

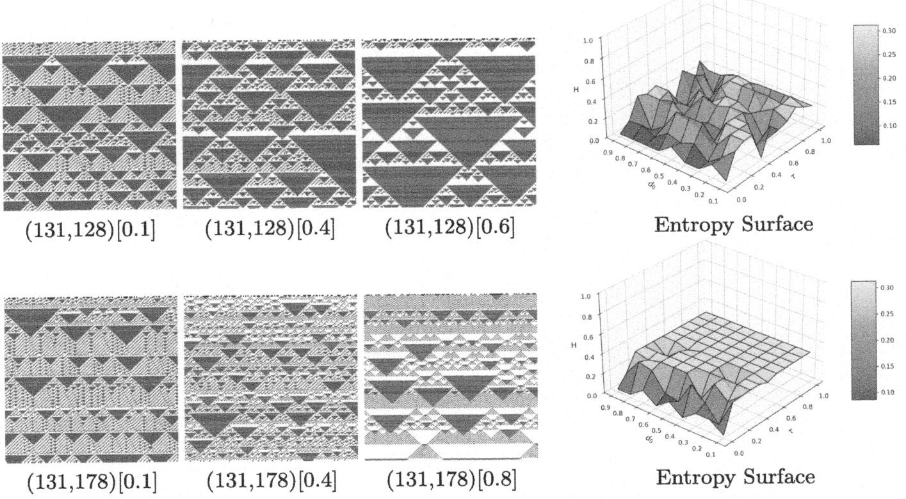

Fig. 3. Evidence space-time pictures and entropy surface for temporally stochastic couples $(131, 128)$ and $(131, 178)$ which show chaotic class dynamics following Table 1.

Next, we discuss chaotic dynamics of temporally stochastic couples $(131, X)$. Here, Fig. 3 reports the space-time pictures and entropy surface for couples $(131, 128)$ (for $\tau = \{0.1, 0.4, 0.6\}$) and $(131, 178)$ (for $\tau = \{0.1, 0.4, 0.8\}$). Note that, according to Table 1, temporal mixture of most of the chaotic rules[9] and ECA 131 shows chaotic dynamics. Moreover, it is not an abnormal situation after considering the assumption: chaotic rule dominates the couple's behaviour with rule 131. However, there are evidences in Table 1 where Wolfram's class I [27] null rules (i.e., 128, 160 and 168) and Wolfram's class II [27] periodic or fixed point rules (i.e., 28, 50, 58, 94, 156 and 178) as a member of $(131, X)$ couple show chaotic dynamics. In other words, two simple rules together show chaotic dynamics under temporally stochastic update, for evidence see Fig. 3. Here, couple $(131, 128)$ shows an uneven entropy surface, and couple $(131, 178)$ reports the supporting high entropy evidence, see Fig. 3. To understand this extreme situation, we discuss the case study of a couple $(131, 50)$ in the next section.

[9] Chaotic ECA rules 18, 22, 30, 60, 90, 122, 126, 146, 150 and complex ECA rules 41, 54, 110.

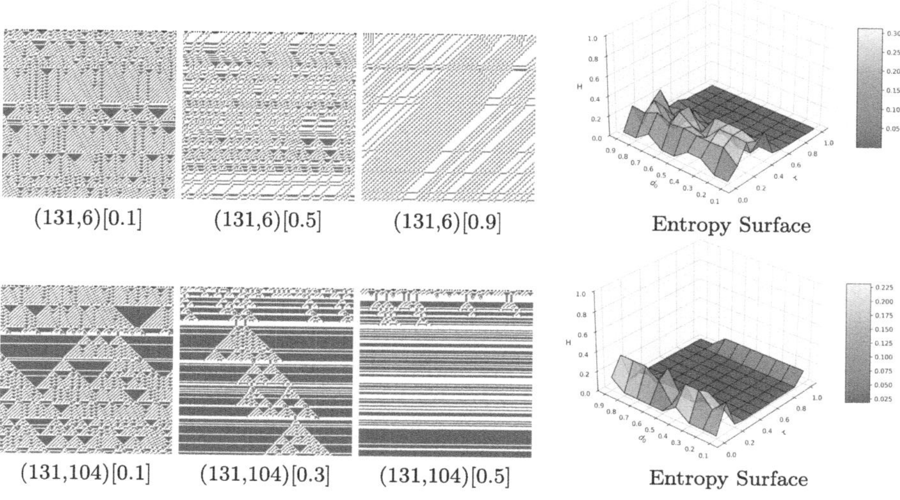

Fig. 4. Evidence space-time pictures and entropy surface for temporally stochastic couples (131, 6) and (131, 104) which show class transition dynamics following Table 1.

Next, we show the evidence of class transition dynamics of $(131, X)$ couples where the systems depict τ dependent dynamics. Under this situation, the system moves from chaotic to simple (chaotic $\leftrightarrow$ simple) dynamics for changing τ parameters. For example, in Fig. 4, couple (131, 6) shows chaotic behaviour for $\tau = 0.1$, however, it reflects simple evolution for $\tau = 0.9$. Here, the entropy surface of the couple $(131, X)$ also captures the class transition, i.e., moves from high entropy to low entropy for changing τ parameter, see Fig. 4. Under this situation, some $(131, X)$ couples also provide the solution of the self-synchronization problem for specific τ value. For example, couple (131, 104) displays chaotic dynamics for $\tau = 0.1$, however, this temporally stochastic rule provides simple dynamics for $\tau = 0.5$ which compute the self-synchronization problem, see Fig. 4. The same is also applicable for ECA 5, 7 and 78. Under the *class transition* category of Table 1, most of the members of couples $(131, X)$ come from Wolfram's class II[10], i.e., two simple rules under temporally stochastic update capable of showing chaotic dynamics for some noise rate τ. However, there is evidence of chaotic members[11] in this class where the chaotic rule as a member of a couple can be able to dominate based on the mixing rate.

[10] ECA rules 5, 6, 7, 9, 14, 15, 19, 23, 27, 29, 38, 46, 51, 73, 77, 78, 104, 132, 134, 142, 152, 154, 164, 184, 200, and 232.

[11] ECA rules 45 and 105.

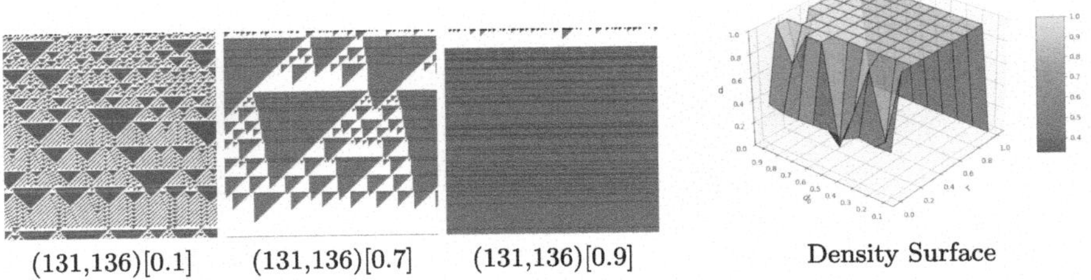

Fig. 5. Evidence space-time pictures and density surface for temporally stochastic couple (131, 136) which show phase transition dynamics following Table 1.

As evidence of phase transition, Fig. 5 reports the dynamics of the temporally stochastic couple (131, 136) for $\tau = \{0.1, 0.7, 0.9\}$. Observe that, for $\tau = 0.1$, the temporally stochastic system reports non all-zero non all-one density. Moreover, the dynamics are chaotic (i.e., *active phase*). Now, for $\tau = 0.9$, rule $(131, 136)$ converges to all-one configuration (i.e., *inactive phase*). Here, the critical value of phase transition is close to $\tau \approx 0.5$, see the density surface in Fig. 5 for evidence. We also discuss the microscopic reason behind this phase transition in the next section.

4 Microscopic Analysis of (131,X) Couples

To understand the microscopic dynamics of $(131, X)$ couples, we first explore the dynamics of ECA 131. According to Section 2, RMTs 7, 5, 4 are passive and RMTs 6, 3, 2, 1, 0 are active for ECA 131. Figure 6(a) displays the space-time picture of ECA 131. According to the space-time dynamics of Fig. 6(a), one can observe the following: (i) it can able to create *large triangles*[12] based on the initial configuration; (ii) After that, it can able to create only *small triangles* which is repetitive (kind of chain) in nature; (iii) There are also event of 'death' of the *small triangle chain*; and (iv) lastly, pattern $(001)^k$ ($k \in \mathbb{N}$) shows stable shift dynamics in ECA 131. Next, we provide the justification of the above events.

[12] Here, we consider a triangle as large if the size of the base is ≥ 5; otherwise, it is a small triangle.

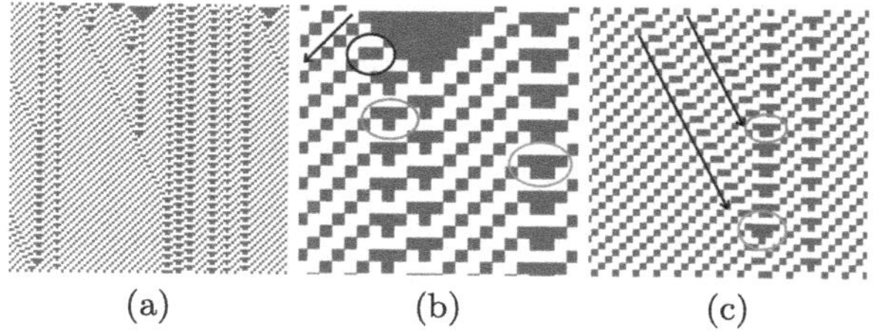

Fig. 6. (a) space-time picture for ECA 131; (b) microscopic dynamics of ECA 131: (i) black arrow depicts the creation and propagation of signal 1, (ii) black circle indicates the creation of signal 2, (iii) red circle indicates the creation of signal 3 or self-reproducing triangle 3, (iv) green circle denotes self-reproducing triangle 4; (c) microscopic dynamics of ECA 131: (i) black arrow depicts the propagation of signal 2, (ii) red circle indicates the creation of self-reproducing triangle 4, (iii) green circle indicates the creation of triangle 5. (Color figure online)

Here, the base of a triangle is consecutive 1 (i.e., $1 \cdots 1$) states. For ECA 131, RMT 0 is active. Therefore, consecutive 0 (i.e., $0 \cdots 0$) states move to consecutive 1 (i.e., $00 \cdots 00 \to d1 \cdots 1d$[13]) states in the next time step. Therefore, if there exists consecutive 0 or 1 *local* configuration in the initial configuration, it can be used as a *base* for triangles (both large and small). Here, RMT 7 is passive, therefore it is not possible to disturb the triangle from the middle. However, RMTs 3 and 6 both are active, therefore, the length of consecutive 1 states decreases from both sides during iteration, and it creates the triangle; for evidence, see the large triangle of base 11 in Fig. 6(b). Now, according to the observation, there exists no large triangle after initial steps which leads to the following question – is it possible to create large length consecutive 0 or 1 states as a part of intermediate configuration? Next, we explore this question.

Firstly, $(001)^k$, $k \in \mathbb{N}$ states follows only shift dynamics (i.e., $001001 \to 010010$) because RMT 4 is passive and RMTs 1 and 2 are active. Here, let us call the pattern 001 as *signal* 1. Now, we discuss the creation of signal 1. If we get a pattern like $d10011 \cdots$ where the pattern 00 is associated with a triangle in the right hand side, then it is possible to create a single 1 because RMTs 1, 2, 3, 6 are active (i.e., $d10011 \cdots$ (RMT sequence 6/2413) $\to d00101 \cdots$). Hereafter, signal 1 propagates in the left and captures the decreasing space of the left triangle. Moreover, the right triangle keeps generating signal 1 at every alternative step. Note that, if a space is occupied by signal 1, then it is not possible to create any other pattern or triangle in that space during iterations. In Fig. 6(b), the black arrow indicates the creation and propagation of signal 1.

Now, in a similar way, if we get a pattern like $d100011 \cdots$ where the pattern 000 is associated with a triangle in the right hand side, then it is possible to create a pattern 0110 (we call *signal* 2). Hereafter, from signal 2, we get signal 1 back. In Fig. 6(b), the black circle indicates the creation of signal 2. Similarly,

[13] Here, 'd' reports don't care about the situation.

If we get a pattern like $d1\underbrace{0\cdots0}_{\text{k zeros}}11\cdots$ where the pattern $\underbrace{0\cdots0}_{\text{k zeros}}$ is associated with a triangle in the right hand side, then it is possible to create a pattern $0\underbrace{1\cdots1}_{\text{k-1 ones}}0$ (we call *signal* $k-1$). However, the maximum possible size of k is four because the right triangle keep generating this kind of signal at every alternative steps. Therefore, apart from signal 1 and 2, we only get the signal 01110 (we call *signal* 3). In Fig. 6(b), the red circle indicates the creation of signal 3. The above discussion explains why it is not possible to create large length consecutive 0 or 1 states ($k \geq 5$) as a part of intermediate configuration.

Here, signal 3 can be able to create small triangle of base 3: $\begin{smallmatrix}1&1&1\\0&1&0\end{smallmatrix}$ (we call *triangle* 3), see Fig. 6(b). Moreover, a triangle (of base) 3 keeps *self-reproducing* itself in a static way because RMTs 0 and 1 are active, i.e., $\begin{smallmatrix}1&1&1\\0&1&0\\0&0&0\\1&1&1\end{smallmatrix}$, see Fig. 6(b). Moreover, for this self-reproduction process, triangle 3 creates space (of size 1) in the left hand side by producing signal 1 in every alternative step and the incoming signal 1 from right hand side also gets synchronized with triangle 3 to create the space of size 1. Additionally, note that, a triangle (of base) 4 also keeps *self-reproducing* itself in a static way, i.e., $\begin{smallmatrix}1&1&1&1\\0&1&1&0\\0&0&0&0\\1&1&1&1\end{smallmatrix}$, see Fig. 6(b) in green circle following the same RMT behaviour. Next, we discuss the creation of triangle 4.

Recall that, signal 2 (pattern 0110) can be able to create a signal 1 in the left hand side. Additionally, if signal 2 encounters a signal 1 coming from the right hand side, it (signal 2) propagates in the right direction. Here, active RMTs 0 and 1 play an important role. Figure 6(c) depicts the propagation of signal 2 (see black arrow). Now, if signal 2 encounters triangle 3, then it is able to create triangle 4 by using active RMT 0, see Fig. 6(c) in the red circle. Again, if signal 2 encounters triangle 4, then it is able to create triangle 5 by using active RMT 0, see Fig. 6(c) in the green circle. However, according to the initial logic, triangle 5 is not able to reproduce itself. And, it will die. Finally, Signal 1 dominates the space-time of ECA 131. The above discussion includes all possibilities. Note that the same analysis is also applicable for equivalent ECA rules 62, 118, and 145. In this context, the works of [7,25] have also discussed these *glider* dynamics of ECA 62.

Next, we discuss the case study of couple (131, 50), where two periodic rules of Wolfram's class II [27] together show chaotic dynamics under the temporally stochastic mixing environment. Recall that ECA 131 is capable of creating triangles of different sizes; however, it is not possible in the intermediate steps due to absence of consecutive 0 or 1 states. Under the temporally stochastic environment, if the noise rule X (here, 50) is able to create consecutive 0 or 1 states after breaking the $(001)^k$ pattern or signal 1, then the default rule 131 is able to produce asymmetric triangles during evolution. Therefore, to create chaos from rule $(131, X)$, we need ECA X with $(001)^k$ pattern breaking capability. Note that pattern 001 is associated with RMTs 1, 2 and 4. Now, if the X rule is associated with following transitions $001 \rightarrow 1$, $010 \rightarrow 1$ and $100 \rightarrow 1$ (i.e., active RMT 1, passive RMT 2 and active RMT 4), then we get consecutive 1 states.

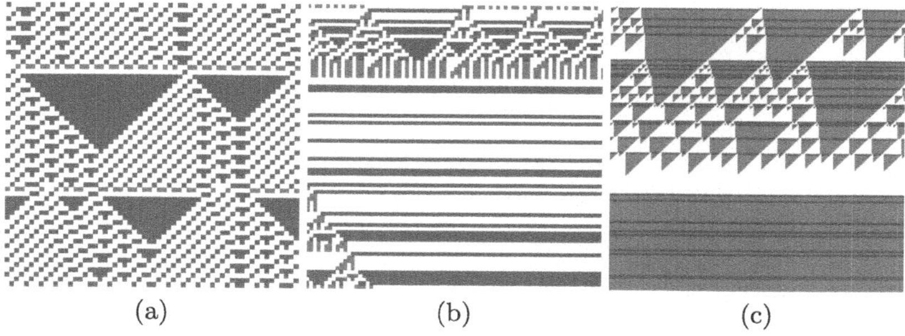

Fig. 7. (a) Microscopic space-time picture for couple (a) $(131, 50)$; (b) $(131, 78)$; and (c) $(131, 136)$.

Moreover, if $001 \rightarrow 1$, $010 \rightarrow 1$, and $100 \rightarrow 0$ is true for rule X, then we get pattern $(011)^k$, and ECA 131 is capable to convert pattern $(011)^k$ into consecutive 0 states (because RMTs 3 and 6 are active and RMT 5 is passive). The similar logic is also applicable for $001 \rightarrow 0$, $010 \rightarrow 1$, $100 \rightarrow 1$; $001 \rightarrow 1$, $010 \rightarrow 0$, $100 \rightarrow 1$; and $001 \rightarrow 0$, $010 \rightarrow 0$, $100 \rightarrow 0$. Therefore, rule X with any of the above properties is capable of breaking the $(001)^k$ pattern and $(131, X)$ with this property is capable of generating chaotic pattern. Here, rule 50 is associated with active RMTs 1 and 4 which depicts the reason behind chaotic behaviour. For evidence, see the space-time picture of Fig. 7(a) where the application of ECA 50 (in red) creates consecutive 1 states. In Table 1, most of the other rules[14] of this chaotic class are associated with this property. Remark that the above condition is one of the necessary conditions only. The next two case studies give an overview of the sufficient condition.

Next, we discuss the class transition dynamics of $(131, 78)$ which is also capable of solving self-synchronization problems based on the τ parameter. Firstly, note that, for ECA 78, RMTs 2 and 4 are passive and RMT 1 is active. Therefore, if we apply noise rule 78 with less probability then the couple $(131, 78)$ is able to show chaotic dynamics following the above argument. Next, observe that ECA 78 (individually) converges to a fixed point which is a combination of pattern 01 and 011 (RMTs 6, 5, 3, 2 are applicable). Therefore, if we apply noise rule 78 with high probability, then the system will get a proper chance for convergence. Now, during this process (after convergence), if we apply ECA 131, then the system moves to all-0 configurations because RMTs 2, 3, 6 are active and RMT 5 is passive. Following this, the system moves to all-1 configuration for ECA 131 because RMT 0 is active. Hereafter, if we apply ECA 78 again, then the system moves to all-0 configuration because RMT 7 is active. It captures the possibility of solving self-synchronization problems. To sum up, if we give ECA 78 repetitive number of chances (where, the number of chances is more than *convergence time* of ECA 78), the temporally stochastic system moves to

[14] Note that, there are some exceptional situations which also can be justified with similar logic with the combination of different RMTs.

periodic blinking state. The above is true for high noise rates. However, if we don't allow ECA 78 to settle down, then the system is capable of generating chaos which justifies the event of class transition for the couple $(131, 78)$, see Fig. 7(b) for evidence.

Next, we discuss the phase transition event for couple $(131, 136)$ where 136 is a null rule. For 136, $001 \rightarrow 0$, $010 \rightarrow 0$, $100 \rightarrow 0$ is applicable; therefore, it is also capable of breaking the $(001)^k$ pattern generated by ECA 131. Moreover, RMTs 7 and 3 are passive for ECA 136. Therefore, 136 disturbs existing triangles (generated by 131) very slowly using only active RMT 6. If we give 136 repetitive chances, then only it is able to create all-0 configuration by destroying the existing triangles. Otherwise, 136 creates local consecutive 0 configurations only which is helpful for 131 in future to create large triangles, see Fig. 7(c). Therefore, if we apply noise rule 136 with less probability, then $(131, 136)$ is also capable of generating chaotic dynamics (or active phase). However, if we give 136 more chances then it is able to destroy all the existing triangles, i.e., all-0 configurations or inactive phases (then, all-1 configuration using ECA 131). The above argument justifies the event of phase transition.

As a closing remark, this study gives an overview of the exciting possibilities of (new) temporally stochastic updating schemes after considering the evidence of $(131, X)$ couples. However, in this study, we have given justification of interesting events after considering the case studies of $(131, 50)$, $(131, 78)$ and $(131, 136)$ only. A proper theoretical understanding of this probabilistic system indicates the future direction.

References

1. Adamatzky, A., Mayne, R.: Actin automata: phenomenology and localizations. Int. J. Bifurcation Chaos **25**(02), 1550030 (2015)
2. Arrighi, P., Schabanel, N., Theyssier, G.: Stochastic cellular automata: correlations, decidability and simulations. Fundam. Inf. **126**(2–3), 121–156 (2013)
3. Bołt, W., Baetens, J.M., De Baets, B.: On the decomposition of stochastic cellular automata. J. Comput. Sci. **11**, 245–257 (2015)
4. Bołt, W., Bołt, A., Wolnik, B., Baetens, J.M., De Baets, B.: A statistical approach to the identification of diploid cellular automata based on incomplete observations. Biosystems **186**, 103976 (2019). Selected papers from the International Conference on the Theory and Practice of Natural Computing 2017
5. Bušić, A., Mairesse, J., Marcovici, I.: Probabilistic cellular automata, invariant measures, and perfect sampling. Adv. Appl. Probab. **45**(4), 960–980 (2013)
6. Chen, B., Chen, F., Martínez, G.J.: Glider collisions in hybrid cellular automaton with memory rule (43,74). Int. J. Bifurcation Chaos **27**(06), 1750082 (2017)
7. Chen, F., Shi, L., Chen, G., Jin, W.: Chaos and gliders in periodic cellular automaton rule 62. J. Cell. Autom. **7**, 01 (2012)
8. Cirillo, E.N.M., Nardi, F.R., Spitoni, C.: Phase transitions in random mixtures of elementary cellular automata. Phys. A: Stat. Mech. Appl. **573**, 125942 (2021)
9. Fatès, N.: Stochastic cellular automata solutions to the density classification problem - when randomness helps computing. Theory Comput. Syst. **53**(2), 223–242 (2013)

10. Fatès, N.: Diploid cellular automata: first experiments on the random mixtures of two elementary rules. In: Dennunzio, A., Formenti, E., Manzoni, L., Porreca, A.E. (eds.) AUTOMATA 2017. LNCS, vol. 10248, pp. 97–108. Springer, Cham (2017). https://doi.org/10.1007/978-3-319-58631-1_8
11. Fatès, N.: A tutorial on elementary cellular automata with fully asynchronous updating -general properties and convergence dynamics. Nat. Comput. **19**, 03 (2020)
12. Fernández, R., Louis, P.-Y., Nardi, F.R.: Overview: PCA models and issues. In: Probabilistic Cellular Automata: Theory, Applications and Future Perspectives, pp. 1–30 (2018)
13. Li, W., Packard, N.H., Langton, C.G.: Transition phenomena in cellular automata rule space. Phys. D **45**(1), 77–94 (1990)
14. Mairesse, J., Marcovici, I.: Around probabilistic cellular automata. Theoret. Comput. Sci. **559**, 42–72 (2014). Non-uniform Cellular Automata
15. Martinez, G.J.: A note on elementary cellular automata classification. J. Cell. Autom. **8**, 233–259 (2013)
16. Martínez, G.J., Aaamatzky, A., Alonso-Sanz, R.: Designing complex dynamics in cellular automata with memory. Int. J. Bifurcation Chaos **23**(10), 1330035 (2013)
17. Park, J.K., Steiglitz, K., Thurston, W.P.: Soliton-like behavior in automata. Phys. D **19**(3), 423–432 (1986)
18. Paul, S.: Self-synchronization of temporally stochastic cellular automata. In: Kamilya, S., Das, S., Formenti, E. (eds.) ASCAT 2025. CCIS, vol. 2499, pp. 1–18. Springer, Cham (2025). https://doi.org/10.1007/978-3-031-94121-4_1
19. Paul, S., Bhattacharjee, K.: Modeling spread of contagious disease by temporally stochastic cellular automata. In: Das, S., Martinez, G.J. (eds.) ASCAT 2023. AISC, vol. 1443, pp. 161–175. Springer, Singapore (2023). https://doi.org/10.1007/978-981-99-0688-8_13
20. Paul, S., Bhattacharjee, K.: Temporally non-uniform cellular automata as pseudo-random number generator. In: Dalui, M., Das, S., Formenti, E. (eds.) ASCAT 2024. CCIS, vol. 2021, pp. 56–71. Springer, Cham (2024). https://doi.org/10.1007/978-3-031-56943-2_5
21. Paul, S., Roy, S., Das, S.: Pattern classification with temporally stochastic cellular automata. In: Manzoni, L., Mariot, L., Roy Chowdhury, D. (eds.) AUTOMATA 2023. LNCS, vol. 14152, pp. 137–152. Springer, Cham (2023). https://doi.org/10.1007/978-3-031-42250-8_10
22. Roy, S., Adak, S.: Effect of family noise in diploid cellular automata. Nat. Comput. (2025)
23. Roy, S., Modi, H., Patel, R., Adak, S.: Are some family members harmful? - a study on diploid cellular automata. In: Bagnoli, F., Baetens, J., Bandini, S., Matteuzzi, T. (eds.) ACRI 2024. LNCS, vol. 14978, pp. 10–21. Springer, Cham (2024). https://doi.org/10.1007/978-3-031-71552-5_2
24. Roy, S., Paul, S., Das, S.: Temporally stochastic cellular automata: Classes and dynamics. Int. J. Bifurcation Chaos **32**(12), 2230029 (2022)
25. Shi, L., Chen, F., Jin, W.: Gliders, collisions and chaos of cellular automata rule 62. In: 2009 International Workshop on Chaos-Fractals Theories and Applications, pp. 221–225 (2009)
26. Vispoel, M., Daly, A.J., Baetens, J.M.: Progress, gaps and obstacles in the classification of cellular automata. Phys. D **432**, 133074 (2022)
27. Wolfram, S.: Cellular Automata and Complexity: Collected Papers, 1st edn. CRC Press (1994)

The Role of Distant Neighbor on Locally Chaotic ECAs: A Study

Bhumika Sikdar(✉)

Department of Information Technology, Indian Institute of Engineering Science and Technology, Howrah 711103, West Bengal, India
bhumikasikdar@gmail.com

Abstract. In this research, we aim to study the dynamical behavior of three locally chaotic elementary cellular automata (ECA) where a few cells are systematically integrated with distant neighbors. A simple model has been proposed where only a few controlled perturbations are sufficient to produce dynamical shifts.

Keywords: ECAs · Distant neighbor · Locally chaotic · Complexity

1 Introduction

Elementary cellular automata rules were classified into four categories by Stephen Wolfram in terms of their increasing complexity [7]. Class I patterns are the least complex, where almost every initial configuration converges to a uniform state. Patterns that settle into an almost periodic structure are in Class II. Class III forms globally chaotic patterns, and Class IV ECAs are dynamically complex. Li and Packard later identified three minimal locally chaotic rules [6] which are periodic as per Wolfram's classification.

The role of a single distant neighbor on the dynamical behavior of 88 minimal ECAs has recently been studied in [3]. This work explores the impact of multiple cells with distant neighbor on locally chaotic cellular automata. All locally chaotic cellular automata with their minimal representation are taken into consideration for the study. For the sake of brevity, in the subsequent sections, they will be addressed as L_{ch}. Their local transition rule tables are tabulated (Table 1), and the space-time diagram evolved from an arbitrary initial configuration is illustrated in Fig. 1

A few earlier studies have implemented the concept of a distant neighbor in different contexts [1,4,5]. Li, in [5], extensively studied a model of 2-state, 3-neighborhood non-local elementary cellular automata. The 3 neighbors are chosen at random. Mean-field parametrization was found beneficial for the identification of critical surfaces separating regions of non-local rule space. The concept of *Robust Universal Computation* has been introduced. It showed that universal

This work is supported by CRG project (File No.: CRG/2023/006799) of SERB, Govt. of India.

H. Raju et al. (Eds.): ASCAT 2026, CCIS 2801, pp. 130–142, 2026.
https://doi.org/10.1007/978-3-032-18612-6_10

computation can be easily found in non-local ECAs irrespective of their dynamics, although it may not necessarily be robust. Another study [4] on long-distance CA (LDCA) found locally chaotic rule 73, a potential candidate for universal computation. This research [1] introduced a hybrid neighborhood topology where, along with local neighbors cells are integrated with distant neighbors at a fixed distance.

Table 1. Local Transition rule tables of L_{ch}

RMT	111 (7)	110 (6)	101 (5)	100 (4)	011 (3)	010 (2)	001 (1)	000 (0)	ECA
f	0	0	0	1	1	0	1	0	**26**
	1	0	0	1	1	0	1	0	**154**
	0	1	0	0	1	0	0	1	**73**

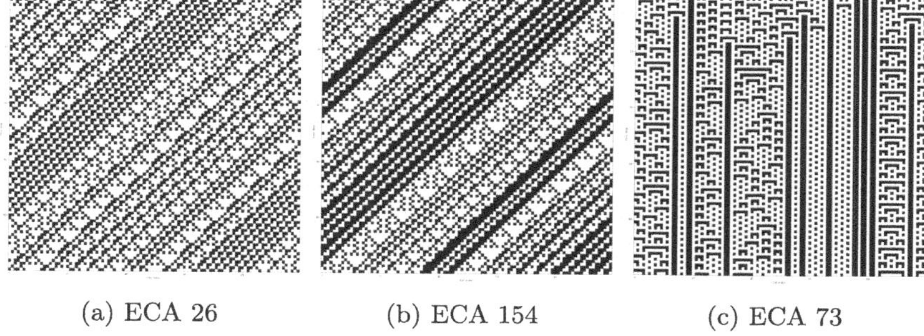

(a) ECA 26 (b) ECA 154 (c) ECA 73

Fig. 1. Space-time evolution diagram of 3 locally chaotic rules, evolved from an arbitrary initial configuration on periodic boundary condition

In this research, we propose a simpler model (see Sect. 2), where a few selected cells are coupled with distant neighbors. Only locally chaotic ECAs are chosen for a focused examination. The number of perturbations used in this study is highly reduced. This study aims to understand complexity and robustness not in terms of an increase in perturbations, but in light of a controlled optimum. Finally, the work provides us insight into certain mechanisms developed by a system to maintain complex dynamics. It is observed that there is a certain limit for a system to become disordered and a limit to the applied perturbations as well.

2 The Model

The work considers elementary cellular automata (ECAs), which are defined on a 1-dimensional lattice $\mathcal{L}$. Each cell of an elementary cellular automaton (ECA)

can assume either state 0 or 1, and update its state following a next state function $f : \{0,1\}^3 \rightarrow \{0,1\}$. During evolution, a cell considers the state of its left, self and right neighbor, and uses f to move to its next state. The rule f can be expressed in tabular form as shown in Table 1. The first row of the table shows all possible combinations of present states of the cell and its left and right neighbor. Each such combination is popularly called as RMT (Rule Min Term) [2]. It is generally represented by its decimal equivalent(the numbers inside parentheses of Table 1). A rule is generally named by the decimal equivalent of 8 next states against the RMTs.

In this work, an ECA rule is used as the primary next function of all the cells. Some of the cells which are elements of a set $A \subset \mathcal{L}$, additionally, utilize another function $g : \{0,1\}^2 \rightarrow \{0,1\}$, which takes the updated state of the cell and the previous state of another cell, to the right, which is a Distant Neighbor (DN) of the former. The distance between a cell $c \in A$ and its corresponding distant neighbor d is such that $d \geq 3$. There are 16 possible functions that can work as g (see Table 2).

Table 2. 16 binary functions (g)

Function	11	10	01	00	Op. name	Popular name
g	0	0	0	0	α_0	–
	0	0	0	1	α_1	NOR
	0	0	1	0	α_2	–
	0	0	1	1	α_3	–
	0	1	0	0	α_4	–
	0	1	0	1	α_5	–
	0	1	1	0	α_6	XOR
	0	1	1	1	α_7	NAND
	1	0	0	0	α_8	AND
	1	0	0	1	α_9	XNOR
	1	0	1	0	α_{10}	–
	1	0	1	1	α_{11}	–
	1	1	0	0	α_{12}	–
	1	1	0	1	α_{13}	–
	1	1	1	0	α_{14}	OR
	1	1	1	1	α_{15}	–

The experimental setup takes an n-length lattice with periodic boundary conditions, partitioned into equal intervals of size m. n/m determines the number cells $c \in A$ and the number of their corresponding DN. In this way, we obtain n/m number of c equally spaced along the length of the lattice n. We will address the original evolution as ECA and that when modified by perturbation will be called $\mathrm{ECA_c}$.

Example 1. Let us take a finite lattice of size 10 whose elements are indexed as *0,1,2...,9*. An initial configuration 1101011100 is taken. When rule 26 is applied alone, the next configuration is 1000010011. However, if two cells with index *0* and *4*, additionally use a distant neighbor, at $d = 3$, each, and AND is applied as g, then the next configuration becomes 1000000011. Similarly, we get the subsequent configurations by applying the same method each time. Figure 2 illustrates a space-time evolution diagram of ECA 26 when all the cells are updated with f (Fig. 2a), and its corresponding $\mathrm{ECA_c}$, some cells c are updated as $g[(f)]$ (Fig. 2b).

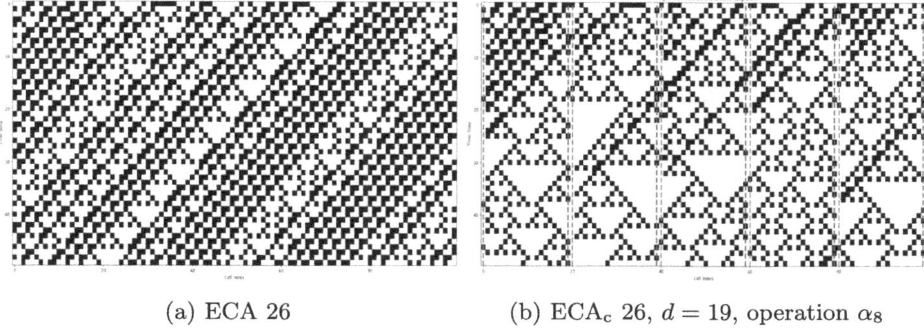

(a) ECA 26 (b) $\mathrm{ECA_c}$ 26, $d = 19$, operation α_8

Fig. 2. (a) The space-time evolution diagram of ECA 26. (b) The space-time evolution after integration with a distant neighbor ($\mathrm{ECA_c}$), $n = 100$, $c = 5$, $m = 20$

3 Dynamical Behavior

A number of experiments are performed to understand the dynamical behavior of the model under study. The experiments are initiated with a n-length lattice, divided into n/m number of equal intervals, and an equal number of cells c, integrated with DN. To keep the setup as simple as possible, the DN are only allowed to move between intervals i.e., $3 \le d \le m$. The dynamical behavior of the L_{ch} for a certain operation g is observed by gradually moving the DN away from the cell c it was coupled with. In this manner, the outcomes regarding all 16 operations are documented. The set of procedures is followed again for different values of m. Different lattice lengths n are also taken into consideration.

For graphical analysis, we use a *density plot* of active cells (cells with state 1) as a function of timesteps. If the number of active cells at a particular timestep is n_a, and ρ_a is the density, then,

$$\rho_a = \frac{n_a}{n}$$

Figure 3 shows such an example. The blue graph represents the ECA, and the red represents ECA_c.

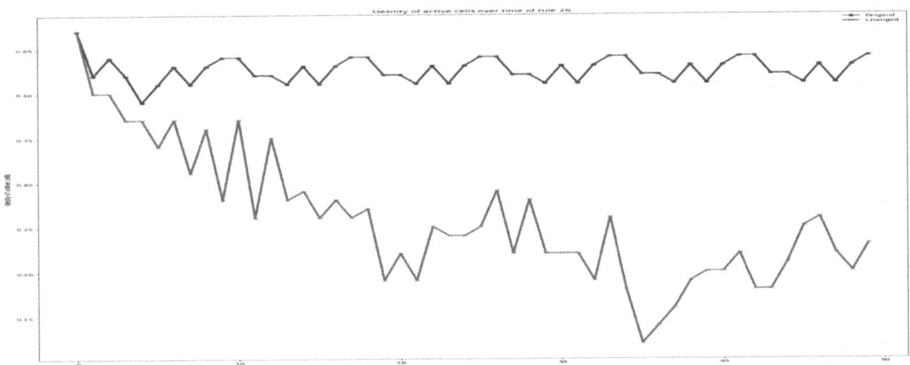

Fig. 3. Example of a *density plot* of ECA and ECA_c 26, $n = 200$, *timesteps*=100

We employ another graphical method to analyse the dynamics. The frequencies of certain RMTs in the ECA_c, are plotted against the number of timesteps. We regard this as an *RMT plot*.

In simulations, n is taken as large as possible, and evolved for longer timesteps. A certain window of the entire space-time diagram is observed for potential outcomes, where the window length $w \ll n$. Due to the excessive amount of time required for compilation, we have resorted to a maximum lattice length $n = 10,000$ and 5000 timesteps, which is $n/2$. For example, Fig. 4 shows an observation window with a lattice window $w = 100$ and a time window between 4900–5000 timesteps.

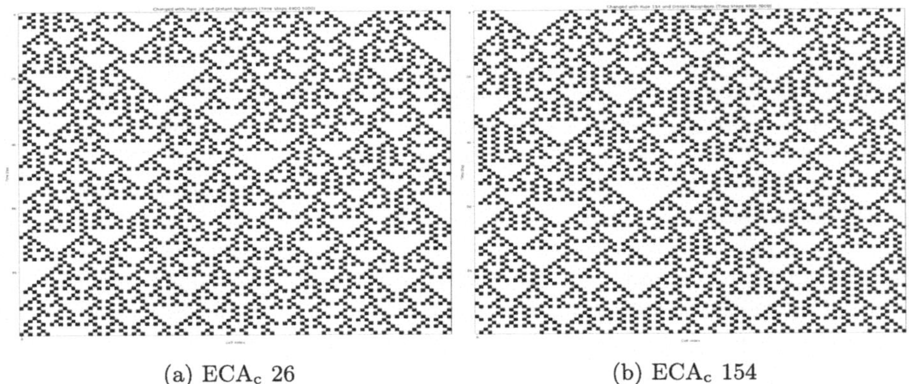

(a) ECA_c 26 (b) ECA_c 154

Fig. 4. Observation window of L_{ch}. $w = 100$, time-window: $4900 - 5000$ *timesteps*. $m = 100$, $d = 99$, α_8

3.1 ECA 73

This ECA is found to be largely robust to the impact of distant neighbors. It retains its locally chaotic dynamics for all the operations being tested. Figure 5 illustrates an example of a 200-length lattice with an arbitrary initial configuration. Here, $m = 20$, $c = DN = 10$. Figure 5a shows the space-time evolution. In the ECA_c (Fig. 5(b-d)), multiple DN has been introduced and integrated with the cell c through the operation α_8 (AND). Three different distant neighbor positions d have been considered and corresponding evolution patterns are shown. In the ECA_c, several local changes can be observed. However, they are almost always bound by some rigid periodic structures.

The *density plot* of ECA and ECA_c 73 is illustrated in Fig. 6. A 1000-length lattice with an interval value of 100 is taken along with the operation AND. Figure 6a and 6b capture the graphical representations for two different values of d. For both $d = 3$ and $d = 99$, the graph of ECA_c does not depart significantly from the corresponding ECA.

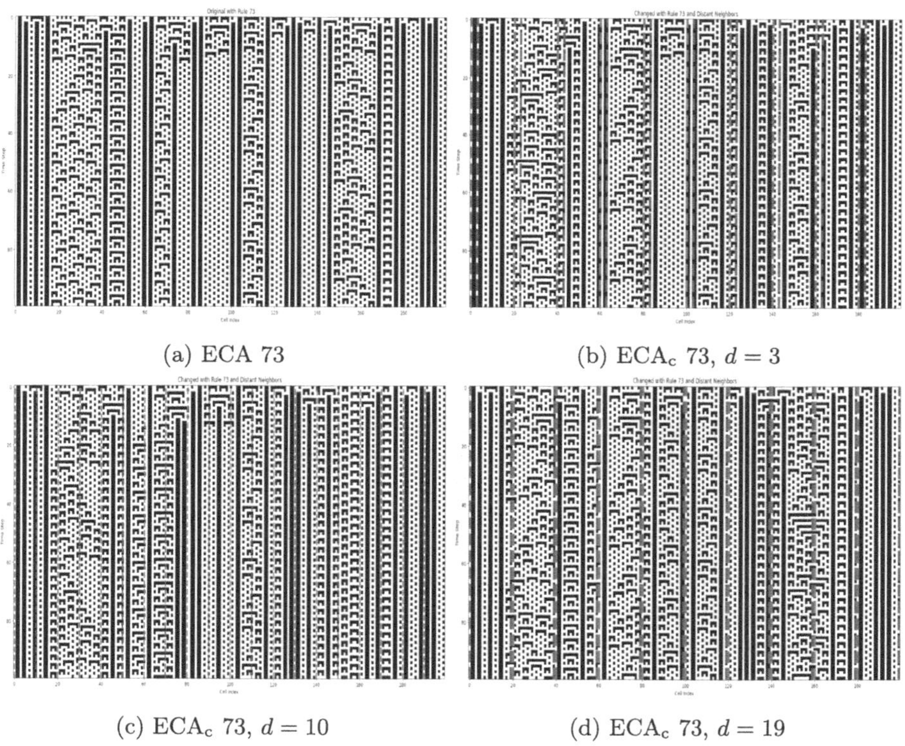

(a) ECA 73

(b) ECA_c 73, $d = 3$

(c) ECA_c 73, $d = 10$

(d) ECA_c 73, $d = 19$

Fig. 5. space-time patterns of ECA and ECA_c 73. $n = 200$, $m = 20$

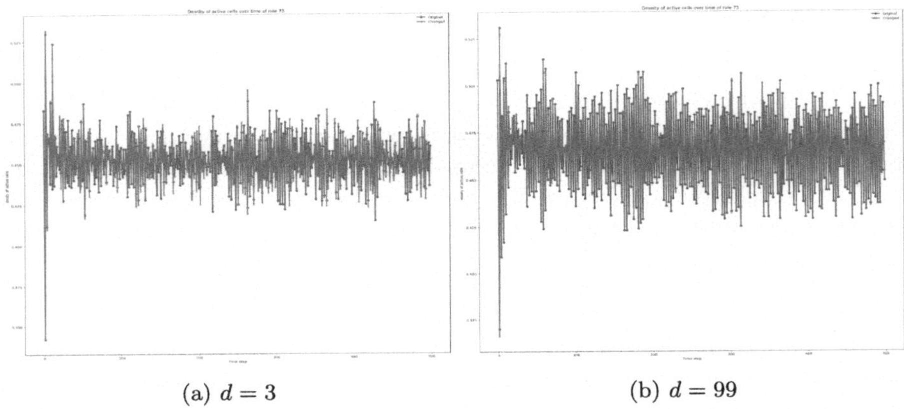

(a) $d = 3$ (b) $d = 99$

Fig. 6. *density plot* of ECA and ECA$_c$ 73, $n = 1000$, $m = 100to$, $timesteps = 500$, α_8

3.2 ECA 26 and 154

These two L_{ch} reveal some interesting properties involving specific operations. Operations α_0, α_{12}, α_{15} are trivial, so they are discarded after initial examination. The rest of the 12 operations are thoroughly studied. In elementary cellular automata, a periodic ECA transitioning to chaos upon perturbation is a rare dynamical behavior. However, we found operation α_8 (AND) highly capable of completely dissolving the globally periodic structures of the *Lch*. A few other operations g are sometimes capable as well, but not as effective. Operation AND requires a minimally invasive distant neighbour integration and an effortless, consistent transition. For certain perturbation conditions, complexity emerges as a result of periodic structures breaking down. Figure 8 shows the original space-time patterns of the L_{ch} (Fig. 8a and 8e) with their dynamics obtained by applying $[g(f)]$ i.e., operation AND (α_8) (Fig. 8). A lattice of size 200 is equally partitioned into 5 intervals of length 20; hence, 5 cells are integrated with DN. The images show the dynamical effects of three different positions of the DN ($d = 3, 20, 39$).

It is apparent from the images that as the DN moves from $d = 3$ to m, the periodic structures start to dissipate. At some point in the middle, large inverted triangles form, which dominate the dynamics for a certain period, eventually breaking them up into smaller ones as d approaches $m - 1$. Although they are approximations, if m is not very small, a complete disappearance of any dominating structures is guaranteed to be observed at $m - 1$.

At $d = m$, every DN is superimposed on a cell c, i.e., every c is a DN of another. Here, we observe locally chaotic patterns (Fig. 7).

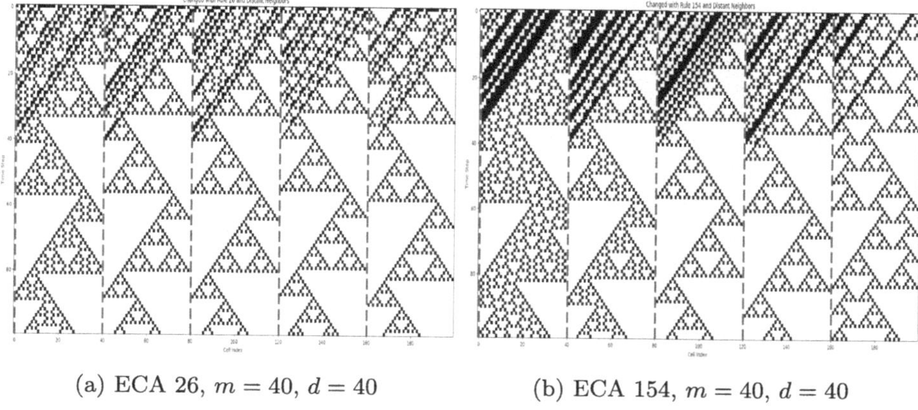

(a) ECA 26, $m = 40$, $d = 40$ (b) ECA 154, $m = 40$, $d = 40$

Fig. 7. ECA_c 26 and 154 for $d = m$

Figure 9 shows ECA_c 26 and its corresponding *density plots*. the ECA_c represents a lattice with $n = 1000$ and $m = 100$. Three different *density plots* correspond to three different positions of DN ($d = 3, 50, 99$). At $d = 3$, the plot of ECA_cn(Fig. 9b), ρ_a drops quite sharply at first, then almost stabilizes ($\rho_a \approx 0.25$). At $d = 50$, the plot (Fig. 9d) has a gradual decrease reaching far lesser values of ρ_a. Here, we can observe greater amplitude variation at lower values of ρ_a, than the graph encountered before. It could imply that after the number of active cells is reduced significantly, the system pushes them back up, so that the system does not converge to all 0 s. The 3rd plot (of ECA_c) has a dip with higher amplitude variation. Then it reaches a baseline at $\rho_a \approx 0.25$.

To get a better understanding of the underlying mechanisms of patterns produced by the operation AND, we look at the plot shown in Fig. 10. Let us call the number of RMTs $\mathcal{N}_{RMT}$. The subfigures represent the spacetime diagrams of ECAs 26 and 154 with an 1000-length lattice for different values of d. The RMT 111 never occur in ECA or ECA_c 26 after the initial timestep (Fig. 10a). On the other hand, in ECA_c 154, $\mathcal{N}_{RMT}(111)$ will eventually go to zero while perturbed. $\mathcal{N}_{RMT}(000)$ on perturbation will initially increase rapidly, then eventually stabilise. however, the quantity will always dominate the rest. In this regard, $\mathcal{N}_{RMT}(010)$ comes next to $\mathcal{N}_{RMT}(000)$. However, $\mathcal{N}_{RMT}(101)$ has a lower frequency, which implies that a configuration ..10101010.. will selectively emerge in the evolution. This particular configuration always produces all zeros for both f and g on AND. What is most surprising is the fact that three RMTs consisting of consecutive 1 s (111,110,011) completely disappear when maximum disorder is achieved (Fig. 10c and 10f). $\mathcal{N}_{RMT}(111)$, $\mathcal{N}_{RMT}(110)$ and $\mathcal{N}_{RMT}(011)$, disappear as $d = m - 1$ is approached. The space-time diagram consisting only of the RMTs with only isolated 1 s (101,010,001,100) is shown in Fig. 4.

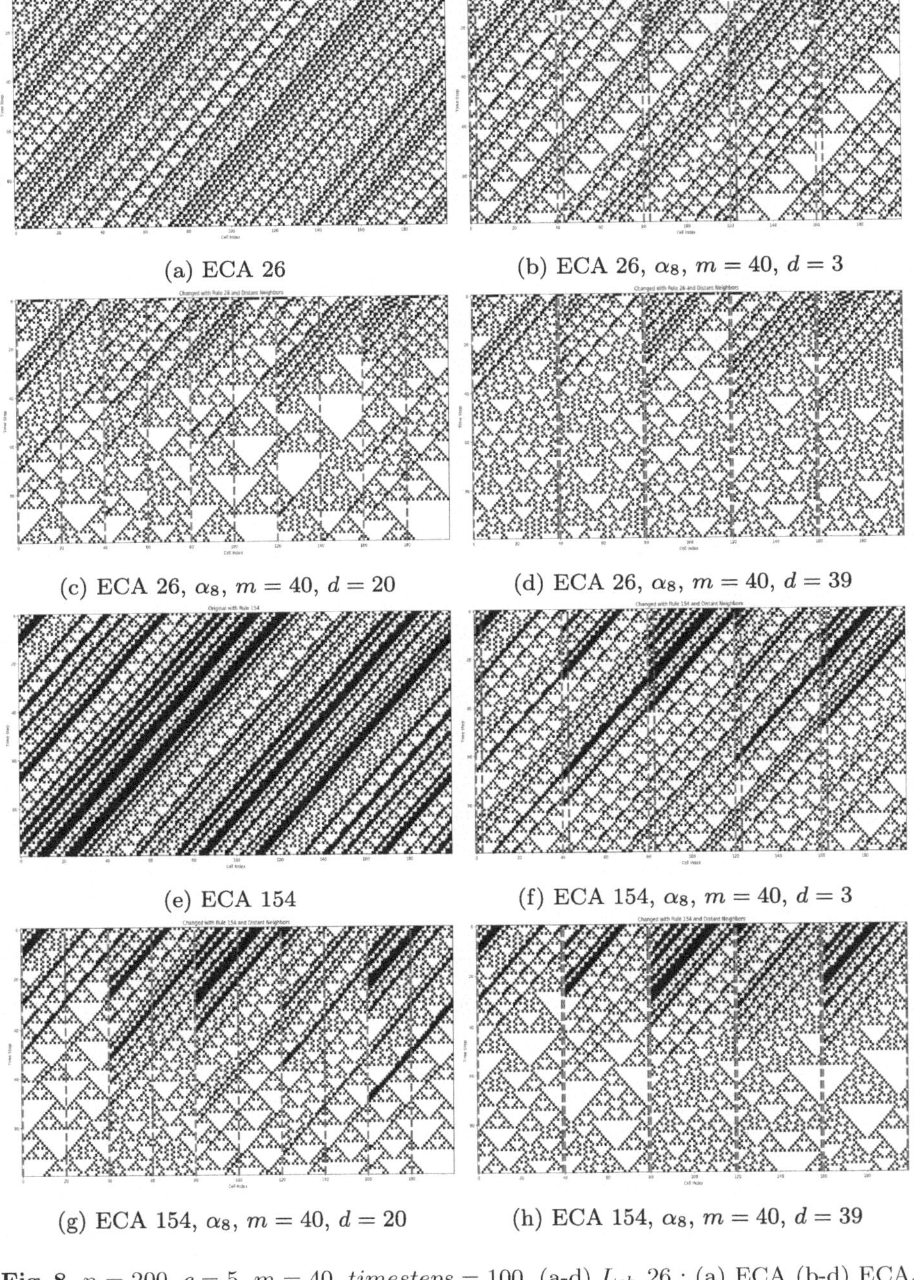

(a) ECA 26

(b) ECA 26, α_8, $m = 40$, $d = 3$

(c) ECA 26, α_8, $m = 40$, $d = 20$

(d) ECA 26, α_8, $m = 40$, $d = 39$

(e) ECA 154

(f) ECA 154, α_8, $m = 40$, $d = 3$

(g) ECA 154, α_8, $m = 40$, $d = 20$

(h) ECA 154, α_8, $m = 40$, $d = 39$

Fig. 8. $n = 200$, $c = 5$, $m = 40$, $timesteps = 100$, (a-d) L_{ch} 26 : (a) ECA (b-d) ECA_c for three different d. (e-h) ECA 154: (a) ECA, (e-g) ECA_c for three different d

All these observations point towards a unique mechanism employed by the system to drive itself towards disorder. In this experiment, a few cells $c \ll n$ are coupled with distant neighbor. It is obvious that a cell has a higher probability of becoming zero or staying zero if AND has been employed. A surge of 0 s in the system can be understood as an increase in $\mathcal{N}_{RMT}(000)$ that populates the system with zeros, and simultaneous elimination of RMTs that populate the system with 1 s. Despite that, we witness the system surviving complete collapse or convergence. Instead, it forms an apparently disordered structure with four out of 8 RMTs granted by the system.

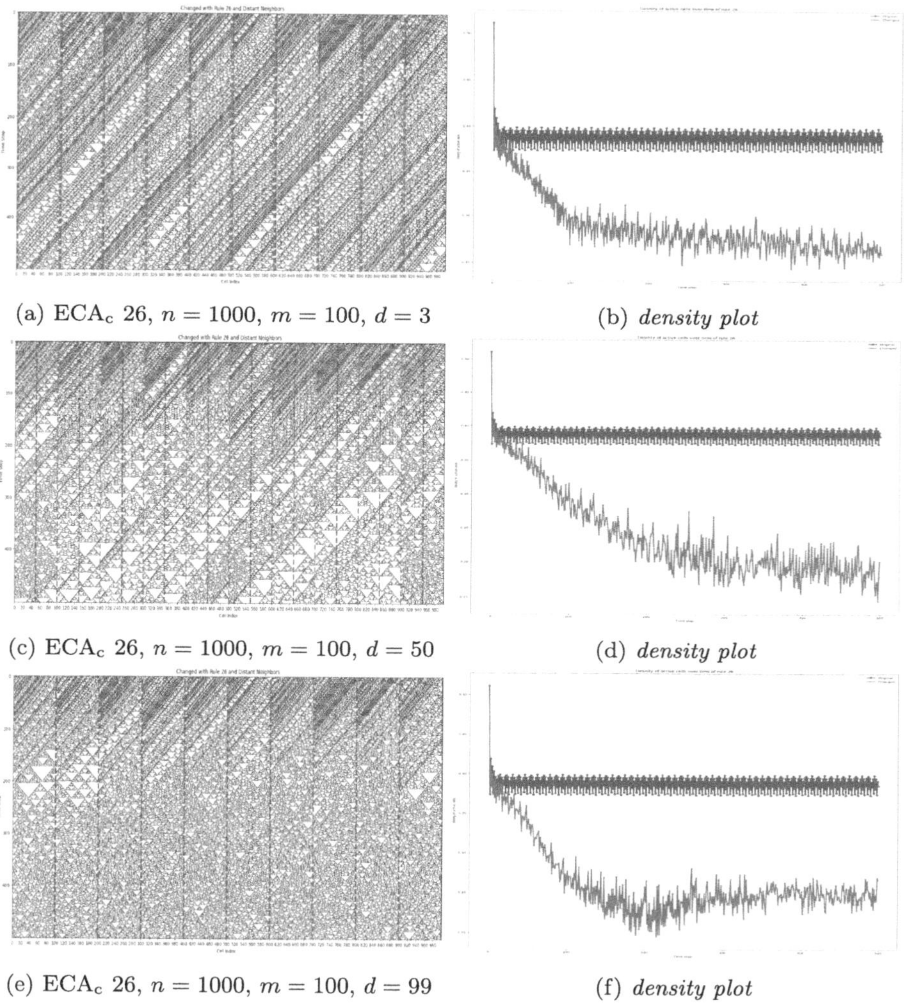

(a) ECA_c 26, $n = 1000$, $m = 100$, $d = 3$ (b) *density plot*

(c) ECA_c 26, $n = 1000$, $m = 100$, $d = 50$ (d) *density plot*

(e) ECA_c 26, $n = 1000$, $m = 100$, $d = 99$ (f) *density plot*

Fig. 9. ECA_c 26, α_8 along with their subsequent *density plots*

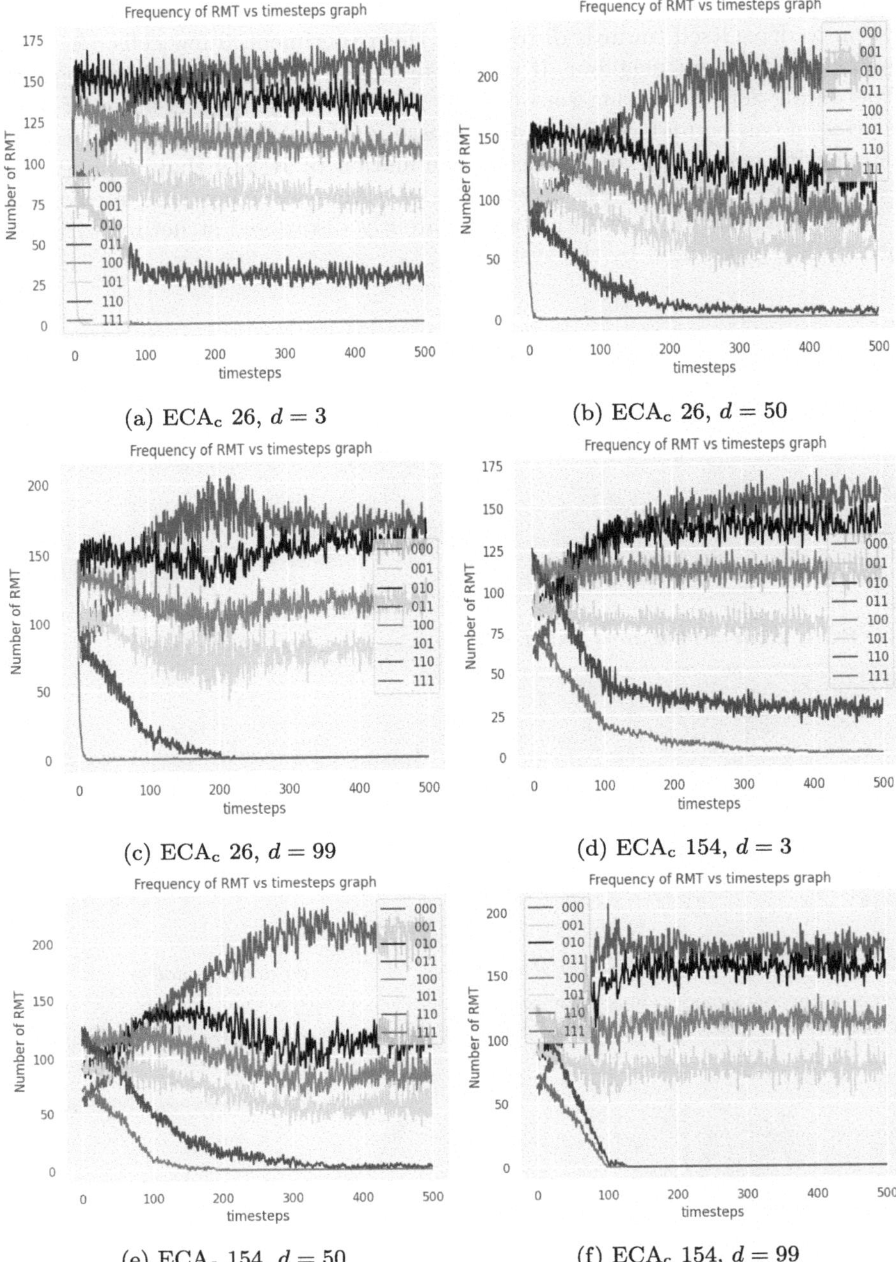

Fig. 10. Fluctuations of RMTs against *timesteps*. $n = 1000$, $m = 100$, α_8

4 Discussion

The experiment has probed into the functionality of three locally chaotic ECAs on integration with distant neighbor. A simple model has been studied, where a few chosen cells have to be systematically placed at equal intervals and integrated with a distant neighbor. In this manner, they are updated twice. The synchronous movements of the distant neighbors have been restricted to their corresponding intervals. All other cells except the chosen few have followed the usual update rules of the ECA. Among the three locally chaotic ECAs examined, 73 has been proven to be globally robust on application of all the 16 mode of perturbations. However, the other two ECAs have showed remarkable alterations to the integration of distant neighbor, especially through the operation AND. In this case, just by shifting each distant neighbor from one end of the interval to the other, one could see a gradual process of globally periodic patterns dissipate completely. This implies that a few systematic perturbations can be sufficient to produce a dynamical shift. We have also been able to record a rare dynamical transition from periodicity to complete disorder. Several operations have required a higher number of cells c to achieve that. However, in the case of AND operation, a few systematic perturbations are enough. The dynamical transitions have provided us with a new insight into how a system apparently seems to approach convergence, but instead forms complete disorder with a few allowable configurations. The future work will aim to incorporate different quantitative and complexity measures, such as entropy, information and Lempel-Ziv complexity, for a deeper and formal understanding of the system's dynamics.

Acknowledgements. We would like to express our sincere gratitude to Dr. Sukanta Das and Dr Kamalika Bhattacharjee for their invaluable support and guidance.

References

1. Andreica, A., Chira, C.: Using a hybrid cellular automata topology and neighborhood in rule discovery. In: Pan, J.-S., Polycarpou, M.M., Woźniak, M., de Carvalho, A.C.P.L.F., Quintián, H., Corchado, E. (eds.) HAIS 2013. LNCS (LNAI), vol. 8073, pp. 669–678. Springer, Heidelberg (2013). https://doi.org/10.1007/978-3-642-40846-5_67
2. Bhattacharjee, K., Naskar, N., Roy, S., Das, S.: A survey of cellular automata: types, dynamics, non-uniformity and applications. Nat. Comput. **19**(2), 433–461 (2020)
3. Bhattacharjee, K., Sikdar, B., Jain, S.: A study on role of distant neighbor over elementary cellular automata. In: Asian Symposium on Cellular Automata Technology, pp. 101–114. Springer, Heidelberg (2025). https://doi.org/10.1007/978-3-031-94121-4_8
4. Kang, L.: Complex behavior in long-distance cellular automata. Complex Syst. **23**(2) (2014)

5. Li, W.: Phenomenology of nonlocal cellular automata. J. Stat. Phys. **68**(5), 829–882 (1992)
6. Li, W., Packard, N., et al.: The structure of the elementary cellular automata rule space. Complex Syst. **4**(3), 281–297 (1990)
7. Wolfram, S.: Statistical mechanics of cellular automata. Rev. Mod. Phys. **55**(3), 601 (1983)

Cellular Automata Based In-Memory Search Operation

Moumita Dutta[1(✉)], Kasturi Ghosh[2], and Biplab K. Sikdar[1]

[1] Computer Science and Technology, Indian Institute of Engineering Science and Technology, Shibpur, West Bengal, India
2024csm014.moumita@students.iiests.ac.in, biplab@cs.iiests.ac.in
[2] Information Technology, University Institute of Technology, Burdwan, West Bengal, India
kasturi.dikpati@gmail.com

Abstract. In a traditional von Neumann machine, execution involves moving operands from memory to the CPU, performing arithmetic or logical operations on them, and then storing the results back in memory. For data centric computation, such memory wall (data movement between CPU and memory) causes hindrance for high speed computation. In this work, we propose an architecture such that the arithmetic – logical operations can be done in-memory with minimal involvement of the CPU. The in-memory computing (IMC) architecture is developed around the cellular automata (CA) framework configured in SRAM without any major change in SRAM structure. The proposed design is then employed to illustrate the effective realization of in-memory search operation.

Keywords: In-memory computation · Cellular Automata · searching · SRAM

1 Introduction

The process of execution on a traditional von Neumann computer system primarily involves continuous interaction between the central processing unit (CPU) and the memory unit. The operands are fetched from memory and the arithmetic/logical operations are performed in the ALU. Then the result is written back to memory. These introduce two critical components of execution time: the processing time of CPU, and the transfer time of data between memory and CPU. Due to the growing demand of faster and more energy-efficient computing system, researchers have been actively investigating techniques to optimize processing time, reduce memory transfer overheads, and enhance system performance.

This Work Is Supported by CRG Project(File No. CRG/2023/006799) of SERB, Govt. of India

H. Raju et al. (Eds.): ASCAT 2026, CCIS 2801, pp. 143–154, 2026.
https://doi.org/10.1007/978-3-032-18612-6_11

Searching a data item in memory is an operation that requires longer time span. As in a conventional computer system, the data item to be searched (key) is stored in the CPU register. Each element of the data set is then brought into the CPU one by one and compared with the key. The search time is, therefore, dominated by the transfer of elements from memory to CPU. The in-memory computing (IMC) [9] is a new computing approach that enables data processing directly at the location of data, saving time and huge amount of energy.

Compute-in-SRAM and analog in-memory units for multiply-accumulate and other arithmetic operations have been demonstrated in [5,11,12]. The recent work in [8] implements elementary cellular automata (CA) in a memristor-based in-memory circuit, enabling logic and pattern-processing directly in the memory. These CA-enabled IMC systems realize memory-centric hardware to support scalable and low-power compute architectures for next-generation systems [6].

The cellular automata (CA), introduced by John von Neumann in 1950's have been found as the tool to perform universal computation. Numerous application domains, including natural evolution [13], physics [4], cryptography and image processing [2], theoretical biology [3] and many more are explored. In this background, this work targets to explore the potential of CA in realizing the IMC. The IMC architecture is developed around the uniform CA framework configured in SRAM without a major change in SRAM's structure. An effective design is then demonstrated while realizing the in-memory searching. It first performs XOR of the key to be searched and the data item, in-memory, and then outcome 'd' of the XOR operation is checked through CA evolution to sense whether the d is all 0s.

2 Cellular Automata

Cellular automaton (CA) [2] is a computational model that can be represented as an autonomous finite state machine, and it evolves in discrete space and time. This makes the CA a powerful model capable of replicating the complex systems of nature and also arithmetic and logical operations.

The CA consists of a grid of cells, where each cell stores a discrete variable at time t, referred to as the present state (PS). It evolves to the next time step $t+1$, to attain its next state (NS). The NS depends on the PS of the cell and the PSs of its immediate neighboring cells. In this work, we refer to 1-dimensional CA, where the cells are organized in an 1-dimensional lattice and each cell having 3 neighbors and can have 2 states, 0 and 1. Such a CA is the *Elementary Cellular Automata* (ECA) [13].

The PS (S_i^t) of the ith cell of an ECA follows the next state function (f_i) for updates. Here,

$$S_i^{(t+1)} = f_i(S_{i-1}^t, S_i^t, S_{i+1}^t)$$

S_{i-1}^t, S_i^t and S_{i+1}^t are the present states of the left neighbor, self and right neighbor of the ith cell at time t. The state of n-cells $S^t = (\ S_1^t,\ S_2^t,\ .\ .\ .\ ,\ S_n^t\ )$ is the PS of the CA (CA configuration at time t). Similarly, the NS of an n-cell CA (configuration at time $t+1$) is

$S^{(t+1)} = (f_1(S_0^t, S_1^t, S_2^t),\ f_2(S_1^t, S_2^t, S_3^t),\ \ldots,\ f_n(S_{n-1}^t, S_n^t, S_{n+1}^t))$

The function f_i: $\{0,1\}^3 \rightarrow \{0,1\}$ is expressed in Table 1. The NSs for the eight (2^3) corresponding PSs converted to decimal equivalent, express the evolution *rule* (R_i) of the cell. In ECA, there can be $2^8 = 256$ rules. Table 1, in its first row, shows all possible combinations of the PSs of the 3 neighbors at time t. The next row refers rule min term (RMT) and the third row shows the NS at time $t+1$ and its decimal equivalent (e.g. 254) as the rule.

Table 1. RMT of the rule 254

PS	111	110	101	100	011	010	001	000	Rule
RMT	(7)	(6)	(5)	(4)	(3)	(2)	(1)	(0)	
NS	1	1	1	1	1	1	1	0	254

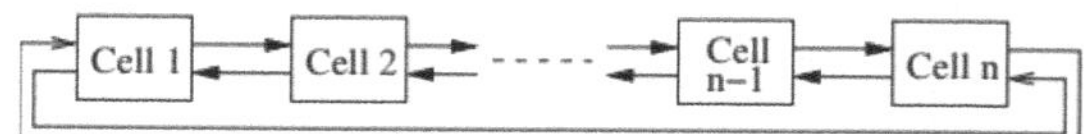

Fig. 1. Periodic boundary n-cell CA.

A CA is periodic boundary if the left (right) neighbor of leftmost (rightmost) terminal cell is the rightmost (leftmost) terminal cell Fig. 1. The PSs S_{i-1}^t, S_i^t and S_{i+1}^t of the $(i-1)^{th}$, i^{th} and $(i+1)^{th}$ cell respectively (3 neighbors of ith cell), as a combination, is called RMT. For example, in Table 1, the second row entry for PS 101 is the fifth RMT (represented as (5)). The NS with respect to the fifth RMT is '1' for rule 254.

A set of states/configuration can form a cycle in a CA and is called attractor. An attractor forms a basin with the states which lead to the attractor. As in Fig. 2, 0→0 and 15→15 are the attractors. The basin with attractor 15, referred to as the 15-basin, has fifteen states. The 0→0 and 15→15 are single length cycle attractor and are called point state attractors. If a CA is having only two point state attractors (no multi-length attractor), then it is a single length cycle two-attractor CA (TACA). The CA configured with rule 254, having state transitions as shown in Fig. 2, is a TACA [10].

3 The Overview of In-Memory Searching

The one dimensional, 3-neighborhood, 2-state periodic boundary CA, introduced in Sect. 2, is considered in this work for realizing in-memory searching within SRAM. The special class of two attractor CA (TACA), having all 0s and all 1s attractors, are employed for the design.

To find (search) the key x within a set of data elements (table), each element is to be compared with x. In conventional von Neumann machine computer, the elements of the table are brought to the CPU, one by one, from memory for comparison with x (involving ALU). In the proposed in-memory searching, the key x is stored in memory (say, at location l). Each memory word (containing a data element) d is then compared with x in-memory. It has two steps - a) XORing (Comparison): First, the x at location l and the data word d are bit-wise XORed and the output (y) of XOR is stored in the cells of a TACA constructed with rule 254 in-memory. If x and d are the same, then y is all 0 s. b) Decision: The TACA loaded with y is evolved for some time steps (say, k) to verify whether all the cells of the TACA are 0. The least significant cell of the TACA (called *checkbit*), after k time steps, indicates the match. For a match between x and d, it is $'0'$ otherwise, it is $'1'$. Figure 3 illustrates the in-memory searching.

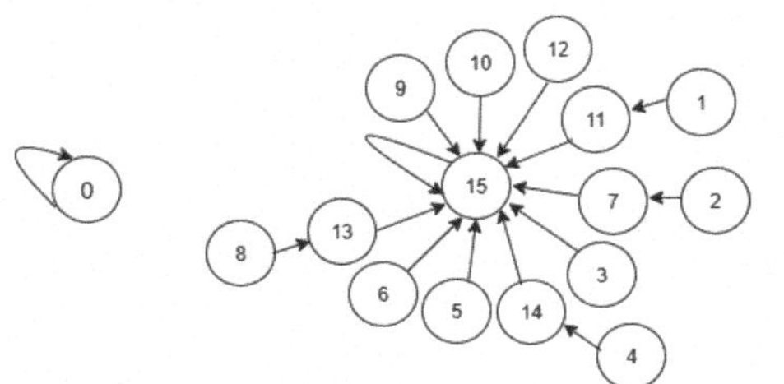

Fig. 2. State transition of 4-cell TACA with rule 254.

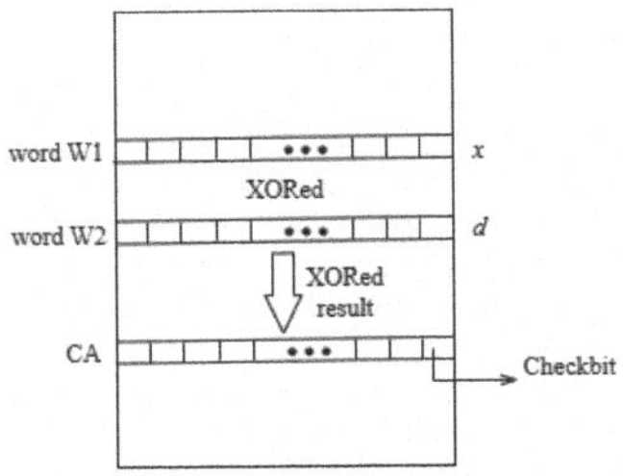

Fig. 3. In-memory searching.

Let consider the 4-bit key x = ABCD and the element d = WXYZ are to be compared. When x and d are XORed, the outcome y of the XOR operation is 0000 if there is a match between x and d otherwise, it is non-zero. The y is now input to the 4-cell TACA constructed with rule 254. It can be observed in Fig. 2 that if y is all 0s, then the CA settles to all 0s attractor otherwise, it settles to all 1s attractor after $k = 2$ time steps (like $k = n/2 = 2$). Now, by sensing the least significant cell, the outcome of the comparison between x and d can be verified. It is 0 for a match and 1 for the mismatch.

Theorem 1. *The least significant cell of an n-cell CA, constructed with rule 254, settles to* 1 *if the CA evolves for atmost $n/2$ time steps, when initialized with any non-zero configuration.*

Proof. Let K_i be the number of 1s in a consecutive chain of 1s (calling it a *sequence*) in the n-cell CA configuration for one time step, where i is the *sequence* number (i = 1, 2, 3, ...) and let d be the number of 0s that separate out two such sequence of 1s, say K_i and K_{i+1}, where K_i is immediate left to K_{i+1} (represented as $K_i \ll K_{i+1}$).

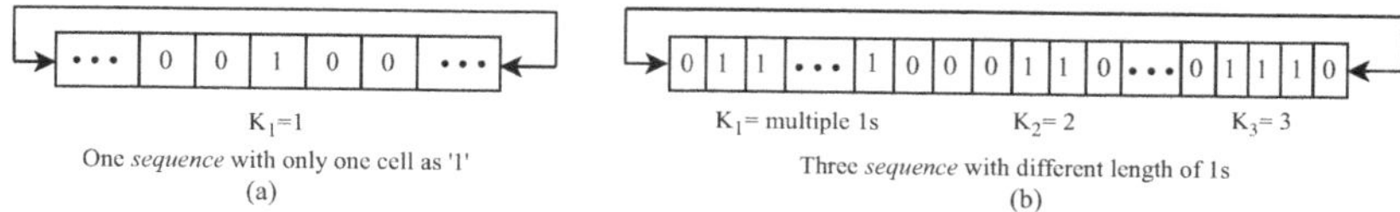

Fig. 4. Characterizing consecutive 1s (*sequence*) in n-cell CA configuration.

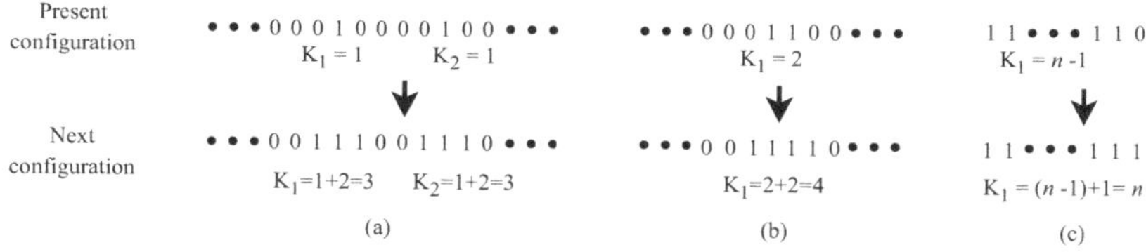

Fig. 5. Patterns of *sequences* of 1s in an n-cell CA with transition from present configuration to next configuration, when K_i+2 or K_i+1.

Figure 4 represents an n-cell CA, where Fig. 4(a) shows presence of only one *sequence* (K_1) with only one cell as '1' and Fig. 4(b) has three *sequences* (K_1, K_2, K_3) within n cells, where

- K_1 « K_2 with $d = 3$ and K_1 = multiple 1s
- K_2 « K_3 with $d \geq 2$ and K_2 = 2 1s
- K_3 « K_1 with $d = 2$ and K_3 = 3 1s

In rule 254 (refer to Table 1), for any combination of present states of neighbors (except 000), the next state (NS) of a cell is 1. So, a *sequence* having single 1 or consecutive 1*s* in the present configuration (except the all 1s configuration) of the n-cell CA, when evolves for one time step, two (or one) more 1*s* are generated in the next configuration. That is, for i^{th} *sequence* having K_i 1*s* in the present configuration, $K_i + 2$ (or $K_i + 1$) 1*s* are generated in the next evolution (shown in Fig. 5). Now, the maximum number of evolutions (e) for which the least significant cell settles to 1 through a function say $f(e)$ (estimates the total number of 1 s after any e^{th} evolution, in the n cells of the CA) can be computed as (here K_i^{e-1} and K_i^e represents the number of 1s in i^{th} *sequence* for any $e - 1$ and e evolution respectively):

Case 1 - The n-cell CA has only one cell as '1' whereas, rest of the $n - 1$ cells are 0s. As in Fig. 6, the initial configuration has only one *sequence* with single 1 ($K_1^0 = 1$). After 1st evolution, the next configuration also has only one *sequence* with $K_1^1 = 1 + 2 = 3$ 1*s*. Similarly, for any $(e-1)^{th}$ evolution, K_1^{e-1} increases by 2 in the eth evolution to get K_1^e. The $f(e)$ for a single evolution is

$$f(e) = K_i^{e-1} + 2 \tag{1}$$

where $i = 1$ and $e = 1, 2, 3, \ldots$ until it is all 1s configuration (with $e = 0$ as initial configuration). After maximum e^{th} evolution, the configuration is all 1s -

that is, $K_1^e = n$ (shown in the last step of Fig. 6). Through recurrent substitution of Eq. 1, computation of K_1^e with the base value of $K_1^0 = 1$ is:
$f(e) = K_1^e$

$K_1^e = K_1^{e-1} + 2 = (K_1^{e-2} + 2) + 2 = (K_1^{e-3} + 2) + 2*2 = (K_1^{e-4} + 2) + 3*2$
$= \ldots\ldots = (K_1^1 + 2) + (e-2)*2 = (K_1^0 + 2) + (e-1)*2$
Therefore,

$$K_1^e = K_1^0 + e * 2 \tag{2}$$

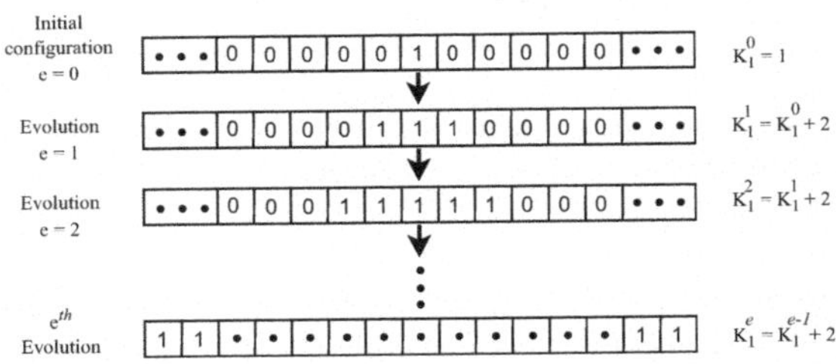

Fig. 6. Evolution of n-cell CA, from a single cell as 1 in an initial configuration to all n cells as 1 s.

Now, for the worst case, the least significant cell (*checkbit*) sets to '1' after an evolution (say e^{th}), when all other $n-1$ cells are already set to '1' from 0 to $(e-1)^{th}$ evolutions, then the value of e in terms of n is:

after e^{th} evolution, $K_1^e = n$
$\implies K_1^0 + e * 2 = n$ [from Equation 2]
$\implies 1 + 2e = n$ [since, $K_1^0 = 1$]
$\implies e = (n-1)/2 \approx n/2$

Case 2 - The n-cell CA has more than one cell as '1' in initial configuration. Now, for a single *sequence* with multiple 1s, maximum e can be estimated through Eq. 1, where K_1^0 is the number of 1s in initial configuration ($e = 0$). Whereas, for multiple *sequences*:

$$f(e) = \sum_{i=1}^{l} K_i^e \tag{3}$$

where l is the total number of *sequences*, $e = 1, 2, 3, \ldots$ with base value $e = 0$, K_i^0 is the number of 1s in initial configuration and

$$K_i^e = \begin{cases} K_i^{e-1} + 1 & \text{if } K_i^{e-1} \text{ and } K_{i+1}^{e-1} \text{ has } d = 1 \\ K_i^{e-1} + 2 & \text{if } K_i^{e-1} \text{ and } K_{i+1}^{e-1} \text{ has } d \geq 2 \\ K_i^{e-1} & \text{if } K_i^{e-1} \text{ and } K_{i+1}^{e-1} \text{ has } d = 0 \end{cases} \tag{4}$$

In the worst case, if the *checkbit* settles to 1 in the eth evolution, when all other $(n-1)$ cells are set to '1' in the previous evolutions, then e is: (a) For single *sequence*, by Eq. 1 and 2

$$K_1^0 + 2 * e = n \text{ [where } K_1^0 > 1 \text{]}$$
$$\implies e = (n - K_1^0)/2$$

which is definitely less than $(n-1)/2$ (since $K_1^0 > 1$).

(b) For multiple *sequences*, following Eq. 3, the cells with state 1 in the next configuration will increase in a rate of summation of all K_i's (i = 1, 2, 3, ..., l) (shown in Fig. 7), resulting in a smaller value of $e < (n-1)/2 \approx n/2$.

Hence, the maximum possible value of e, for all the cases of initial configuration, is atmost $n/2$.

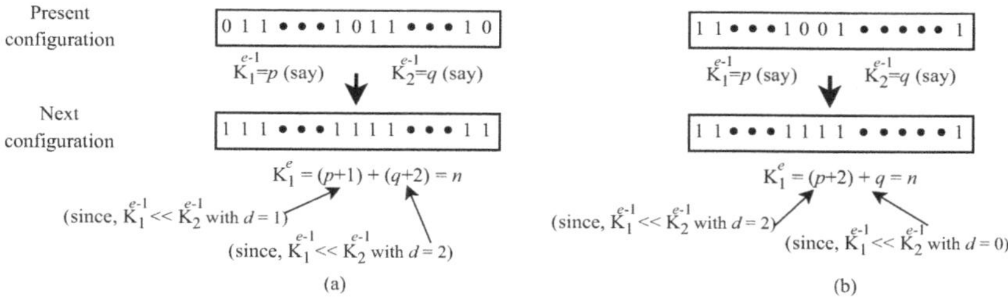

Fig. 7. Configuration of an n-cell CA with *sequences* of multiple 1s from $(e-1)^{th}$ to e^{th} evolution.

4 The Design in SRAM

This section reports implementation of proposed in-memory searching scheme in SRAM. Figure 8 shows the abstract model of the realization, where each *Cell-i* of the memory word Wi is XORed with *Cell-i* of Wj and fed to the CA constructed for the design. The following subsections detail out the realization of in-memory XOR operation and the design of CA in-memory with 8T SRAM cells for comparison.

4.1 8T SRAM Cell

An 8-transistor SRAM (8T SRAM) cell, as proposed in [1], is shown in Fig. 9. It is choosen for the current design for its decoupled read-write operations (seperate bit-lines for read and write). The 8T SRAM cells give an optimized bit-cell structure that facilitates in-memory computing for logic operations within the memory array. Such a cell consists of 6 basic transistors (6T), as that of the 6T SRAM cell (for storing a bit and performing write operation using the write

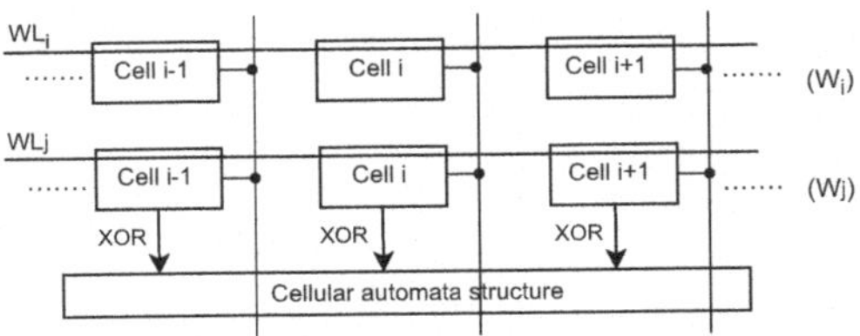

Fig. 8. Realization of in-memory comparison in SRAM.

word-line (WWL) and write bit-line (WBL) and write bit-line bar (WBLB)). The two additional transistors (M1 and M2) are for the read port (through read bit-line (RBL) and the read word-line (RWL) of Fig. 9). For in-memory logic operation, multiple word-lines can be enabled and the outcome of computation can be accessed through memory read bit-line [9].

To read the stored data within an SRAM cell, RWL is activated while WWL remains deactivated, and RBL is initially precharged to high. If Q is high, transistor M_1 is turned ON and RBL is discharged otherwise, RBL remains in precharge state. When the RWL of the cell is activated, the read is performed. As because of decoupled read ports, multiple read wordline activation does not allow a cell to flip (which occurs in 6T SRAM cell, called read disturb). This isolated read mechanism of 8T SRAM structure enables realization of NOR, NAND and XOR logic within the memory array [1].

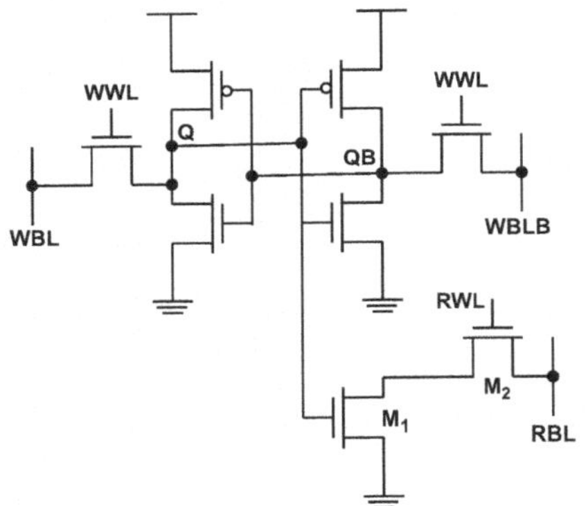

Fig. 9. A standard 8T SRAM cell [1].

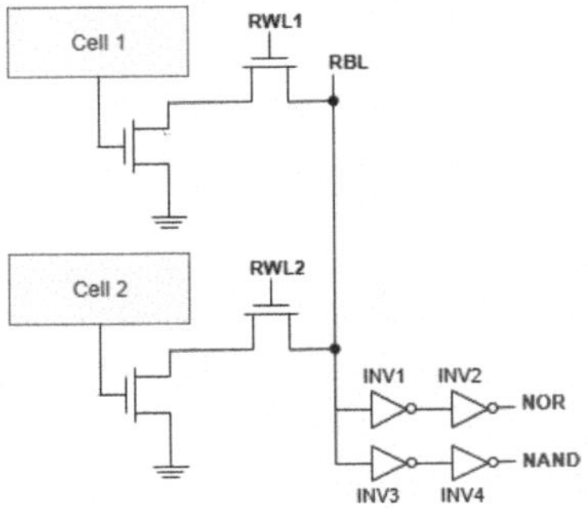

Fig. 10. Schematic of 8T-SRAM for realizing NOR and NAND [7].

4.2 Realization of Logic Operations in SRAM

In SRAM array, with 8T cells, the NOR, NAND and XOR operations can be realized as described in [1,7,9].

1. *Two-input NOR logic* has output '1' only when both the inputs are '0' otherwise, it is '0'. To realize NOR operation between Cell1 and Cell2 of Fig. 10, the RBL is initially precharged to high. Let RWL1 and RWL2 are activated simultaneously. If both the Cell1 and Cell2 contain '0' (that is, Q = 0 for both the cells), then RBL is not discharged - that is, remains in high state. For any other value (01/10/11) of Cell1 and Cell2, the RBL is discharged. A gated inverter (INV1) is connected to the RBL so that the inverted output is set to low only when the RBL is high. The cascaded inverter (INV2 of Fig. 10) output is the NOR of Cell1 and Cell2.

2. *Two-input NAND logic* has output '0' only when both the inputs are '1' otherwise, it is '1'. For NAND realization, as in [9], we need to activate the two RWLs (RWL1 and RWL2 of Cell1 and Cell2 respectively). The precharged RBL gets discharged to 0V if any one of the operands (Cell1/Cell2) Q is '1'. The fall latency (rate of discharge) of the precharged RBL to 0V depends on whether a single operand's (Cell1 or Cell2) Q is '1' or both the operands' (Cell1 and Cell2) Qs are '1' simultaneously. For the later case, the rate of discharge of precharged RBL is much faster. This different fall latency of RBL is considered to set the signal timing of RWLs such that the RBL is not discharged fully for the inputs '01/10'. According to the difference of voltage levels on RBL for '01/10' and '11', the inverter INV3 is triggered in such a way that it goes high only for input '11'. Hence INV4 reflects the outcome of NAND operation (Fig. 10).

3. The realization of *2-input XOR logic* is shown in the Fig. 11. Here, SL1 is connected to VDD while SL2 is grounded, forming a voltage divider. Both the RWL1 and RWL2 are initially held at ground, and the read bit-line (RDBL) is precharged to V_{pre}. The transistors M_2 and M_4 are then switched ON and a voltage divider is formed with RDBL (shown in Fig. 11(b)). For the stored values of Cell1 and Cell2 as

- '00' or '11', the RDBL remains at V_{pre} (as M_2 and M_4 are OFF),
- '01', the RDBL is discharged to ground (M_4 is ON while M_1 is OFF),
- '10', the RDBL is charged to VDD (M_1 is ON while M_4 is OFF).

The XOR output is finally obtained by sensing the RDBL using two skewed inverters [1] [7]. The INV2 is skewed such that it goes high only when the RDBL is much lower than V_{pre} - that is, close to ground (for cell values '01'). The INV1 is skewed such that it goes low only when the RDBL is much higher than the V_{pre}, approximately to VDD (for cell values '10'). Hence by ORing (in the peripheral circuit of SRAM) the outputs of INV2 and INV3 of Fig. 11(a), we can get the outcome of XOR operation.

When an inverter INV2 or INV3 is high, it indicates that the data stored in the cells are unequal ('01 or '10'). Whereas, when both the INV2 and INV3 are low, it indicates that the data stored in the cells are equal ('00' or '11').

4.3 Realization of CA in-Memory

The read-compute-store (RCS) scheme in 8T SRAM is carried out through the decoupled read and write wordlines [1]. The Boolean operations NOR, NAND

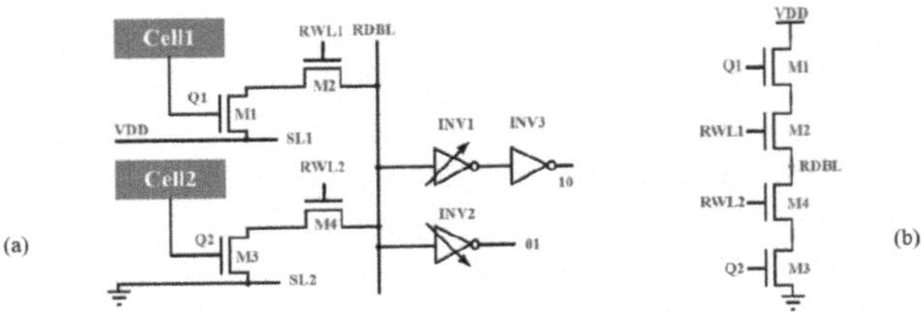

Fig. 11. Schematic of 8T-SRAM realizing XOR [7].

and XOR can be computed through the RBL for multiple cells in a column. As the data is being read (source read) from RBL, for activated RWLs of two cells, simultaneously the WWL of a third cell can be enabled to perform write operation (destination write).

In Fig. 12(a), the read lines (RWL1 of Cell1 and RWL2 of Cell2, belonging to two different words) are activated, and the compute block, which is the abstract form of the skewed inverters of Fig. 10 and Fig. 11, computes the result of boolean logic operation. The decoupled read and write ports of 8T cell facilitate the simultaneous activation of the third wordline. This is the write wordline say, WWL3 of Cell3 (belonging to any third word). The output from the compute block can be selected through a multiplexer and input to the write drivers for directly storing the boolean result in the cell corresponding to WWL3 on the same cycle. This reflects the RCS scheme as shown in Fig. 12(a). The schematic of implementation of RCS in the memory array for three words, storing NOR operation (on operands A and B) result to the third memory word (through WWL3), is shown in Fig. 12(b). A copy operation is also possible through the RCS. It is done by activating the RWL of source word row and WWL of the destination word row. The data stored in the bit cells of source word, through the respective RBL, goes to the sense amplifier (SA), and the output of the SA is the input of the RCS block, to be stored in the destination word [1].

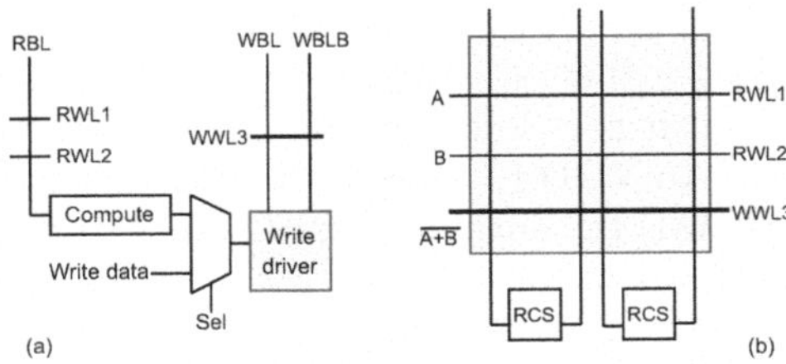

Fig. 12. (a) 'Read-Compute-Store' (RCS) scheme, (b) RCS blocks in memory array [1].

To realize cellular automata (CA) within the memory, four memory words are used, CA1, CA2, L and R, as shown in Fig. 13. The CA1 stores the initial configuration of the CA. The other two words L and R store the neighbors (left and right) of a cell of CA1. Initially the contents of L and R are the copy of the word CA1. Now, to get the corresponding left (right) neighbors for the cells of CA1, the word L (R) is operated with a single circular right (left) shift. That is, a cell of CA1 ($CA1_i$) and its corresponding left (L_i) and right (R_i) neighbors are in a column within the memory array. These are the PSs of the $Cell_i$ and its neighbors. The boolean logic for *rule* (say XOR) is applied on the three cells (L_i, $CA1_i$, R_i) state to find NS of $Cell_i$. The NS is then stored in the respective cell ($CA2_i$) of the fourth memory word CA2. This effectively gives the first evolution of the CA for a time step t. For the next evolution, at time step $t+1$, the CA rule is applied on the cells of CA2 ($CA2_i$) and its corresponding left (L_i) and right (R_i) neighbors, where L and R is the copy of CA2 with a respective circular right and left shift. The CA2, L and R are then considered for the next evolution. The NS thus generated is stored in CA1, and the next evolution is done on CA1. This process is continued for the desired number of evolution steps. For the current design, we need to construct CA1 and CA2 with rule 254 - that is,

$$CA_i^{t+1} = L_i^t + CA_i^t + R_i^t$$

which is an OR operation of the three cells in a column of the memory array. The OR operation is realized at the skewed inverter (INV1) of Fig. 10, which is also represented in the compute block of Fig. 12(a).

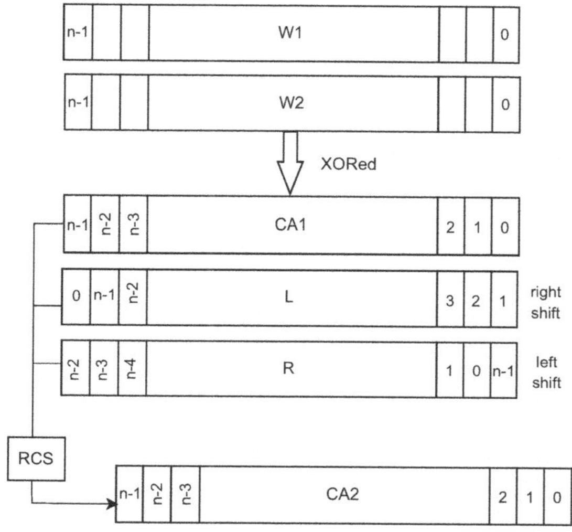

Fig. 13. Schematic of CA within memory for in-memory computation.

5 Conclusion

This work employs an architecture for the in-memory computation (IMC). The design is developed around the 3-neighborhood, 2-state, one-dimensional periodic boundary cellular automata (CA). Such an in-memory computing architecture, defined for in-memory search in SRAM, can effectively be extended for other in-memory operations that are conventionally done by the CPU. The design proposed is a proof of concept. Further evaluation in terms of CA evolution time, latency breakdown, total search time and area overhead are the future scope.

Acknowledgments. - We would like to express our gratitude towards Indian Summer School on Cellular Automata 2025, for the support, advice and guidance.

References

1. Agrawal, A., Jaiswal, A., Lee, C., Roy, K.: X-sram: Enabling in-memory Boolean computations in CMOS static random access memories. IEEE Trans. Circuits Syst. I Regul. Pap. **65**(12), 4219–4232 (2018)
2. Chaudhuri, P.P., Chowdhury, D.R., Nandi, S., Chattopadhyay, S.: Additive cellular automata: theory and applications, Volume 1, vol. 1. John Wiley & Sons (1997)
3. Ermentrout, G.B., Edelstein-Keshet, L.: Cellular automata approaches to biological modeling. J. Theor. Biol. **160**(1), 97–133 (1993)
4. Grinstein, G., Jayaprakash, C., He, Y.: Statistical mechanics of probabilistic cellular automata. Phys. Rev. Lett. **55**(23), 2527 (1985)
5. Khwa, W.S., et al.: A mixed-precision memristor and SRAM compute-in-memory AI processor. Nature, 1–7 (2025)
6. Kleyko, D., Frady, E.P., Sommer, F.T.: Cellular automata can reduce memory requirements of collective-state computing. IEEE Trans. Neural Netw. Learn. Syst. **33**(6), 2701–2713 (2021)
7. Lin, Z., et al.: A review on SRAM-based computing in-memory: circuits, functions, and applications. J. Semicond. **43**(3), 031401 (2022)
8. Liu, Y., et al.: Cellular automata imbedded memristor-based recirculated logic in-memory computing. Nat. Commun. **14**(1), 2695 (2023)
9. Mohammad, B., Halawani, Y.: In-memory computing hardware accelerators for data-intensive applications. Springer (2023)
10. Saha, M., Sikdar, B.K.: A self testable hardware for memory. In: 2013 IEEE International Conference on Circuits and Systems (ICCAS), pp. 45–50. IEEE (2013)
11. Shin, H.J., Jo, S.H.: Low-power 8t SRAM compute-in-memory macro for edge AI processors. Appl. Sci. **14**(23), 10924 (2024)
12. Wang, J., Wang, X., Eckert, C., Subramaniyan, A., Das, R., Blaauw, D., Sylvester, D.: A 28-nm compute SRAM with bit-serial logic/arithmetic operations for programmable in-memory vector computing. IEEE J. Solid-State Circuits **55**(1), 76–86 (2019)
13. Wolfram, S.: Statistical mechanics of cellular automata. Rev. Mod. Phys. **55**(3), 601 (1983)

Multi-length Attractors First Degree Decimal Cellular Automata as Classifier

Vicky Vikrant[1(✉)], C. J. Baby[2], and Kamalika Bhattacharjee[3]

[1] School of Computer Science Engineering and Technology, Bennett University, Greater Noida 201310, Uttar Pradesh, India
vicky.vikrant.cse@gmail.com
[2] Department of Computer Science and Engineering, National Institute of Technology, Tiruchirappalli 620015, Tamil Nadu, India
[3] Department of Information Technology, Indian Institute of Engineering Science and Technology, Shibpur 711103, West Bengal, India

Abstract. This work introduces a data classification algorithm based on *multi-length attractors* of decimal First Degree Cellular Automata (FDCAs). The proposed model uses the basins of multi-length attractors FDCAs to form distinct attractor configuration spaces that correspond to different data classes. Since only the information of the labeled attractors are enough to classify any new data instance, this method drastically improves the time and space requirements of existing cycle-based classification approaches whereas keeping the benefits of the traditional attractor based approaches. Some selection criteria are identified to select a set of non-chaotic candidate CAs which always generate only multi-length attractors. Experimental results demonstrate that multi-length attractors FDCAs as models achieve excellent classification accuracy with execution time comparable to conventional machine learning methods, thereby validating their potential as efficient data classifiers.

Keywords: First Degree Cellular Automaton (FDCA) · Decimal CA · Cycle Structure · Multi-Length Attractor · Classification

1 Introduction

A Cellular Automaton (CA), originally introduced by John von Neumann to study self-reproduction, is a discrete dynamical system where simple local interactions among cells can give rise to complex global behavior. Later, Stephen Wolfram extended CA theory and demonstrated its effectiveness in modeling a wide range of natural phenomena. Over the years, CA has emerged as an important model in natural computing and has been applied to several domains, including data classification.

Several variants of cellular automata (CAs) have been explored for classification tasks. These include Multiple-Attractor Cellular Automata (MACAs) [3], Generalized Multiple-Attractor Cellular Automata (GMACAs) [4], Asynchronous Cellular Automata with fixed-point attractors [2], and Temporally

H. Raju et al. (Eds.): ASCAT 2026, CCIS 2801, pp. 155–168, 2026.
https://doi.org/10.1007/978-3-032-18612-6_12

Stochastic Cellular Automata [6]. In most of these studies, *elementary cellular automata* operating in the binary domain have been used. Consequently, a data encoding process is required to convert decimal or other non-binary data into binary form. However, this encoding often alters the intrinsic properties of the original features, which may reduce the classification performance compared to traditional machine learning techniques [14].

To overcome this limitation, it is more natural to use a CA model that can directly handle decimal-valued data. A *decimal cellular automaton*, where each cell can take any state from 0 to 9, eliminates the need for such encoding. However, a decimal CA with an m-neighborhood dependency results in 10^{10^m} possible rules, making exhaustive characterization infeasible. To address this, some recent works have concentrated on a smaller but powerful subclass of decimal CAs known as *First Degree Cellular Automaton (FDCA)* [12]. FDCAs are defined by nearest-neighbor interactions and are compactly represented by only eight parameters, reducing the rule space to 10^8. This makes their study practical while allowing direct use of decimal numbers as input, which is a major advantage over binary CA-based classifiers. Taking that lead, this paper also concentrates on decimal FDCAs for the classification so that there is no need of encoding and the subsequent data-loss.

Traditional CA-based classification approaches are mainly driven by the concept of an *attractor*. In these methods, data instances are mapped to CA configurations that evolve under specific rules until they reach an attractor. Each attractor corresponds to a class label, and all configurations converging to that attractor are assigned the same class. This approach requires very less space to store the trained model since only the attractor information is required for deciding the class of a new test data. However, since this process relies on a single configuration, it often limits the richness of the classification structure and induces bias. Also, depending on the length of the chain to reach the attractor, the testing time can be high.

On the other hand, some recent works have introduced reversible cellular automata as model of data classification [10,13] where the cycles of a reversible CA are used as the trained classifier. In these cases, majority of configurations of a cycle are instrumental in deciding the class label of a test data. However, a major limitation of these approaches is their high space requirement; because to obtain a fully labeled cyclic space, all d^n configurations of the d-state n-cell CA is needed to be explored and stored as the trained model. So, the storage complexity becomes significant. Moreover, in these methods, all cycles in the cyclic space are labeled during the training phase itself, which further increases the overall time complexity of training.

To address the problems of both of these approaches, this work introduces the concept of *multi-length attractors* as the basis for grouping data into classes. In a multi-length attractors CA, the configuration space is partitioned into several attractor basins, each consisting of a cycle along with some chains of configurations that evolve toward that cycle. When a data instance—represented as a configuration—is evolved under the FDCA rule, it eventually converges to one of the multi-length cycles within these basins. The class label for each basin is then determined by the majority class of the training instances whose configurations converge to the corresponding multi-length cycle.

The main advantage of this approach is that only the labeled multi-length cycles to which the training data converge need to be stored, significantly reducing storage requirements. Likewise, during training, it is sufficient to evolve only the configurations corresponding to the training instances, rather than the entire configuration space, which greatly improves time efficiency. In the testing phase, each input configuration is evolved until it reaches one of the labeled multi-length cycles, and the class associated with that cycle is assigned as the predicted label. So, this approach avoid the limitations of the traditional attractor based approaches as well as the storage complexity of the reversible CA based approaches while keeping the benefits of both the approaches making it the most efficient classification model in terms of both time and space requirement.

Further details on FDCA and the use of multi-length attractors FDCA as a classifier are provided in Sects. 2 and 3. To perform data classification using multi-length attractors FDCA, it is essential to identify suitable multi-length rules that satisfy the necessary conditions for effective classification; the procedure for selecting such rules is described in Sect. 4. The proposed classification algorithm is presented in Sect. 5, with its implementation and performance analysis are discussed in Sect. 6.

2 First Degree Cellular Automata

First Degree Cellular Automata (FDCAs), represent a class of cellular automata characterized by a large state space per cell but a fixed three neighborhood dependency. Each variable in the governing rule appears with degree one, hence the term *first degree.*

Definition 1. *A* ***first degree cellular automaton rule*** *uses local transition function* $R : S^3 \to S$ *expressible as a polynomial:*

$$R(x, y, z) = (c_0xyz + c_1xy + c_2xz + c_3yz + c_4x + c_5y + c_6z + c_7) \bmod d,$$

where $x, y, z \in S$, *and coefficients* $c_0, c_1, \ldots, c_7 \in \mathbb{Z}_d$.

A cellular automaton following the above rule is termed a *first degree cellular automaton* [1]. Such a rule is uniquely defined by the constants $c_0, c_1, \ldots, c_7$, which serve as its defining parameters. The rule space can be efficiently represented through these parameters. In this study, we consider the case $d = 10$, representing a decimal first degree CA under the periodic boundary condition.

Let G_n denote the global transition function of an n-cell CA, and let C_n represent the set of all possible n-cell configurations. The evolution of the CA can be visualized using a *configuration transition diagram.* For an irreversible CA, its the configuration transition diagram contains some configurations with more than one predecessor. In that case, the configuration space is partitioned into several *basins*, where each basin consists of a single cycle, called *attractor*, together with one or more *chains* of configurations that eventually converge to that cycle. If any cycle is of length 1, it is called a *fixed-point attractor*. If any of these cycles contain more than one configuration, the CA is called a *multi-length attractor cellular automaton*.

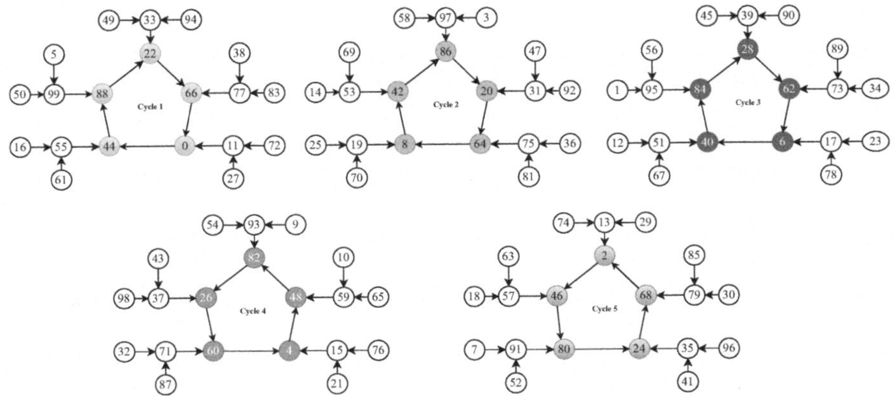

Fig. 1. Five Basins of 10-State 2-Cell multi-length attractors FDCA $\langle 0, 0, 5, 0, 5, 1, 5, 4\rangle$

Example 1. For instance, consider a 2-cell FDCA with the rule vector $R = \langle 0, 0, 5, 0, 5, 1, 5, 4\rangle$ over a decimal state space ($d = 10$). The corresponding parameter values are $c_0 = 0$, $c_1 = 0$, $c_2 = 5$, $c_3 = 0$, $c_4 = 5$, $c_5 = 1$, $c_6 = 5$, and $c_7 = 4$. Now, consider a two-cell configuration 98 in the Cycle 4 of the transition diagram (see Fig. 1), where the state of cell 0 is 9 and the state of cell 1 is 8. Under periodic boundary conditions, the local rule $R(x, y, z) = (c_0xyz + c_1xy + c_2xz + c_3yz + c_4x + c_5y + c_6z + c_7) \bmod d$ is applied to cell 0 as $R(8, 9, 8) = (5xz+5x+y+5z+4) \bmod 10 = (5{\cdot}8{\cdot}8+5{\cdot}8+9+5{\cdot}8+4) \bmod 10 = 3$. Similarly, applying the rule to cell 1 yields $R(9, 8, 9) = 7$. Hence, the next configuration corresponding to the current configuration 98 is 37. For this FDCA, its configuration space contains five basins. Figure 1 illustrates transition diagram containing five basins. Here each of the basins contains a cycle of length 5.

3 Classification with Multi-length Attractors

In this work, we use decimal first degree multi-length attractors cellular automata under periodic boundary condition. A key benefit of using decimal CAs is that they directly support numerical datasets, eliminating the need for additional encoding processes. Each object in the dataset is transformed into a configuration by concatenating its feature values into a single decimal string. This string then serves as the configuration for the decimal CA. As a result, the overall computation becomes more efficient, particularly by reducing the preprocessing overhead typically associated with data encoding.

Data points are classified into two or more groups depending on their labels. We may accomplish this by utilizing the basin of multi-length attractors FDCA rules. The dataset is initially divided into training and testing segments. During the training phase, the CA basins are classified into distinct classes based on the labels in the training data. In some cases, a test instance may evolve to a cycle within a basin that was not assigned a label during the training phase. In such situations, the label of the test instance is determined based on the similarity between its configuration and the median configurations of the labeled cycles.

Table 1. Hypothetical sample dataset used for classification

Data ID	A1	A2	Target Configuration	Class
1	7	7	77	1
2	1	1	11	1
3	8	8	88	1
4	7	5	75	2
5	8	6	86	2
6	8	1	81	2
7	0	5	05	1
8	1	4	14	2
9	4	1	41	1

Let us consider the hypothetical dataset shown in Table 1 to demonstrate the classification process using a multi-length attractors FDCA. The dataset comprises two attributes A1 and A2 and a corresponding class label. The values of the two attributes are concatenated to form a *Target Configuration*. Suppose that during the training phase, data instances with IDs 1, 2, and 3, having target configurations 77, 11, and 88 respectively, belong to *Class 1*. Similarly, instances with IDs 4, 5, and 6, whose target configurations are 75, 86, and 81, are assigned to *Class 2* and constitute the training data.

Example 2. Consider the basins for a 2-cell FDCA rule $\langle 0, 0, 5, 0, 5, 1, 5, 4 \rangle$, as illustrated in Fig. 1. When the CA is evolved from the configurations 77, 11, and 88, all three configurations eventually converge to the same multi-length cycle (Cycle 1: $22 \mapsto 66 \mapsto 0 \mapsto 44 \mapsto 88 \mapsto 22$). Since these configurations correspond to training instances belonging to *Class 1*, the basin associated with Cycle 1 is labeled as *Class 1*. Similarly, the configurations 75, 86, and 81 converge to another multi-length cycle (Cycle 2: $86 \mapsto 20 \mapsto 64 \mapsto 8 \mapsto 42 \mapsto 86$), and the majority of these configurations belong to *Class 2*. Therefore, the basin corresponding to Cycle 2 is labeled as *Class 2*. In both cases, it is sufficient to store only the multi-length cycle along with its assigned label. When a new input is given during the testing phase, its label is predicted by determining to which basin it belongs, and the cycle information learned during training.

Consider the test instance with data ID 7, whose target configuration is 05. To classify this instance, the FDCA is first evolved from the configuration 05 to determine the multi-length cycle to which it converges. In this case, the configuration reaches Cycle 1, and therefore the test instance is classified as belonging to *Class 1*. In the same manner, the label of the instance with data ID 8 and target configuration 14 can be predicted as *Class 2*. Now, consider the test data ID 9 with target configuration 41. This configuration evolves under CA rule $R = \langle 0, 0, 5, 0, 5, 1, 5, 4 \rangle$ to reach an unlabeled basin cycle given by $2 \mapsto 46 \mapsto 80 \mapsto 24 \mapsto 68 \mapsto 2$ (see Fig. 1), which has a median value of 46. The nearest labeled cycle to this unlabeled cycle is Cycle 1, defined as $22 \mapsto 66 \mapsto 0 \mapsto 44 \mapsto 88 \mapsto 22$, with a median value of 44 and an associated class label 1. Therefore, the predicted class label for this test data instance is 1.

This example illustrates that an FDCA can effectively perform data classification by exploiting its cyclic attractor structure. Each multi-length cycle functions as a representative of a specific class, allowing new data instances to be classified based on the cycle to which they converge. This cyclic-space-based mapping provides a natural, deterministic mechanism for classification using multi-length attractors FDCA.

4 Selection of Proper Rules

In a multi-length attractors CA, the configuration space is partitioned into a collection of basins of attractors, where each basin consists of a cycle and one or more chains of configurations that converge to that cycle. When such multi-length attractors CAs are used as a tool for data classification, the distribution of configurations within these basins becomes important. Specifically, the way in which input patterns are mapped to different basins plays a significant role in the testing time. A balanced and well-separated basin structure results in good classification performance. So, our target is to have CAs where all cycles are multi-lengths with a limited number of such cycles. To obtain such multi-length attractors CAs, we apply specific selection criteria to identify those most suitable for data classification. The step-by-step procedure used for this selection is outlined below (see Fig. 2 for an overview).

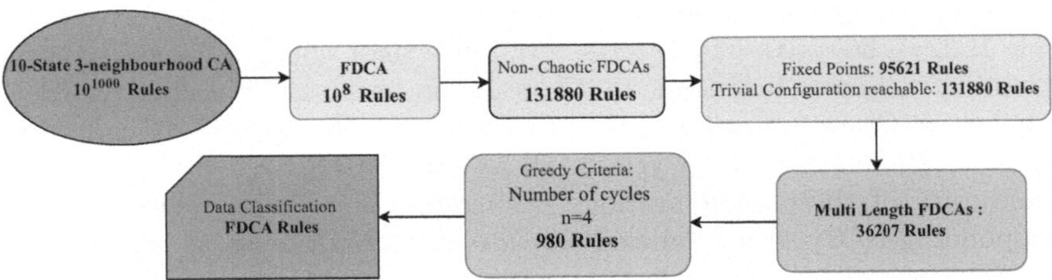

Fig. 2. Criteria for selecting proper FDCA rules for Data classification

4.1 Finding Multi-Length Attractors FDCAs

To identify suitable multi-length attractors FDCAs for classification, it is essential to focus on rules that generate attractors of balanced basin structure. Extremely large cycles, typically produced by chaotic CAs, do not align with the objective of this study, as they lead to unpredictable and computationally expensive behavior. Similarly, in a chaotic CA, unrelated configurations may become part of the same basin, resulting poor grouping of data unsuitable for classification. Therefore, the selection process emphasizes on *non-chaotic* FDCAs which are more likely to yield fixed-point or multi-length attractors. The primary goal is to identify FDCAs that consistently produce a mix of fixed-point and small-to-medium-length cycles, irrespective of the automaton size. The following steps are applied to find the initial set of multi-length attractors CA:

1. **Identification of Non-Chaotic FDCAs Using the P Parameter:** P Parameter, proposed by Kamilya and Das (2018) [7] is a good way to identify and evaluate to the degree of chaos in any CA. As proposed in Ref. [7], a CA can be considered non-chaotic if its parameter value is $P = (l, r)$ where $l < 0.75$ or $r < 0.5$. Rules with lower P values are preferred, as they represent non-chaotic dynamics that lead to stable and interpretable attractor formation. So, in the first step, we identify the non-chaotic decimal FDCAs, likely to produce smaller cycles, yielding a total of 131880 rules satisfying this criterion.
2. **Determine FDCA with only Multi-Length Attractors:** Here, we take a greedy approach to identify a subset of FDCAs which will always produce multi-length attractors for any n. For this, we first identify which of the FDCA rules generated in Step 1 can generate fixed-point attractors for some n. Those 95621 CAs are not desirable. Next, as the greedy approach, we determine whether the trivial configurations, like 0^n, 1^n, etc. have any predecessor in the CA. Since, we are considering periodic boundary condition, so these CAs are more likely to have multi-length cycles. However, all the 131880 non-chaotic CAs have reachable trivial configurations. Now, some of these CAs may have these trivial configurations as fixed-point attractors. So, we remove the 95621 CAs with fixed-point attractors from this set, getting 36259 FDCAs which will always generate multi-length attractors.

Further, since, we do not want reversible CAs, any reversible CAs present are identified and removed. As a result, we obtain a total of 36207 FDCA rules having only multi-length cycles. Since this number is too large for practical use in data classification, we apply additional filtering criteria based on the number of cycles present in the configuration space of each FDCA.

4.2 Filtering Based on Cycle Structure

A large number of basins in an FDCA is not suitable for data classification, as the multi-length cycles within these basins are treated as representatives of the classes in the dataset. Therefore, from the initial set of 36,207 multi-length attractors CAs, we further filter the rules based on the ratio between the number

Table 2. Sample multi-length attractors FDCA parameters selected for data Classification

0,0,5,5,2,5,0,9	0,0,0,5,0,2,0,7	0,0,0,5,0,7,5,3	0,0,5,0,5,1,5,4	0,0,5,0,5,7,0,7
0,0,0,5,0,3,5,3	0,0,0,5,0,2,0,9	0,0,0,5,0,7,5,5	0,0,5,0,5,1,5,6	0,0,5,0,5,7,0,9
0,0,5,0,1,5,0,5	0,0,0,5,0,3,5,1	0,0,0,5,0,7,5,7	0,0,5,0,5,1,5,8	0,0,5,0,6,5,5,5
0,0,0,5,0,1,0,2	0,0,0,5,0,3,5,5	0,0,0,5,0,7,5,9	0,0,5,0,5,3,0,1	0,0,5,0,7,5,0,1
0,5,0,5,0,6,5,1	0,0,0,5,0,3,5,7	0,0,0,5,0,8,0,1	0,0,5,0,5,3,0,3	0,0,5,0,7,5,0,3
0,0,0,5,0,1,0,4	0,0,0,5,0,3,5,9	0,0,0,5,0,8,0,3	0,0,5,0,5,3,0,5	0,0,5,0,7,5,0,5
0,0,0,5,0,1,0,6	0,0,0,5,0,6,5,1	0,0,0,5,0,8,0,5	0,0,5,0,5,3,0,7	0,0,5,0,7,5,0,7
0,0,0,5,0,1,0,8	0,0,0,5,0,6,5,3	0,0,0,5,0,8,0,7	0,0,5,0,5,3,0,9	0,0,5,0,7,5,0,9
0,0,0,5,0,2,0,1	0,0,0,5,0,6,5,7	0,0,0,5,0,8,0,9	0,0,5,0,5,7,0,1	0,0,5,0,8,5,5,1
0,0,0,5,0,2,0,3	0,0,0,5,0,6,5,9	0,0,0,5,2,0,0,1	0,0,5,0,5,7,0,3	0,0,5,0,8,5,5,3
0,0,0,5,0,2,0,5	0,0,0,5,0,7,5,1	0,0,0,5,2,0,0,3	0,0,5,0,5,7,0,5	0,0,5,0,8,5,5,5

of cycles and the total number of configurations in the CA. Our selection criterion is defined as follows:

Greedy Criterion: *Select those rules for which the number of multi-length cycles lies between $l_1\%$ and $l_2\%$ of the total configuration space 10^n, where n denotes the number of cells in the FDCA.*

In this work, we apply this filtering criterion for cell length $n = 4$, using $l_1 = 5\%$ and $l_2 = 10\%$ and obtain 980 FDCA rules that satisfy our criterion. Representative examples from the set of 980 rules are presented in Table 2. For larger lattice sizes, additional criterion may be applied for further filtering. In this work, however, we work with these 980 FDCAs.

5 The Proposed Classification Algorithm

Classification in machine learning refers to the process of categorizing data instances into distinct classes based on their feature characteristics. In the proposed approach, we perform multi-label classification by using the basins of multi-length attractors FDCAs.

For each of the configurations corresponding to the training data instances, the FDCA is evolved. Since we use only multi-length attractors FDCAs, each configuration from a training instance eventually converges to the multi-length cycles within a basin. The class label for each basin is assigned according to the majority label of the training instances whose configurations reach the corresponding cycle. To address storage limitations, we store only the cycles and their associated labels, rather than storing the entire basin structure, which significantly reduces memory requirements. During testing, each input configuration is evolved until it reaches a labeled multi-length cycle, and the label of that cycle is then used to classify the test instance.

In the case of certain outlier data instances, there is a possibility that the evolved configuration may converge to an unlabeled multi-length cycle belonging to a basin that was not assigned a class label during training. For such situations, we determine the label by measuring the similarity of the test configuration to the existing labeled multi-length cycles. Specifically, we compute the distance between the test configuration and the median configuration of each stored labeled cycle, and assign the label of the nearest cycle (see Example 2). This strategy ensures classification even when the configuration does not directly fall into a labeled basin. Algorithm 1 shows the step-by-step training and testing process.

Algorithm 1. CA-Based Data Classification using Multi-length Attractor FDCA

Require: CA rule R, Dataset D.
Ensure: Trained model

Split D into training set $Train$ and test set $Test$
Identify attribute and class columns in D
Normalize all attribute values to digit strings of uniform length
for each sample $(x, label)$ in $Train$ **do** ▷ — Training Phase —
    Concatenate attributes of x to form pattern string PS
    $cycle_i \leftarrow$ Multi-length cycle reached from configuration x under rule R
end for
$cycle_labels \leftarrow$Assign label to each $cycle$ base on majority voting.
▷ — Testing Phase —
Concatenate attributes of x to form PS
$test\_cycle \leftarrow$ Multi-length cycle reached from configuration x under rule R
if $test\_cycle \in cycle\_labels$ **then**
    $pred \leftarrow cycle\_labels[test\_cycle]$ ▷ Exact cycle match found
else
    $pred \leftarrow$ label of nearest cycle ▷ Based on closest attractor by median similarity
end if

6 Implementation and Performance Analysis

To ensure compatibility with the configuration space of FDCAs, we represent each dataset instance as a decimal string in our implementation. For our experimental analysis, we use the set of 980 rules selected at Sect. 4.2 (sample rules are presented in Table 2). For ease of use, reproducibility, and adaptability to different datasets, the complete system is created as a Python module and uploaded in GitHub. [1]

[1] GitHub repository: https://github.com/kamalikaB/Classification.

Table 3. Dataset Details Used for Experiments

Dataset	Dataset	Instances	Attributes	Training	Test
DS 1	Haberman	306	4	360	90
DS 2	Balloons	16	5	14	4
DS 3	Monk 1	432	7	345	87
DS 4	Monk 2	432	7	464	116
DS 5	Monk 3	432	7	364	92
DS 7	Gender	66	5	54	12

6.1 Datasets and Preprocessing

Due to computational constraints, we restrict our dataset selection to cases in which each data instance could be represented by a CA of size 1 to 9. Consequently, this limitation reduced the number of real-world datasets available for evaluating the efficiency of our algorithm. Datasets containing non-integer values are preprocessed to satisfy the CA cell length requirements. Additionally, variations in attribute lengths across datasets are normalized before processing to maintain uniformity in representation.

Table 3 provides the details of the datasets used in this study, where n denotes the cell length of each CA. To evaluate the performance of the proposed classification algorithm, each dataset is divided into two randomly generated subsets: 20% of the data are used for testing, while the remaining 80% are used for training. The algorithm is validated using six different datasets obtained from Ref. [9] and the classification accuracy is calculated for each dataset.

Table 4. Performance of multi-length attractors FDCA for Classification

Dataset	c_0	c_1	c_2	c_3	c_4	c_5	c_6	c_7	Accuracy
Haberman	0	0	5	5	2	5	0	9	0.72
Balloons	0	0	0	5	0	3	5	3	1.00
Monk1	0	0	0	5	0	6	5	1	1.00
Monk2	0	0	5	0	1	5	0	5	0.87
Monk3	0	0	0	5	0	1	0	2	0.89
Gender	0	5	0	5	0	6	5	1	0.84

Table 5. Performance of different models in classification

Dataset	Measure	Multi length	Reversible	SVM	NB	DT	LR	KNN	MLP
Haberman	Accuracy	0.721	0.966	0.661	0.710	0.645	0.710	0.677	0.677
	Precision	0.617	1.000	0.695	0.741	0.739	0.710	0.750	0.722
	Recall	0.589	0.929	0.932	0.909	0.773	1.000	0.818	0.886
	F1-Score	0.595	0.963	0.796	0.816	0.756	0.830	0.783	0.796
Balloons	Accuracy	**1.000**	1.000	0.750	0.750	1.000	0.500	1.000	1.000
	Precision	1.000	1.000	0.667	1.000	1.000	0.500	1.000	1.000
	Recall	1.000	1.000	1.000	0.500	1.000	1.000	1.000	1.000
	F1-Score	1.000	1.000	0.800	0.667	1.000	0.667	1.000	1.000
Monk 1	Accuracy	**1.000**	1.000	0.724	0.724	0.885	0.724	0.897	1.000
	Precision	1.000	1.000	0.732	0.732	0.867	0.732	0.886	1.000
	Recall	1.000	1.000	0.698	0.698	0.907	0.698	0.907	1.000
	F1-Score	1.000	1.000	0.714	0.714	0.886	0.714	0.897	1.000
Monk 2	Accuracy	0.872	0.922	0.621	0.534	0.957	0.644	0.638	0.681
	Precision	0.844	0.982	0.573	0.512	0.932	0.579	0.592	0.644
	Recall	0.864	1.000	0.839	0.534	0.982	0.982	0.804	0.750
	F1-Score	0.852	0.991	0.681	0.603	0.957	0.728	0.682	0.694
Monk 3	Accuracy	**0.895**	0.688	0.826	0.954	1.000	0.839	0.920	1.000
	Precision	0.895	0.955	0.864	0.955	1.000	0.826	0.911	1.000
	Recall	0.895	1.000	0.839	0.955	1.000	0.864	0.932	1.000
	F1-Score	0.895	0.977	0.844	0.955	1.000	0.844	0.921	1.000
Gender	Accuracy	**0.846**	0.714	0.214	0.429	0.643	0.500	0.643	0.500
	Precision	0.875	1.000	0.167	0.455	0.600	0.500	0.625	0.500
	Recall	0.857	0.500	0.143	0.714	0.857	1.000	0.714	0.571
	F1-Score	0.845	0.667	0.154	0.556	0.706	0.667	0.667	0.533

6.2 Performance Analysis Measures

Performance indicators are critical in determining the efficacy of classification models and the quality of their learning processes. Numerous measures have been offered in the literature to aid in more informed evaluations, whether for general use or specialized to specific application areas [5]. In this work, we use four commonly used performance evaluation metrics like accuracy, precision, recall, and F-score, to compare our technique to other machine learning models. The mathematical definitions for these metrics are provided below.

$$\text{Accuracy} = \frac{\text{Number of correct predictions}}{\text{Total predictions}}$$

$$\text{Recall} = \frac{\text{Correctly classified positives}}{\text{Total positives}}$$

$$\text{Precision} = \frac{\text{Correctly classified positives}}{\text{Total predicted as positives}}$$

$$\text{F-measure} = \frac{2 \times \text{Recall} \times \text{Precision}}{\text{Recall} + \text{Precision}}$$

The classification accuracies for the different datasets, along with their corresponding decimal CA constants, are summarized in Table 4.

6.3 Performance Comparison with Existing ML Models

We use conventional machine learning benchmark metrics to assess the performance of the proposed approach on a set of real-world datasets. We compare the accuracy, precision, recall, and F-measure (F-score) of the proposed first Degree CA-based classification algorithm with several standard machine learning (ML) algorithms [8], including Decision Trees (DT), Naive Bayes (NB), Logistic Regression (LR), K-Nearest Neighbors (KNN), Multi-Layer Perceptron (MLP), and Support Vector Machine (SVM). In the case of the decimal CA classifier, preprocessing steps are applied to adapt the data to the input requirements of the CA framework. We also compare our results with existing reversible FDCA-based data classification models [10]. A detailed analysis of the results is presented in Table 5. As observed, the multi-length CA-based classifier achieves competitive or superior performance compared to conventional ML models as well as the existing FDCA based model.

Finally, we compare the execution time of our model with that of the reversible FDCA-based classification model as well as one traditional machine learning method, the KNN classifier (see Table 6 and Fig. 3). The result shows that our classification model with multi-length attractors is very fast in comparison to the reversible FDCA based classifier and comparable to the state-of-the-art ML classifiers. This provides evidence to our claim that our proposed algorithm is the most efficient one for classifying data with CA exploiting the benefits of CAs to the fullest.

Table 6. Execution time of CA-based classification models.

Dataset	Multi-length FDCA(sec)	Reversible FDCA(sec)	KNN (sec)
Haberman	0.167	9.811	0.035
Balloons	0.016	0.063	0.505
Monk 1	0.432	49.72	0.033
Monk 2	0.173	11.37	0.033
Monk 3	0.182	52.09	0.059
Gender	0.042	0.099	0.037

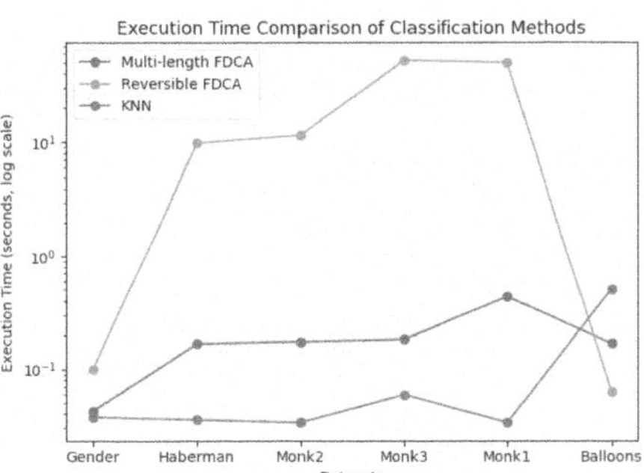

Fig. 3. Execution time comparison between the CA-based classification models.

7 Conclusion

This work introduces a data classification algorithm based on the basins of multi-length attractors decimal first degree CA. Using the target configurations generated by concatenating the attributes and the selected CA rules for data classification, a *majority voting* mechanism is used to determine the label of each basin. The labeled multi-length cycles generated through this process have then been used to predict labels for unknown data instances. Experimental evaluations on real-world datasets demonstrate that the proposed classification method performs very well when compared to existing machine learning classification techniques with speed almost comparable to these methods. However, a key limitation arises when using *decimal cellular automata* with ten states: once the number of cells exceeds six, the exponential increase in configuration space leads to slower computation. Likewise, since the number of FDCA rules used here is large, additional filtering based on the distribution of configurations in the chains and multi-length cycles may obtain a more refined set of rules for data classification. These two limitations can be considered as potential avenues for future enhancement of this classification approach.

Acknowledgment. The authors are grateful to Prof. Sukanta Das, Prof. Biplab K Sikdar and Dr. Sukanya Mukherjee for their valuable comments and suggestions during ASCAT 2025 which has initiated this work.

References

1. Bhattacharjee, K., Vikrant, V.: Study of first degree cellular automata for randomness. J. Cellular Automata **17** (2023)
2. Sethi, B., Das, S.: Modeling of asynchronous cellular automata with fixed-point attractors for pattern classification. In: 2013 International Conference on High Performance Computing & Simulation (HPCS), pp. 311–317. IEEE (2013)
3. Maji, P., Shaw, C., Ganguly, N., Sikdar, B.K., Chaudhuri, P.P.: Theory and application of cellular automata for pattern classification. Fund. Inf. **58**(3–4), 321–354 (2003)
4. Ganguly, N., Maji, P., Sikdar, B.K., Chaudhuri, P.P.: Design and characterization of cellular automata based associative memory for pattern recognition. IEEE Trans. Syst. Man Cybern. Part B (Cybern.) **34**(1), 672–678 (2004)
5. Ferri, C., Hernández-Orallo, J., Modroiu, R.: An experimental comparison of performance measures for classification. Pattern Recogn. Lett. **30**(1), 27–38 (2009)
6. Paul, S., Roy, S., Das, S.: Pattern classification with temporally stochastic cellular automata. In: International Workshop on Cellular Automata and Discrete Complex Systems, pp. 137–152. Springer, Cham (2023). https://doi.org/10.1007/978-3-031-42250-8_10
7. Kamilya, S., Das, S.: A study of chaos in cellular automata. Int. J. Bifurcat. Chaos **28**(03), 1830008 (2018)
8. Alauthman, M., et al.: Enhancing small medical dataset classification performance using GAN. Informatics **10**(1), 28 (2023)

9. Kelly, M., Longjohn, R., Nottingham, K.: The UCI Machine Learning Repository. https://archive.ics.uci.edu
10. Baby, C.J., Bhattacharjee, K.: Reversible decimal first degree cellular automata for data classification. In: International Conference on Cellular Automata for Research and Industry. Springer, Cham (2024). https://doi.org/10.1007/978-3-031-71552-5_13
11. Bhattacharjee, K.: First degree cellular automata as pseudo-random number generators. In: Asian Symposium on Cellular Automata Technology. Springer, Singapore (2022). https://doi.org/10.1007/978-981-19-0542-1_10
12. Baby, C.J., Bhattacharjee, K. Data classification using reversible decimal First Degree cellular automata. Natural Comput. 1–20 (2025)
13. Baby, C.J., Bhattacharjee, K.: Information propagation based data classification with reversible non-uniform elementary cellular automata. In: Asian Symposium on Cellular Automata Technology (ASCAT 2025). Communications in Computer and Information Science, vol. 2499. Springer, Cham (2025). https://doi.org/10.1007/978-3-031-94121-4_15
14. Narodia Parth, P., Bhattacharjee, K.: Gödel number based encoding technique for effective clustering. In: International Conference on Pattern Recognition and Machine Intelligence. Springer, Cham (2023). https://doi.org/10.1007/978-3-031-45170-6_6

Dynamics of Non-convergent Skewed Elementary Cellular Automata

Souvik Roy[1(✉)], Shreena Patel[1], Dhyey Patel[1], Hetvi Raval[1], and Sumit Adak[2]

[1] School of Engineering and Applied Science, Ahmedabad University, Gujarat, India
{souvik.roy,shreena.p,dhyey.p7,hetvi.r1}@ahduni.edu.in, svkr89@gmail.com
[2] DTU Compute, Technical University of Denmark, Kgs. Lyngby, Denmark
suad@dtu.dk

Abstract. This study explores the notion of skewed asynchronous cellular automata (ACA) which violates the atomicity property of fully ACA. In an early work, Roy et al. [18] have identified the convergent elementary cellular automata (ECA) rules under skewed update. In this direction, this study explores the remaining 28 non-convergent skewed ECA rules. To understand the dynamics of these rules, this study considers space-time pictures, communication class structure and empirical tests. First, we classify the non-convergent skewed systems following space-time dynamics. Hereafter, we record the number of recurrent configurations and transient configurations for these rules following finite size experiments which identify the reversible skewed systems. Moreover, the communication class dynamics of these rules also show connection with different non-trivial OEIS sequences (like, *A*001644, *A*001608, *A*001639 etc.). According to the results, ECA 90 acts as a good randomness enhancer following the Dieharder tests. Therefore, we theoretically analyse the communication class of ECA 90 which shows recurrent or reversible phenomenon under skewed update.

Keywords: Elementary Cellular Automata (ECA) · Skewed ACA · Communication class · Reversibility · ECA 90

1 Introduction

The study of asynchronous cellular automata (ACA) has gained a lot of attention in the recent past which breaks the centralized concept of global clock in cellular automata (CA) [1–3,6]. The CA research community has proposed different versions of asynchronous updates to capture independence from the global clock where fully asynchronous update is the most studied after considering the simplicity [9,15]. In fully ACA, we choose one cell at each time step for update following uniform distribution. In a different view, two neighbouring cells are restricted to update together which captures the model of concurrent and distributed systems (also known as *atomicity property* [5]). In the direction of computing with fully asynchronous noise, in an early work, Sethi et al. [21]

H. Raju et al. (Eds.): ASCAT 2026, CCIS 2801, pp. 169–181, 2026.
https://doi.org/10.1007/978-3-032-18612-6_13

have proposed an efficient pattern classifier[1] after comparing with benchmark algorithms (like, C4.5, Bayesian) following the theoretical analysis of convergent fully elementary ACA [21]. Fully ACA[2] with eccentric cloud properties have also performed well in the computational task of density classification[3] [16]. From the theoretical point of view, the ACA research community also completely analysed the convergence [10,22], convergence time [10,12], and recurrence (or reversibility) [17] property of this fully asynchronous system for elementary CA (ECA) rules. Apart from the fully asynchronous CA, α-asynchronous [4], asynchronism with *information loss* [14] and m-asynchronous [7] updates have also gained popularity after considering the literature [9].

In this direction, in a recent work [18], Roy et al. have observed the dynamics of elementary CA systems in the absence of *atomicity property* (or after breaking the notion of fully asynchronous update). In other words, the works of Ref. [18] have introduced the notion of skewed asynchronous update where it chooses two cells for update at each time step and the chosen two cells should be neighbours of each other [18]. Intuitively, it breaks the restriction of fully asynchronous systems. Moreover, this proposed skewed asynchronous update is also close to physical (burst error, particle systems [1]) and societal [18] complex systems in the context of modelling. Salvi et al. [20] have also proposed a classification of convergent skewed ECA systems after considering the notion of eccentric, partially eccentric and deterministic cloud properties. In the context of computational ability of these skewed systems, Fatès has proposed a four-neighbourhood rule[4] to compute well-known parity problems[5] [11].

According to the initial results of Ref. [18], fully versus skewed update shows following remarkable change in dynamics – (a) ECA rules 26, 38, 54, 58, 90, 122, 134, and 150 display abrupt phase transition for changing updating schemes, i.e., if we change updating method from fully ACA to skewed ACA; (b) ECA rules 6 and 22 also report abrupt phase transition dynamics for changing updating schemes where the CA size is odd; (c) Lastly, ECA 105 displays class transition [18] dynamics for changing updating schemes where the CA size is divisible by four. For evidence, Fig. 1 depicts the abrupt phase change dynamics for ECA 26 and 54. Roy et al. [18] have also identified the convergent ECA rules following the experimental simulation based approach. According to the results of [18], 52 (resp. 8) ECA rules, out of 88, show convergent (resp. CA size dependent convergent) dynamics under skewed asynchronous update. In this article, the rest 28 non-convergent ECA rules are our point of interest.

---

[1] Elementary CA rules 4, 12, 36 and 68 [21].

[2] Elementary CA rules 170, 178 and 184 [16].

[3] In the density classification problem, if the number of 1's in the initial configuration is more, then the system should converge to all-1; otherwise, the destination should be an all-0 configuration [8].

[4] if $a \neq b$ and $c \neq d$, then $f(a,b,c,d) = (1-b, 1-c)$; else $f(a,b,c,d) = (b,c)$ [11].

[5] In the parity problem, if the number of 1's in the initial configuration is even, then the system should converges to all-0; otherwise, the destination should be an all-1 configuration [11].

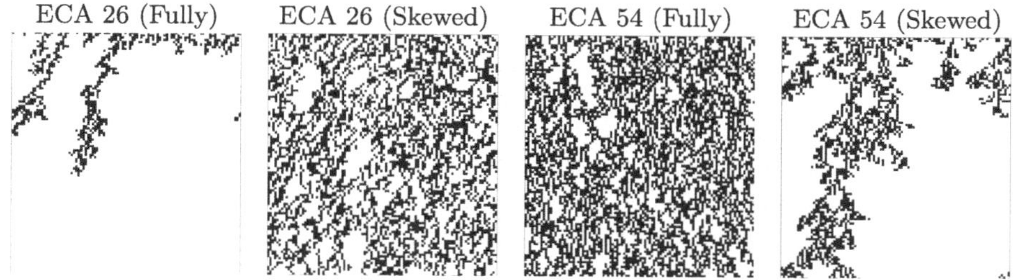

Fig. 1. Space-time pictures of ECA 26 and 54 which report abrupt phase change dynamics for changing updating schemes, i.e., fully ACA to skewed ACA.

In this study, we explore the 28 minimal non-convergent rules under skewed asynchronous update following space-time dynamics, structure of communication class, and randomness enhancing capabilities. We first classify the non-convergent skewed ACA rules according to space-time pictures which classifies the system into – *periodic with noise*, *noisy*, and *chaos with noise* classes. Hereafter, we compute the number of recurrent configuration, number of transient configurations, and number of communication class for these rules to understand their cycle structure (here, we call communication class) following a finite CA size experimental setup. The above two experiments give us an idea about randomness enhancing capabilities of these non-convergent rules. Next, we validate the experimental results after considering the Dieharder [13] empirical test. According to the empirical test results, ECA 90 under skewed update acts as a good randomness enhancer. To understand the reason behind that, we theoretically analyse the communication class dynamics of ECA 90 which shows recurrent (reversible) phenomenon (i.e., number of recurrent configuration is 2^n).

2 Elementary CA and Skewed Asynchronous Update

Here, we consider two-state three-neighbourhood elementary CA (ECA) where the cells (denoted as $\mathcal{L} = \mathbb{Z}/n\mathbb{Z}$) are organized as a one-dimensional ring [24]. A cell is associated with a state ({0,1}) at every time step $t \in \mathbb{N}$. A configuration is the collection of states at a given time. Here, the system follows the local transition function $f : \{0,1\}^3 \to \{0,1\}$. For example, ECA 90 can be written as follows:

(a,b,c) (RMT)	111 (7)	110 (6)	101 (5)	100 (4)	011 (3)	010 (2)	001 (1)	000 (0)	Rule
$f(a,b,c)$	0	1	0	1	1	0	1	0	90

Here, we use the *rule mean term* (RMT) for the naming purpose of the neighbourhood triplet (a, b, c); and we use classical *decimal code* for the naming purpose of the 'rule'. If $f(a, b, c)= b$, then the RMT is *passive*; otherwise, we call it as *active*. Note that, there are 256 rules in this elementary cellular systems, however, we only consider 88 because the rest are equivalent [24].

Next, we discuss the notion of fully and skewed ACA systems. For both of the systems, during computation, the model chooses a cell (uniformly and randomly) at each time step. Let us consider that $(U_t)_{t\in\mathbb{N}} \in \mathcal{L}^{\mathbb{N}}$ represents the sequence of chosen cells. Now, for fully ACA, we apply the local rule for only the chosen cell; and, for skewed ACA, we apply the local rule for the chosen cell and its right neighbouring cell. Formally, let us consider a_i^t denotes the state of cell i at time t. Therefore, for fully ACA, we can able to define the next state at time $t+1$ as

$$a_i^{t+1} = \begin{cases} f(a_{i-1}^t, a_i^t, a_{i+1}^t) & \text{if } i = U_t \\ a_i^t & \text{otherwise.} \end{cases}$$

Similarly, for skewed ACA, we can able to define the next state as :

$$a_i^{t+1} = \begin{cases} f(a_{i-1}^t, a_i^t, a_{i+1}^t) & \text{if } i = U_t \text{ or } i = U_t + 1 \\ a_i^t & \text{otherwise.} \end{cases}$$

For evidence, if we start from an initial configuration '101100' and follow elementary CA 90, the system shows following evolution under fully update: $\begin{smallmatrix} 1&0&1&1&\mathbf{0}&0 \\ 1&0&1&\mathbf{1}&1&0 \\ 1&\mathbf{0}&1&0&1&0 \\ 1&0&1&0&1&0 \end{smallmatrix}$ and following evolution under skewed update: $\begin{smallmatrix} 1&0&1&1&\mathbf{0}&\mathbf{0} \\ 1&0&1&\mathbf{1}&\mathbf{1}&1 \\ 1&\mathbf{0}&\mathbf{1}&0&0&1 \\ 1&0&0&0&0&1 \end{smallmatrix}$. Here, the updated cells are marked in bold. If a configuration is associated with only passive RMTs, we call the configuration as a *point attractor* [10,12]. Next, we call a configuration (say, a) as *recurrent* if the configuration is associated with 'internal return' property; that is, $\forall b$ configuration reachable from a, there exists a path from b to a using one or more steps [17]. Otherwise, the configuration can be viewed as a *transient* configuration [17]. For an ECA system, if all configurations are recurrent, the overall system is also recurrent or *reversible*. If all the recurrent configurations of an ECA system is a point attractor, then the system is *convergent* [10,12].

3 Non-convergent Elementary Skewed ACA Systems

In this section, we explore the 28 minimal non-convergent rules under skewed asynchronous update following space-time dynamics, structure of communication class, and randomness enhancing capabilities. Recall that, Roy et al. [18] have identified the convergent elementary skewed ACA rules following the experimental simulation based approach. According to the results of [18], 52[6] (resp. 8[7]) ECA rules, out of 88, show convergent (resp. CA size dependent convergent) dynamics under skewed asynchronous update. Therefore, we write the following remark.

[6] ECAs 0, 2, 4, 5, 6, 8, 10, 12, 13, 18, 22, 24, 32, 34, 36, 38, 40, 42, 44, 50, 54, 56, 72, 74, 76, 77, 78, 94, 104, 106, 122, 128, 130, 132, 136, 138, 140, 146, 150, 152, 154, 160, 162, 164, 168, 170, 172, 178, 184, 200, 204 and 232 show convergent dynamics under skewed asynchronous update.

[7] ECAs 7, 14, 15, 23, 30, 37, 45 and 105 show size dependent convergent dynamics under skewed asynchronous update.

Remark 1 *Elementary CA rules* 1, 3, 9, 11, 19, 25, 26, 27, 28, 29, 33, 35, 41, 43, 46, 51, 57, 58, 60, 62, 73, 90, 108, 122, 129, 137, 142, 156 *show non-convergent dynamics under skewed asynchronous updating scheme [18].*

Here, we ask the following questions regarding the dynamics of these 28 non-convergent rules: `What can be said about recurrent or reversibility properties of these rules?` If any rule is neither reversible nor convergent, then `what about the multi-length attractor dynamics of the corresponding rule.` To understand these questions, we need to know the number of recurrent, transient configurations, and number of communication classes for a rule. Moreover, from a space-time dynamics perspective, `is it possible to capture the dynamics of these rules using classical Wolfram's` [24] `classification?` If not, what we are able to say about the space-time dynamics of these rules. However, space-time dynamics observation is a qualitative approach of observation. To make it quantitative, we also consider empirical tests which capture the randomness of the space-time pictures. In this direction, next, we explore the space-time pictures of these 28 non-convergent skewed ACA rules.

3.1 Space-Time Dynamics of Skewed Systems

Here, to capture the space-time dynamics of the skewed ACA systems, we consider ring size $n = 100$, and plot the evolution of the system for 100 time steps. To validate the results, we also evolve the system for 1000 time steps where the ring size is 1000. Moreover, for an ECA rule, we run the (same) experiment for 50 times to capture the ambiguity. Note that, Fig. 2 depicts the example space-time pictures for ring size 100 where states 0 and 1 are respectively denoted by white and black boxes.

According to these qualitative experimental results, here, we classify the dynamics of 28 non-convergent skewed ACA rules into the following three classes: *periodic with noise*, *noisy*, and *chaos with noise*. Firstly, remark that it is not possible classify this asynchronous systems following classical Wolfram's classification [24]. Table 1 depicts the classification of these 28 non-convergent rules. According to Table 1, periodic with noise (resp. noisy, chaos with noise) class contains 7 (resp. 8 and 13) ECA rules.

Table 1. Classification of 28 non-convergent ECA rules based on space-time diagrams.

Periodic with noise	28	29	46	73	108	142	156	
noisy	1	3	9	11	25	33	129	137
chaos with noise	19	26	27	35	41	43	51	57
	58	60	62	90	122			

Figure 2(a) shows the evidence of periodic with noise class for ECA rules 28, 142, and 156. For example, ECA 142 shows periodic 'left shift' dynamics,

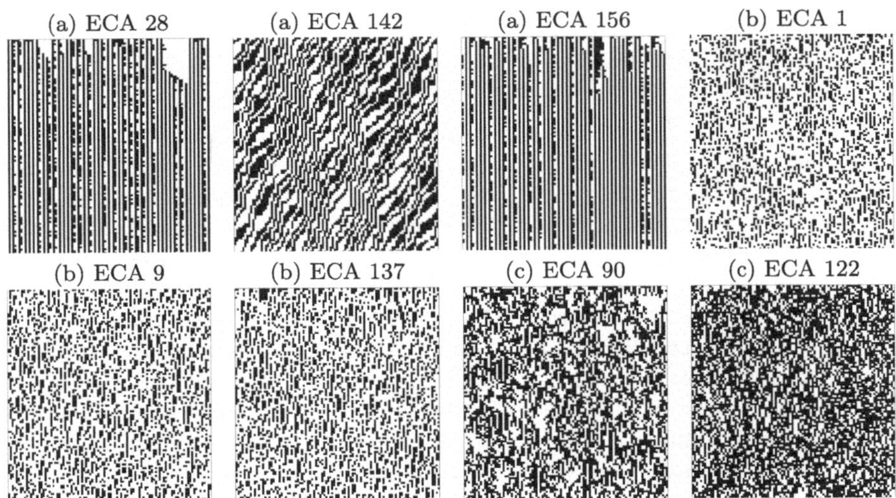

Fig. 2. Space-time pictures of (a) periodic with noise class (ECA rules 28, 142, and 156), (b) noisy class (ECA rules 1, 9, and 137) and (c) chaos with noise class (ECA rules 90 and 122).

however, the periodicity is not uniform here. Next, Fig. 2(b) reports the evidence of noisy class for ECA rules 1, 9 and 137. For these ECA rules, the system creates many small size state 1s blocks and it dies quickly; however, these 1s blocks do not interact with each other. Therefore, space-time looks like only noise without any pattern. That is, the creation and death event of state 1s block is purely random. Lastly, Fig. 2(c) depicts the evidence of chaos with noise class for ECA rules 90 and 122. Here, for ECA 90, the system creates non-uniform triangles with noisy structure. If one observes carefully the difference between the space-time picture of ECA 9 and 90, for ECA 9, there is no information flow; however, the same is not true for ECA 90.

Remark that this classification purely follows qualitative observations. In the literature of CA classification problems, CA researchers have used the notion of transfer entropy, Kolmogorov entropy, hamming distance etc. [23] to identify the class of a CA rule. Therefore, studying the complexity of these skewed rules following these quantitative parameters indicates the `open` direction.

3.2 Communication Class Dynamics of Skewed Systems

Next, we explore the communication class dynamics of these 28 non-convergent skewed ACA rules following a finite size experiment.

In this experiment, we consider ring size $n \in [4, 12]$. Now, we compute the predecessor-successor relationship of all configurations. For example, for rule 90, the successor of 0011 configuration are $\begin{smallmatrix}0&0&1&1\\1&1&1&1\end{smallmatrix}$, $\begin{smallmatrix}0&0&1&1\\0&1&1&1\end{smallmatrix}$, $\begin{smallmatrix}0&0&1&1\\0&0&1&1\end{smallmatrix}$, and $\begin{smallmatrix}0&0&1&1\\1&0&1&1\end{smallmatrix}$. Next, to compute the communication graph, we consider the configurations as the nodes, and predecessor-successor relationships as edges. For example, in the situation of the above evidence, we consider the following edges: 0011 → 1111, 0011 → 0111, 0011 → 0011 (self-loop), and 0011 → 1011. For example, Fig. 3 depicts the communication graph for ECA 90 under skewed update after considering ring size 4. Next, we compute the number of recurrent configurations, number of transient configurations, and number of communication classes using trivial graph theoretic notions[8].

As a result, Table 2 reports the communication class dynamics of the 28 non-convergent skewed rules. According to the results of Table 2, remark the following: (a) ECA 19, 41, 43, 46, 51, 57, 58, 60, 62, 90, and 142 show recurrent or reversible dynamics under skewed asynchronous update where the number of recurrent configuration is 2^n; (b) Moreover, ECA 26 shows *semi-reversible* dynamics where the skewed system is reversible for $n \notin 3\mathbb{N}$; (c) ECA 3, 11, 27 and 35 are associated with $2^n - 1$ recurrent configurations; (d) ECA 28, 29, 73, 108, 142, 156 are associated with large number of communication classes (mostly, exponential) which indicates small cycle sizes. Note that, we have classified these rules as periodic with noise class which justifies these communication class dynamics; (e) Lastly these communication class dynamics show connection with following On-Line Encyclopedia of Integer Sequences (OEIS): *A001644*[9], *A290612*[10], *A072328*[11], *A001608*[12], *A000295*[13], *A001639*[14], *A130509*[15].

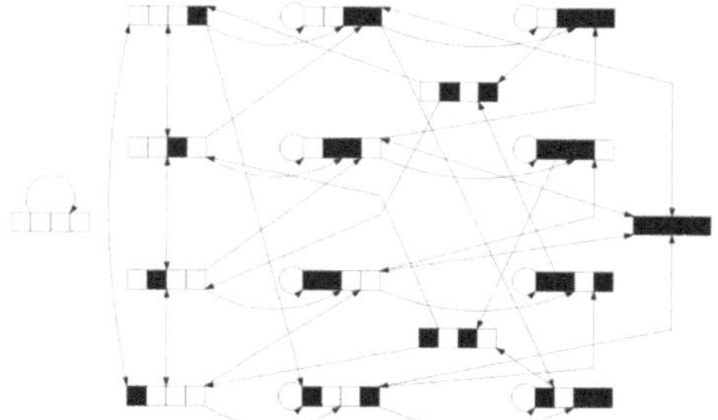

Fig. 3. Communication graph of ECA 90 for ring size 4.

[8] strongly connected components.
[9] $a(n) = a(n-1) + a(n-2) + a(n-3), a(0) = 3, a(1) = 1, a(2) = 3$.
[10] $a(n) = a(n-1) + a(n-2) - a(n-4)$.
[11] $a(n+1) = 2 \times a(n-2) + a(n-1), a(0) = 3, a(1) = 0, a(2) = 2$.
[12] Perrin Sequence (or Ondrej Such sequence, $a(n) = a(n-2) + a(n-3), a(0) = 3, a(1) = 0, a(2) = 2$.
[13] $a(n) = 2^n - n - 1$.
[14] A Fielder sequence, $a(n) = a(n-1) + a(n-3) + a(n-4) + a(n-5)$.
[15] $a(n+3) = 3 + a(n), a(1) = 3, a(2) = 1, a(3) = 2$.

Table 2. Communication class dynamics (number of recurrent configurations, number of transient configurations, and number of communication classes) of 28 non-convergent ECA rules.

ECA	No of recurrent configurations	No of transient configurations	No of communication classes
ECA 1	$11, 21, 39, 71, 131, \cdots$ (*A*001644)	$5, 11, 25, 57, 125 \cdots$	1
ECA 3	2^{n-1}	1	1
ECA 9	$11, 21, 39, 71, 131, \cdots$ (*A*001644)	$5, 11, 25, 57, 125 \cdots$	1
ECA 11	2^{n-1}	1	1
ECA 19	2^n	0	1
ECA 25	2^{n-1}	1	1
ECA 26	2^n (for $n \notin 3\mathbb{N}$) or $2^n - 3$ (for $n \in 3\mathbb{N}$)	0 (for $n \notin 3\mathbb{N}$) or 3 (for $n \in 3\mathbb{N}$)	2
ECA 27	2^{n-1}	1	1
ECA 28	$3, 11, 15, 15, 35, \cdots$	$13, 21, 49, 113, 221, \cdots$	$3, 6, 6, 8, 11, \cdots$ (*A*290612)
ECA 29	$2, 10, 14, 14, 34, \cdots$ (*A*072328)	$14, 22, 50, 114, 222, \cdots$	$2, 5, 5, 7, 10, \cdots$ (*A*001608)
ECA 33	$11, 26, 57, 120, 247, \cdots$ (*A*000295)	$5, 6, 7, 8, 9, \cdots$ (*A*000027)	1
ECA 35	2^{n-1}	1	1
ECA 41	2^n	0	1
ECA 43	2^n	0	1
ECA 46	2^n	0	2
ECA 51	2^n	0	2
ECA 57	2^n	0	1
ECA 58	2^n	0	2
ECA 60	2^n	0	2
ECA 62	2^n	0	2
ECA 73	$4, 10, 3, 7, 20, \cdots$	$12, 22, 61, 121, 236, \cdots$	$4, 5, 3, 7, 12, \cdots$
ECA 90	2^n	0	2 (for $n \notin 3\mathbb{N}$) or 5 (for $n \in 3\mathbb{N}$)
ECA 108	$9, 21, 28, 43, 81, \cdots$	$7, 11, 36, 85, 175, \cdots$	$9, 16, 22, 36, 65, \cdots$ (*A*001639)
ECA 122	$12, 27, 55, 121, 248, \cdots$	$4, 5, 9, 7, 8, \cdots$ (*A*130509)	2
ECA 129	$12, 22, 40, 72, 132, \cdots$	$4, 10, 24, 56, 124, \cdots$	2
ECA 137	$12, 22, 40, 72, 132, \cdots$	$4, 10, 24, 56, 124, \cdots$	2
ECA 142	2^n	0	$5, 4, 6, 5, 7, \cdots$ (*A*028242)
ECA 156	$4, 12, 16, 16, 36, \cdots$	$12, 20, 48, 112, 120, \cdots$	$4, 7, 7, 9, 12, \cdots$

However, here, we report the results based on observations of small ring sizes. Moreover, the experimental setup is not applicable for large ring sizes. In a recent work, Roy et al. [17] have theorized the communication class dynamics of fully ACA after considering the notion of active and passive RMTs. Therefore, building proper theory behind these communication class dynamics indicates the open direction of this research. Moreover, Roy et al. [19] have recently proposed the notion of s-skewed ACA (for skewed ACA $s = 2$). Therefore, the change of communication class number sequence for changing value of s also gives the future open direction.

3.3 Randomness Enhancing Capabilities of Skewed Systems

Next, to understand the randomness or noise in the space-time pictures, we explore the randomness enhancing capability of these 27[16] non-convergent skewed ECA rules.

Here, the model uses a source of randomness to select the sequence of chosen cells $(U_t)_{t\in\mathbb{N}} \in \mathcal{L}^{\mathbb{N}}$. In this study, we use *rand* (following the GCC Library) for creating the sequence $(U_t)_{t\in\mathbb{N}} \in \mathcal{L}^{\mathbb{N}}$. Therefore, we call the skewed system a *randomness enhancer* (not a *pseudo-random number generator*).

Table 3. Dieharder battery test results for 27 non-convergent skewed ECA rules.

ECA	ECA 1	ECA 3	ECA 9	ECA 11	ECA 19	ECA 25	ECA 26	ECA 27	ECA 28
Avg (Passed)	8	10	7	12	14	7	14	13	1
Avg (Weak)	1	0	1	1	1	1	3	2	2
ECA	ECA 29	ECA 33	ECA 35	ECA 41	ECA 43	ECA 46	ECA 57	ECA 58	ECA 60
Avg (Passed)	0	9	12	21	17	2	48	19	12
Avg (Weak)	2	5	1	3	2	0	4	1	1
ECA	ECA 62	ECA 73	ECA 90	ECA 108	ECA 122	ECA 129	ECA 137	ECA 142	ECA 156
Avg (Passed)	11	0	**89**	0	36	6	5	0	0
Avg (Weak)	1	0	**6**	0	5	1	0	1	1

Now, to check the randomness enhancing capability of skewed systems, we use the Dieharder battery of tests (introduced by George Marsaglia [13]). In this experiment, we consider a ring size of 64 and run the system for 32 iterations to obtain the next configuration (since we update only two cells per iteration). Moreover, we consider 50 initial configurations. Here, the testbed contains 114 tests where the output p-value between 0.025–0.975 indicates the passing criteria. Table 3 gives the statistics of these 27 non-convergent rules after considering the Dieharder tests. According to Table 3, ECA 90 under skewed update provides competitive results (89 passed). Note that the target of this study is not to identify a good pseudo-random number generator (or enhancer); therefore,

[16] Here, we have not considered the inversion rule ECA 51, where the local transition function is not dependent on neighbouring cells, i.e., $f(a, b, c) = 1 - b$.

the experimental setup is not associated with other empirical tests, like NIST. Moreover, we have considered only ring size 64 and a small experimental setup. However, the randomness enhancing capability of ECA 90 provides interesting possibilities that contain the `open` direction of this study. Lastly, the results of Table 3 provide a strong connection with the space-time results of Table 1, which justifies the qualitative results. In this context, we theoretically explore the communication class dynamics of ECA 90 under skewed update in the next section.

4 ECA 90 Under Skewed Update

Theorem 1 *ECA* 90 *is reversible under skewed asynchronous update where the number of recurrent configurations is* 2^n *and number of communication classes is* 2 *(for* $n \in 3\mathbb{N}$*) or* 5 *(for* $n \notin 3\mathbb{N}$*).*

Sketch of Proof: Firstly, observe that, RMTs 1, 2, 4, 7 are active and RMTs 0, 3, 5, 6 are passive for ECA 90. Here, we denote 0^n, 1^n, $(011)^{\frac{n}{3}}$ configurations as **0**, **1**, **011** respectively. To prove this theorem, we write the following properties of ECA 90 under skewed update.

Lemma 2 *Configuration **0** creates an isolated communication class for skewed ECA* 90.

RMT 0 is passive; therefore, configuration **0** is self-reachable. Note that, configurations $\cdots 00100 \cdots$ and $\cdots 001100 \cdots$ are the predecessors of configuration **0**. To reach configuration **0** from the predecessors, we need either active RMT 2, passive RMT 1 or 4 (for configuration $\cdots 00100 \cdots$), or active RMTs 3 and 6 (for configuration $\cdots 001100 \cdots$). None of these is true for ECA 90. Therefore, no other configuration can reach configuration **0**. Hence, configuration **0** creates an isolated communication class.

Lemma 3 *Configuration **011** creates an isolated communication class for skewed ECA* 90.

RMTs 3, 5, 6 are passive; therefore, configuration **011** is self-reachable. Here, the predecessor patterns of pattern 011 are $0\underbrace{00}$, $0\underbrace{01}$, $0\underbrace{10}$, $\underbrace{10}1$, $\underbrace{11}1$. However, none of these predecessor patterns can reach 011 under the skewed update for ECA 90 after considering the following argument: passive RMT 0 (for pattern $0\underbrace{00}$), active RMT 2 (for patterns $0\underbrace{01}$, $0\underbrace{10}$), passive RMT 5 (for pattern $\underbrace{10}1$), active RMT 7 (for pattern $\underbrace{11}1$). Therefore, no other configuration can reach configuration **011**. Hence, configuration **011** creates an isolated communication class.

Following a similar argument, we also write the following lemma.

Lemma 4 *Configurations **101** and **110** create an isolated communication class for skewed ECA* 90.

Let us consider that set of all configurations is denoted by ε_n.

Lemma 5 *Configurations of set* $\varepsilon_n \setminus$ {***0,011,101,110***} *creates a communication class for skewed ECA* 90.

To establish this fact, we show the following two properties:

- Any configuration of the set $\varepsilon_n \setminus$ {**0,1,011,101,110**} can reach configuration **1**.
- Configuration **1** can reach any configuration of the set $\varepsilon_n \setminus$ {**0,1,011,101,110**}.
- Note that, if the above two arguments are true, then configuration **1** is also reachable from itself.

Firstly, we establish the fact: any configuration $a \in \varepsilon_n \setminus$ {**0,1,011,101,110**} can reach configuration **1**.

To reach configuration **1** from configuration a, we need to convert state(s) 0 into state(s) 1. For a block of 0 states, ECA 90 is able to decrease the size of the block of 0 states both from left and right hand side because RMTs 1 and 4 are active and RMT 0 is passive, see the following update sequence: $\cdots 0\underbrace{00}1$ and $1\underbrace{00}0\cdots$. If the block of 0 states is associated with size two, then active RMTs 1 and 4 are applicable, see the following update sequence $1\underbrace{00}1$. Lastly, for the isolated 0 state, we can increase the size of block 0 by using active RMTs 2 and 7, see the following example $\begin{smallmatrix} 1 & \mathbf{1} & \mathbf{1} & 0 & 1 \\ 1 & \mathbf{0} & \mathbf{1} & 0 & 1 \\ 1 & 0 & 0 & 0 & 1 \end{smallmatrix}$. Hereafter, we can apply the previous technique. Note that this technique will not work for a block 0 with neighbouring 011 pattern. If the configuration is associated with only the 011 pattern, then it is a point attractor (we have discussed earlier). If the configuration is associated with any other pattern (except 011), we follow this technique starting from that pattern. The skewed ACA system is an absorbing Markov chain; therefore, it is sufficient to show one update pattern.

Next, to reach configuration a from configuration **1**, we need to construct blocks of 0 states of different sizes. To create a block 0 of size two, we can directly apply active RMT 7, see the following update sequence $1\underbrace{11}1$. To create a block 0 of size four, we first create block 0 of size two; hereafter, we use active RMTs 7 and 2, see the following example $\begin{smallmatrix} 1 & 0 & 0 & \mathbf{1} & \mathbf{1} & 1 \\ 1 & 0 & 0 & \mathbf{1} & \mathbf{0} & 1 \\ 1 & 0 & 0 & 0 & 0 & 1 \end{smallmatrix}$. Following this technique, we can construct an even-length block 0 where the block 0 size is greater than four. Now, to construct block 0 of size one, we first create block 0 of size two; hereafter, active RMT 1 or 4 and passive RMT 3 or 6 take on the responsibility, see the following update sequences $10\underbrace{01}1$ and $1\underbrace{10}01$. However, following this procedure, we are not able to create configuration **011** because in this technique, we also use the neighbouring block 1 of size more than two. Next, following the same technique, we are able to create block 0 of size three from block 0 of size four, and so on.

From the above two arguments, it is possible to claim that configuration **1** is also reachable from itself. □

Therefore, if the CA size is divisible by three, ECA 90 under skewed update creates 5 communication classes, where **0**, **011**, **101**, **110** are isolated communication classes, and the rest $2^n - 4$ configurations are associated with the other (single) communication class. However, for $n \notin 3\mathbb{N}$, there is no existence of configurations **011**, **101**, **110**. Therefore, it contains 2 communication classes of sizes 1 and $2^n - 1$.

Concluding Remark. Finally, as a concluding remark, this study has only theoretically explored ECA 90 under skewed update. For the rest 27 non-convergent rules, the experimental results have already provided good insight about the skewed systems dynamics. The formal theoretical exploitation of the rest 27 non-convergent skewed rules show the open direction of this study.

References

1. Adachi, S., Peper, F., Lee, J.: Computation by asynchronously updating cellular automata. J. Stat. Phys. **114**, 261–289 (2004)
2. Bandini, S., Bonomi, A., Vizzari, G.: An analysis of different types and effects of asynchronicity in cellular automata update schemes. Nat. Comput. **11**(2), 277–287 (2012)
3. Bersini, H., Detours, V.: Asynchrony induces stability in cellular automata based models. In: Proceedings of Artificial Life IV, pp. 382–387. MIT Press (1994)
4. Bouré, O., Fatès, N., Chevrier, V.: Probing robustness of cellular automata through variations of asynchronous updating. Nat. Comput. **11**(4), 553–564 (2012)
5. Cori, R., Metivier, Y., Zielonka, W.: Asynchronous mappings and asynchronous cellular automata. Inf. Comput. **106**(2), 159–202 (1993)
6. Dennunzio, A., Formenti, E., Manzoni, L.: Computing issues of asynchronous ca. Fund. Inform. **120**(04), 165–180 (2012)
7. Dennunzio, A., Formenti, E., Manzoni, L., Mauri, G.: M-asynchronous cellular automata: from fairness to quasi-fairness. Nat. Comput. **12**(4), 561–572 (2013)
8. Fatès, N.: Stochastic cellular automata solutions to the density classification problem - when randomness helps computing. Theory Comput. Syst. **53**(2), 223–242 (2013)
9. Fatès, N.: Guided tour of asynchronous cellular automata. J. Cell. Autom. **9**(5–6), 387–416 (2014)
10. Fatès, N.: A tutorial on Elementary cellular automata with fully asynchronous updating - General properties and convergence dynamics. Nat. Comput. **19**(1), 179–197 (2020)
11. Fatès, N.: Asynchronous cellular systems that solve the parity problem. In: Gadouleau, M., Castillo-Ramirez, A. (eds.) Cellular Automata and Discrete Complex Systems, pp. 133–145. Springer, Cham (2024). https://doi.org/10.1007/978-3-031-65887-7_9
12. Fatès, N., Thierry, E., Morvan, M., Schabanel, N.: Fully asynchronous behavior of double-quiescent elementary cellular automata. Theor. Comput. Sci. **362**(1), 1–16 (2006)
13. . Marsaglia, G: DIEHARD: A battery of tests of randomness (1996)
14. Roy, S.: A study on delay-sensitive cellular automata. XXPhys. A **515**, 600–616 (2019)

15. Roy, S.: Distributed Computing on Cellular Automata with Applications to Societal Problems. PhD thesis, Indian Institute of Engineering Science and Technology, Shibpur (2021)
16. Roy, S., Das, S.: Clouds in the basins of fully asynchronous cellular automata. Adv. Complex Syst. **25**(08), 2250013 (2022)
17. Roy, S., Fatès, N., Das, S.: Reversibility of elementary cellular automata with fully asynchronous updating: An analysis of the rules with partial recurrence. Theor. Comput. Sci. **1011**, 114721 (2024)
18. Roy, S., Gautam, V.K., Das, S.: A note on skew-asynchronous cellular automata. In: Gadouleau, M., Castillo-Ramirez, A. (eds.) Cellular Automata and Discrete Complex Systems, pp. 146–158. Springer, Cham (2024)
19. Roy, S., Shelat, S., Salvi, S.: Revisiting the notion of independence in asynchronous cellular automata – a study on s-skewed automata. Int. J. Bifurcat. Chaos **35**, 25501 (2025)
20. Salvi, S., Shelat, S., Adak, S., Roy, S.: The "distributed ghost" with independence - a study on computational ability of skewed asynchronous cellular automata. Northeast J. Complex Syst. **7**, 03 (2025)
21. Sethi, B., Roy, S., Das, S.: Asynchronous cellular automata and pattern classification. Complexity **21**(S1), 370–386 (2016)
22. Sethi, B., Roy, S., Das, S.: Convergence of asynchronous cellular automata: does size matter? J. Cell. Autom. **13**(5–6), 527–542 (2018)
23. Vispoel, M., Daly, A.J., Baetens, J.M.: Progress, gaps and obstacles in the classification of cellular automata. Physica D **432**, 133074 (2022)
24. Wolfram, S.: Cellular Automata and Complexity: Collected Papers, 1st edn. CRC Press, Boca Raton (1994)

Fuzzy Cellular Automata Representation of a Logistic Map for a Population with Fuzzy Initial Value and Fuzzy Growth Rate

Sreeya Ghosh[1](✉), Abhilash Purohit[2], and Sumita Basu[3]

[1] Swami Vivekananda Centre for Multi-Disciplinary Research in Basic Sciences and Social Sciences (Unit of RKMVCC Rahara), RKMSVAHCC, Kolkata, India
sreeya135@gmail.com

[2] Institute of Mathematics and Applications, Bhubaneswar, India
purohitabhilash854@gmail.com

[3] Department of Mathematics, Bethune College, Kolkata, India
sumi_basu05@yahoo.co.in

Abstract. A dynamical system can be represented by a cellular automata (CA). A logistic map is an example of a non-linear discrete dynamical system. Dynamics of population growth may be modelled by using a logistic map. In real world situations very often variables representing population growth are fuzzy. We have modelled fuzzy CA to represent such growth trends.

Keywords: Fuzzy Cellular Automata · Logistic Map · Triangular Fuzzy Number

1 Introduction

Dynamical systems may be linear or non-linear. Analysis of the nature of trajectory of a non-linear system is a challenging problem. Evolution patterns of one-dimensional dynamical systems represented by using homogeneous CA [7] as well as hybrid or non-uniform CA [8,9,13] are well-known.

The logistic differential equation was first published by *Pierre Verhulst* in 1845 to find the population of an organism. *Robert May* in 1976 proposed *logistic map* by discretizing the logistic equation. So logistic map modelling population growth is a non-linear difference equation. *Ibrahim* in [11], studied the chaotic behaviour of the logistic map with respect to CA Rule 90 of Wolfram [16]. *Stariolo and Tsallis* in [14] have investigated logistic map dynamics which is governed by deterministic CA. Again in [10] a correspondence between a tent map and CA evolution has been indicated. *Kumar et al.* in [12], have proposed an image encryption model based on logistic map and CA. The forecasting of chaotic logistic map using fuzzy time series have been dealt in [1] by *Alves*.

H. Raju et al. (Eds.): ASCAT 2026, CCIS 2801, pp. 182–193, 2026.
https://doi.org/10.1007/978-3-032-18612-6_14

Studying a logistic map on the basis of fuzzy values of a population is not commonly practised.

This paper attempts designing a fuzzy CA [6] model to depict population growth trend of an organism on the basis of a logistic map. Fuzzy initial value of the population and fuzzy growth rate parameter in the range between 0 and 3 has been considered. The fuzzy phase points over discrete time steps have been calculated using the concept of triangular fuzzy numbers over some α-cut intervals. The different α levels indicate the different density levels of the population.

For formulating, the growth of the population, trapezoidal fuzzy numbers could also be used. However, in this work a triangular fuzzy number which is a specific instance of a trapezoidal fuzzy number has been used to indicate the peak value of the density proportion of the population with respect to its maximum capacity at any particular time instant.

The paper is arranged as follows. In Sect. 2, the basic concepts of logistic map, triangular fuzzy number and fuzzy algebraic operations on α-cut intervals have been stated. Also fuzzy cellular automata with the framework of CA modelling of discrete dynamical systems has been described. Evolution pattern of non-linear dynamical system represented by a logistic map, has been analysed. Then in Sect. 3, fuzzy CA models of population growth in terms of fuzzification of the initial value and growth rate on the basis of the logistic map linearized around fixed points of the system have been designed.

2 Preliminaries

2.1 Logistic Map

The logistic map [1,3,15] is a mathematical equation in discrete form that depicts how complex forms can arise from simple non-linear dynamical systems. It is used to model population growth expressed by the difference equation $x_{t+1} = rx_t(1 - x_t)$ where $x_t \geq 0$ is a measure of population at t^{th} generation and $r \geq 0$ is the growth rate.

Values of the parameter r is varied in the range [0,4] and x_t remains bounded on [0,1].

Characteristics observed in the population for different values of r (see [17]) are such that the population will :

- Die out just after the initial state for $r = 0$
- Die out eventually for $0 < r < 1$,
- Approach the value $1 - 1/r$ for $1 < r \leq 2$,
- Fluctuate around the value $1 - 1/r$ before approaching it for $2 < r < 3$.

Beyond $r = 3$ oscillatory and chaotic behaviour of the trajectory can be observed.

2.2 Triangular Fuzzy Number, α-Cut

A fuzzy set is a subset of a collection of elements or objects having degrees of membership between 0 and 1. A fuzzy number is a subclass of a fuzzy set over real line.

Definition 1 ***Triangular fuzzy number*** *is a fuzzy number represented with three points as* $\widetilde{A} = (a_1/a_2/a_3)$ *having membership function*

$$\mu_{\widetilde{A}}(x) = \begin{cases} 0 & if\ x < a_1, \\ \frac{x-a_1}{a_2-a_1} & if\ a_1 \leq x \leq a_2 \\ \frac{a_3-x}{a_3-a_2} & if\ a_2 \leq x \leq a_3 \\ 0 & x > a_3 \end{cases} \tag{1}$$

Definition 2 *Given a fuzzy set* $\widetilde{A}$ *in* S *and any real number* $\alpha \in [0,1]$*, then the* ***alpha cut*** ($\alpha - cut$)*, denoted by* $\widetilde{A}[\alpha]$ *is the crisp set* $\widetilde{A}[\alpha] = \{x \in \widetilde{A} \mid \mu_{\widetilde{A}}(x) > \alpha\}$.

Thus on setting the left and right reference functions of $\widetilde{A}$ *as* $\alpha = \frac{x-a_1}{a_2-a_1}$ *and* $\alpha = \frac{a_3-x}{a_3-a_2}$*, it follows that*

$$\widetilde{A}[\alpha] = [a_1 + (a_2 - a_1)\alpha, a_3 - (a_3 - a_2)\alpha] \tag{2}$$

Fuzzy Algebraic Operations on α-cut Intervals

Let $\tilde{A}[\alpha] = [a_1(\alpha), a_2(\alpha)]$ and $\tilde{B}[\alpha] = [b_1(\alpha), b_2(\alpha)]$ be α-cut intervals for some $\alpha \in (0,1]$ of two fuzzy numbers $\tilde{A}$ and $\tilde{B}$. Then fuzzy algebraic operations(see [2]) between them are as follows :

- **Addition**: $\tilde{A}[\alpha] + \tilde{B}[\alpha] = [a_1(\alpha) + b_1(\alpha), a_2(\alpha) + b_2(\alpha)]$
- **Subtraction**: $\tilde{A}[\alpha] - \tilde{B}[\alpha] = [a_1(\alpha) - b_2(\alpha), a_2(\alpha) - b_1(\alpha)]$
- **Multiplication**: $\tilde{A}[\alpha] * \tilde{B}[\alpha] = [min(c_1(\alpha)), max(c_1(\alpha))]$ where, $c_1(\alpha) = \{a_1(\alpha) * b_1(\alpha), a_1(\alpha) * b_2(\alpha), a_2(\alpha) * b_1(\alpha), a_2(\alpha) * b_2(\alpha)\}$
- **Division** : $\tilde{A}[\alpha]/\tilde{B}[\alpha] = [min(c_2(\alpha)), max(c_2(\alpha))]$ where, $c_2(\alpha) = \{a_1(\alpha)/b_1(\alpha), a_1(\alpha)/b_2(\alpha), a_2(\alpha)/b_1(\alpha), a_2(\alpha)/b_2(\alpha)\}$

2.3 Fuzzy Cellular Automaton

A fuzzy CA [6] is a generalization of Boolean CA defined as follows :

Definition 3 *A one-dimensional* ***fuzzy CA (FCA)*** *is a one-dimensional CA where the local transition function is a fuzzy transition function. So the formal definition is as follows:*

An ***FCA*** *(denoted by* $\mathcal{C}^{\mathcal{F}}$*) is a four-tuple* $(\tilde{Q}, \tilde{Q^{\mathbb{Z}}}, \tilde{f}, \tilde{C_0})$*, where,*

- $\tilde{Q} \subset [0,1]$ *is the state set*
- $\tilde{Q^{\mathbb{Z}}}$ *is the set of all configurations*

- $\tilde{f}$ *is the local transition function which is fuzzy in nature such that if* r *be the radius of the neighbourhood then* $\tilde{f} : \widetilde{Q^{2r+1}} \rightarrow \tilde{Q}$
- $\tilde{C}_0$ *is the initial configuration which is a fuzzy number*

The local transition function $\tilde{f}$ is a fuzzification of Boolean function which is crisp in nature. *Disjunctive Normal Form-fuzzification* of the transition function of a classical Boolean CA gives a **Fuzzy transition function**.

Here we have replaced $(a \wedge b)$ by $min(a, b)$, $(a \vee b)$ by $max(a, b)$ and $\neg a$ by $(1 - a)$(see [4]).

For example, the DNF for RULE 200 of Wolfram is
$(c_{i-1} \wedge c_i \wedge c_{i+1})_{(111)} \vee (c_{i-1} \wedge c_i \wedge \neg c_{i+1})_{(110)} \vee (\neg c_{i-1} \wedge c_i \wedge c_{i+1})_{(011)}$

Thus fuzzification of this DNF gives the fuzzy transition rule $\tilde{f}_{200}$ as $max\{min(c_{i-1}, c_i, c_{i+1})_{(111)}, min(c_{i-1}, c_i, 1-c_{i+1})_{(110)}, min(1+c_{i-1}, c_i, c_{i+1})_{(011)}\}$

2.4 Non-Linear Dynamical System Represented Through Logistic Map

Let us consider the logistic map representing population growth of some organism in a breeding tank of the form :

$$x_{t+1} = rx_t(1 - x_t) \tag{3}$$

where $x_t \in [0, 1]$ is considered to be the ratio of the existing population to the maximum possible population at the tth generation and $r \geq 0$ is the growth rate. The fixed points of the system obtained from $x^* = rx^*(1 - x^*)$ are $x^* = 0$ for $r \geq 0$ and $x^* = 1 - 1/r$ for $r \geq 1$.

In general we cannot analyse the stability of the fixed points of a non-linear system. So we linearize the system around fixed point x^* by Taylor expansion.

1. The linearized form of the given system around $x^* = 0$ is :

$$x_{t+1} = rx_t \tag{4}$$

2. The linearized form of the given system around $x^* = 1 - 1/r$ is :

$$x_{t+1} = (2 - r)x_t + \frac{(r - 1)^2}{r} \tag{5}$$

Here, $x^* = 1 - 1/r$ is in the range of allowable x only if $r \geq 1 \Rightarrow 2 - r \leq 1$.

3 FCA Model for Population Growth Formulated by a Logistic Map

The dynamical system depicting population growth is non-linear in nature and can be suitably represented by the logistic map. In this section, the variables of the logistic map are assumed to be imprecise. This implies that the variables of the linearized system are also imprecise and we consider them to be fuzzy(triangular fuzzy numbers). So the density of a population at any time t is fuzzy and is denoted by $\tilde{x}_t$. The growth trend of this population is modelled using FCA having exactly one active cell at a time. Fuzzy growth rate between 0 and 3 around the fixed points with fuzzy initial population value has been considered.

Example 1 *Let us consider the logistic map representing growth population of an organism in a breeding tank of the form $x_{t+1} = rx_t(1 - x_t)$ where $x_t \in [0, 1]$ is considered to be the ratio of the existing population to the maximum possible population at the tth generation and $r \geq 0$ is the growth rate.*

As discussed in subsection(2.4), the linearized form of the system around the crisp fixed points $x^* = 0$ and $x^* = 1 - 1/r$ are given by equations (4) and (5) respectively. In general, getting an exact count of the organisms present in a breeding tank may not be feasible. So a fuzzy value can be considered for either of the following :

- the initial proportion of the organisms present to its maximum capacity denoted by the triangular fuzzy number $\tilde{x}_0$
- the growth rate denoted by $\tilde{r}$
- both the above two quantities

Here, we consider a fuzzy value for the initial proportion of the organisms present to its maximum capacity as well as the growth rate. Then the proportion of the population at time t is denoted by $\tilde{x}_t$.

We further assume that $\tilde{x}_t[\alpha]$ is the density of the proportion of the organisms present at any time t to its maximum capacity for some $\alpha \in [0.1]$. Precision of this situation is:

- highest if $\alpha = 1$ ($\tilde{x}_t[1]$ becomes identical to crisp value x_t)
- high if $0.75 \leq \alpha < 1$
- medium if $0.50 \leq \alpha < 0.75$
- low if $0.25 \leq \alpha < 0.50$
- negligible if $0 < \alpha < 0.25$

1. The fuzzified system of the linearized equation (4) for $\alpha \in [0, 1]$ is :

$$\tilde{x}_{t+1}[\alpha] = \tilde{r}[\alpha]\tilde{x}_t[\alpha] \tag{6}$$

2. The fuzzified system of the linearized equation (5) for $\alpha \in [0, 1]$ is :

$$\tilde{x}_{t+1}[\alpha] = (2 - \tilde{r}[\alpha])\tilde{x}_t[\alpha] + \frac{(\tilde{r}[\alpha] - 1)^2}{\tilde{r}[\alpha]} \tag{7}$$

Now, *Rule 2* of an elementary CA(ECA) having exactly one cell in the active state is equivalent to *Rule 170* which is a 1-place left shift transition function (see [7]). Likewise, *Rule 16* has 1-place right shift transitions.

Proposition 1 *The fuzzy transition function corresponding to Rule 2 of an ECA with exactly one active cell which is a left-shift transition function, will also be a left-shift function.*

Proof. Let us consider Rule 2 of an ECA with exactly one active cell which is a 1-place left-shift function. Then, fuzzification using triangular fuzzy number around any i^{th} cell, for $i \in \mathbb{Z}$, gives $\tilde{x}_i = (a_1(i)/a_2(i)/a_3(i))$ to be a band of values such that $a_1(i) \leq a_2(i) \leq a_3(i)$.

Let the local transition function of an ECA for the ith cell be denoted by $f(x_{i-1}, x_i, x_{i+1})$ such that at any time $t > 0$, $f(x_{i-1}^t, x_i^t, x_{i+1}^t) = x_i^{t+1}$. The corresponding fuzzy transition function at time t will be $\tilde{f}(\tilde{x}_{i-1}^t, \tilde{x}_i^t, \tilde{x}_{i+1}^t)$.

DNF for Rule 2 is $(\neg c_{i-1} \wedge \neg c_i \wedge c_{i+1})$ which on fuzzification gives the fuzzy transition function $\tilde{f}$ as $min(1 - c_{i-1}, 1 - c_i, c_{i+1})$.

Therefore, at any time $t > 0$, $\forall i \in \mathbb{Z}$, Rule 2 gives, $f(x_{i-1}^t, x_i^t, x_{i+1}^t) = x_i^{t+1} \leq x_i^t$, while the corresponding fuzzy transition function gives, $\tilde{f}(\tilde{x}_{i-1}^t, \tilde{x}_i^t, \tilde{x}_{i+1}^t) = \tilde{x}_i^{t+1} \leq \tilde{x}_i{}^t$.

Hence the proposition.

FCA models can be designed to represent fuzzy finite difference equations as discussed in Section(1.4) of [6]. Following the same method, we have presented here FCA models corresponding to equations (6) and (7) for different values of the fuzzy growth rate $\tilde{r}$ in the range between 0 and 3. The α-cut levels(see definition 2) are calculated with respect to the initial value $\tilde{x}_0 = (0.4/0.5/0.6)$. If time increases vertically down then the FCA generated for $\tilde{x}_t[\alpha]$ intervals at $\alpha = 0.25, 0.50, 0.75$ and 1 are as follows:

1. FCA model of trajectory of the fuzzy equation (6) :
   - Case 1 : $0 < r < 1$; let $\tilde{r} = (0.6/0.7/0.8)$
     The fuzzy dynamical system converges monotonically to the crisp fixed point $x^* = 0$ for time $t \to \infty$ as given in Table 1.
     The trajectory follows a left-shift and the FCA generated will follow fuzzified **RULE 2** of Wolfram code as shown in Fig. 1.
   - Case 2 : $r > 1$; let $\tilde{r} = (1.2/1.3/1.4)$
     The fuzzy dynamical system diverges monotonically away from the crisp fixed point $x^* = 0$ for time $t \to \infty$ as given in Table 2.
     The trajectory follows a right-shift and the FCA generated will follow fuzzified **RULE 16** of Wolfram code as shown in Fig. 2.
2. FCA model of trajectory of the fuzzy equation (7) :
   - Case 1 : $1 < r < 2$; let $\tilde{r} = (1.4/1.5/1.6)$
     The fuzzy dynamical system converges monotonically to the crisp fixed point $x^* = 1 - 1/1.5 = 0.333$ for time $t \to \infty$ as given in Table 3.
     The trajectory follows a left-shift and the CA generated will follow fuzzified **RULE 2** of Wolfram code as shown in Fig. 3.

- Case 2 : $2 < r < 3$; let $\tilde{r} = (2.7/2.8/2.9)$
The fuzzy dynamical system converges in oscillations to the crisp fixed point $x^* = 1 - 1/2.8 = 0.6296$ for time $t \to \infty$ as given in Table 4. Corresponding to the trajectory of the fuzzy dynamical system in this case, we get a non-uniform FCA whose cells follow fuzzified **RULE 16** of Wolfram at time $t = 0, 2, 4, 6, ...$ and fuzzified **RULE 2** of Wolfram at time $t = 1, 3, 5, 7, ...$ as shown in Fig. 4.

Table 1. Trajectory around $x^* = 0$ for $\tilde{x}_0 = (0.4/0.5/0.6)$ and $r = (0.6/0.7/0.8)$

t	$\tilde{x}_t[0.25]$	$\tilde{x}_t[0.5]$	$\tilde{x}_t[0.75]$	$\tilde{x}_t[1] = x_t$
0	[0.425000, 0.575000]	[0.450000, 0.550000]	[0.475000, 0.525000]	0.500000
1	[0.265625, 0.445625]	[0.292500, 0.412500]	[0.320625, 0.380625]	0.350000
2	[0.166016, 0.345359]	[0.190125, 0.309375]	[0.216422, 0.275953]	0.245000
3	[0.103760, 0.267654]	[0.123581, 0.232031]	[0.146085, 0.200066]	0.171500
⋮	⋮	⋮	⋮	⋮
10	[0.003865, 0.044945]	[0.006058, 0.030972]	[0.009327, 0.021064]	0.014124
⋮	⋮	⋮	⋮	⋮
15	[0.000369, 0.012566]	[0.000703, 0.007350]	[0.001307, 0.004219]	0.002374
⋮	⋮	⋮	⋮	⋮
20	[0.000035, 0.003513]	[0.000082, 0.001744]	[0.000183, 0.000845]	0.000399
⋮	⋮	⋮	⋮	⋮
25	[0.000003, 0.000982]	[0.000009, 0.000414]	[0.000026, 0.000169]	0.000067
⋮	⋮	⋮	⋮	⋮
30	[0.000000, 0.000275]	[0.000001, 0.000098]	[0.000004, 0.000034]	0.000011
⋮	⋮	⋮	⋮	⋮
40	[0.000000, 0.000021]	[0.000000, 0.000006]	[0.000000, 0.000001]	0.000000

Table 2. Trajectory around $x^* = 0$ for $\tilde{x}_0 = (0.4/0.5/0.6)$ and $r = (1.2/1.3/1.4)$

t	$\tilde{x}_t[0.25]$	$\tilde{x}_t[0.5]$	$\tilde{x}_t[0.75]$	$\tilde{x}_t[1] = x_t$
0	[0.425000, 0.575000]	[0.450000, 0.550000]	[0.475000, 0.52500]	0.500000
1	[0.520625, 0.790625]	[0.562500, 0.742500]	[0.605625, 0.695625]	0.650000
2	[0.637766, 1.087109]	[0.703125, 1.002375]	[0.772172, 0.921703]	0.845000
3	[0.781263, 1.494775]	[0.878906, 1.353206]	[0.984519, 1.221257]	1.098500
⋮	⋮	⋮	⋮	⋮
10	[3.234073, 13.889763]	[6.203632, 7.582217]	[5.392566, 8.756249]	6.892925
⋮	⋮	⋮	⋮	⋮
15	[8.921343, 68.266607]	[12.789769, 49.587158]	[18.169642, 35.759997]	25.592947
⋮	⋮	⋮	⋮	⋮
20	[24.609948, 335.522623]	[39.031278, 222.350474]	[61.220561, 146.041696]	95.024819
⋮	⋮	⋮	⋮	⋮
25	[67.887708, 1649.055601]	[119.114008, 997.026960]	[206.275776, 596.425578]	352.820501
⋮	⋮	⋮	⋮	⋮
30	[187.271458, 8104.921067]	[363.507105, 4470.702226]	[695.022962, 2435.766491]	1309.997822
⋮	⋮	⋮	⋮	⋮
40	[1425.057795, 195783.357214]	[3385.423730, 89890.424074]	[7890.435979, 40625.099186]	18059.432404

Table 3. Trajectory around $x^* = 0.3333$ for $\tilde{x}_0 = (0.4/0.5/0.6)$ and $r = (1.4/1.5/1.6)$

t	$\tilde{x}_t[0.25]$	$\tilde{x}_t[0.5]$	$\tilde{x}_t[0.75]$	$\tilde{x}_t[1] = x_t$
0	[0.425000, 0.575000]	[0.450000, 0.550000]	[0.475000, 0.525000]	0.500000
1	[0.295308, 0.562643]	[0.333145, 0.511121]	[0.373576, 0.462489]	0.416667
2	[0.240188, 0.555537]	[0.280560, 0.489737]	[0.325399, 0.429671]	0.375000
3	[0.216763, 0.551451]	[0.256897, 0.477976]	[0.302516, 0.412442]	0.354167
⋮	⋮	⋮	⋮	⋮
10	[0.199491, 0.546039]	[0.237609, 0.463820]	[0.281924, 0.393608]	0.333496
⋮	⋮	⋮	⋮	⋮
15	[0.199448, 0.545931]	[0.237538, 0.463613]	[0.281814, 0.393407]	0.333338
⋮	⋮	⋮	⋮	⋮
20	[0.199448, 0.545924]	[0.237537, 0.463602]	[0.281814, 0.393407]	0.333333
⋮	⋮	⋮	⋮	⋮
25	[0.199448, 0.545924]	[0.237537, 0.463602]	[0.281811, 0.393399]	0.333333
⋮	⋮	⋮	⋮	⋮
30	[0.199448, 0.545924]	[0.237537, 0.463602]	[0.281811, 0.393399]	0.333333
⋮	⋮	⋮	⋮	⋮
40	[0.199448, 0.545924]	[0.237537, 0.463602]	[0.281811, 0.393399]	0.333333

Table 4. Trajectory around $x^* = 0.6296$ for $\tilde{x}_0 = (0.4/0.5/0.6)$ and $r = (2.7/2.8/2.9)$

t	$\tilde{x}_t[0.25]$	$\tilde{x}_t[0.5]$	$\tilde{x}_t[0.75]$	$\tilde{x}_t[1] = x_t$
0	[0.425000, 0.575000]	[0.450000, 0.550000]	[0.475000, 0.525000]	0.500000
1	[0.531875, 0.982013]	[0.607061, 0.907045]	[0.682140, 0.832100]	0.757143
2	[0.175739, 0.904528]	[0.303573, 0.789249]	[0.428783, 0.671566]	0.551429
3	[0.243538, 1.162727]	[0.403699, 1.016866]	[0.561223, 0.867919]	0.716000
⋮	⋮	⋮	⋮	⋮
10	[-0.214215, 1.409396]	[0.088595, 1.140505]	[0.360535, 0.887996]	0.627518
⋮	⋮	⋮	⋮	⋮
15	[-0.281934, 1.545033]	[0.070069, 1.197034]	[0.361487, 0.927085]	0.647883
⋮	⋮	⋮	⋮	⋮
20	[-0.351503, 1.579766]	[0.050541, 1.202681]	[0.348305, 0.926783]	0.641210
⋮	⋮	⋮	⋮	⋮
25	[-0.369318, 1.615449]	[0.048590, 1.208633]	[0.348406, 0.930960]	0.643397
⋮	⋮	⋮	⋮	⋮
30	[-0.387620, 1.624587]	[0.046534, 1.209228]	[0.346998, 0.930927]	0.642680
⋮	⋮	⋮	⋮	⋮
40	[-0.397122, 1.636378]	[0.046112, 1.209917]	[0.346858, 0.931370]	0.642838

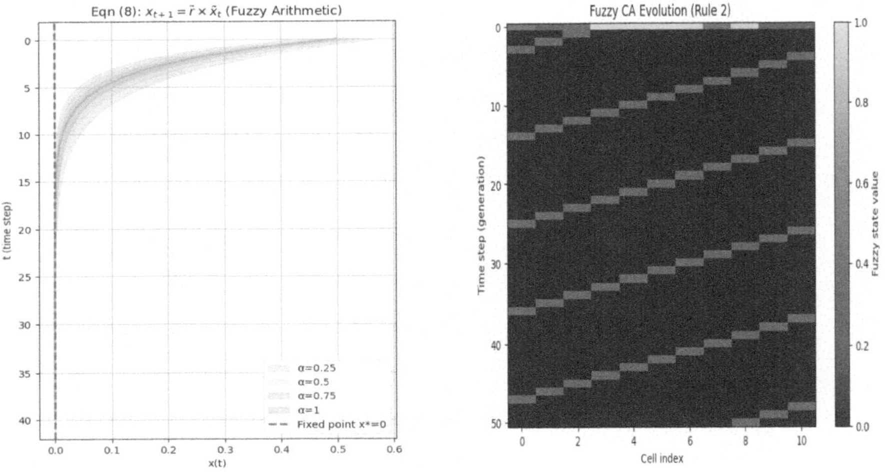

(a)

(b) FCA following fuzzy Rule 2

Fig. 1. FCA model of the trajectory around $x^* = 0$ for $r = (0.6/0.7/0.8)$.

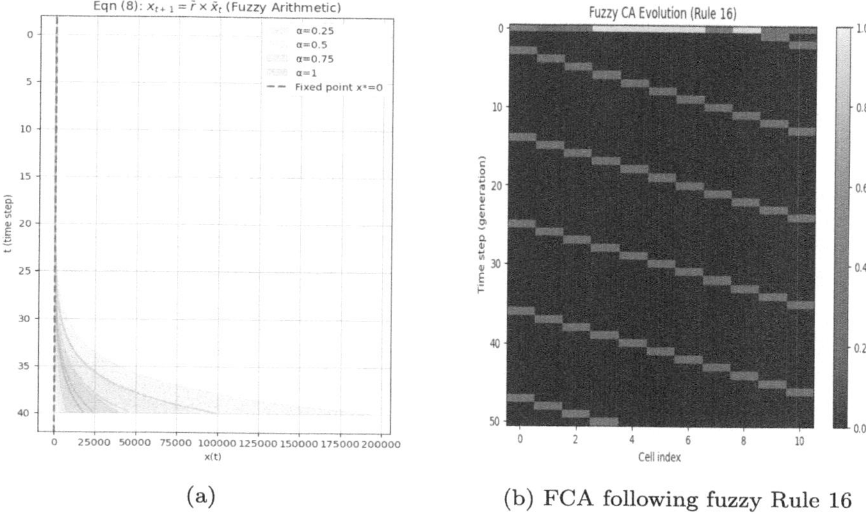

(a) (b) FCA following fuzzy Rule 16

Fig. 2. FCA model of the trajectory around $x^* = 0$ for $r = (1.2/1.3/1.4)$.

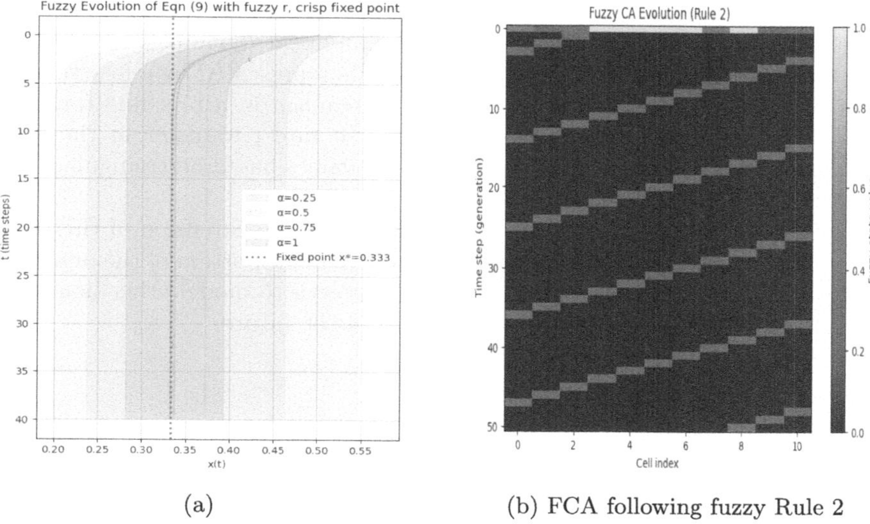

(a) (b) FCA following fuzzy Rule 2

Fig. 3. FCA model of the trajectory around $x^* = 0.3333$ for $r = (1.4/1.5/1.6)$.

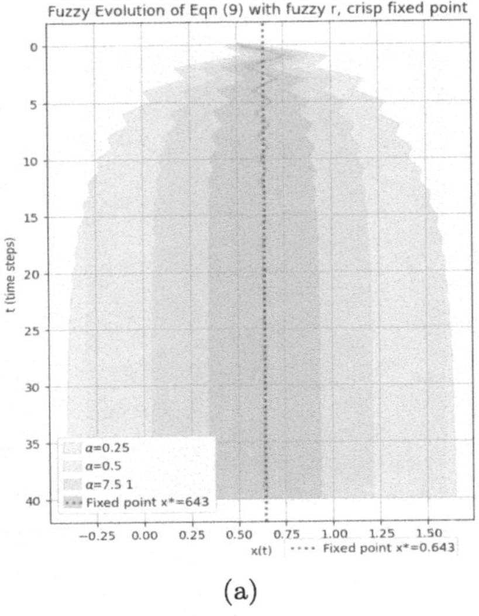

(a)

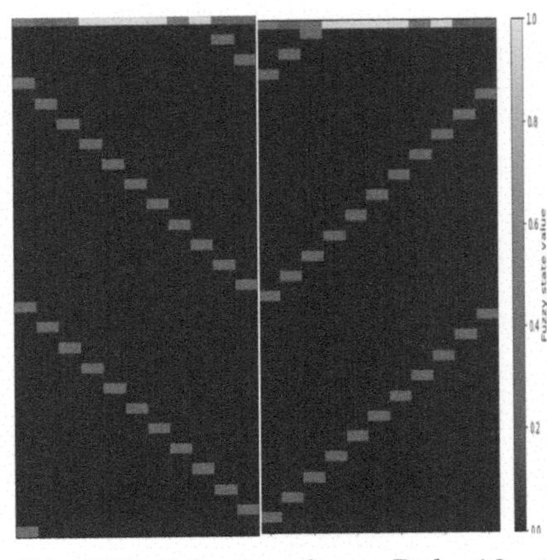

(b) FCA following fuzzy Rule 16 at even time steps 0,2,4,6... and fuzzy Rule 2 at odd time steps 1,3,5,..

Fig. 4. FCA model of the trajectory around $x^* = 0.6296$ for $r = (2.7/2.8/2.9)$.

Conclusion

In this paper we have discussed the nature of the trajectories of a fuzzified system linearized around the crisp fixed points of a non-linear dynamical system. The dynamical system depicting population growth is represented by a logistic map.

FCA models of population growth of an organism in a breeding tank represented by a logistic map having fuzzy growth rate parameter in the range between 0 and 3 and starting with fuzzy initial values based on triangular fuzzy numbers have been designed.

Trapezoidal fuzzy numbers may be considered as an extension of this work, to obtain a range of population density with highest proportion of the organism.

Moreover, the formulation for population growth to model other non-linear and chaotic trajectories may also be explored in the future.

References

1. Alves, LVR: Chaotic Logistic Map Forecast using Fuzzy Time Series, arXiv2103.07550 [math.DS] (2021). https://doi.org/10.48550/arXiv.2103.07550
2. Buckley, J.: Fuzzy Probability and Statistics. Springer-Verlag, Berlin Heidelberg ISBN: 978-3-540-30841-6 (2006). https://doi.org/10.1007/3-540-33190-5
3. Chen, S., et al.: Logistic map : stability and entrance to chaos. J. Phys. Conf. Ser. **2014** 012009 (2021). https://doi.org/10.1088/1742-6596/2014/1/012009
4. Dutta, P., et al.: Fuzzy Arithmetic with and without using α-cut method: a comparative study. Int. J. Latest Trends Comput. **2**(1), 99–107 (2011) (E-ISSN: 2045-5364)
5. Galor, O. : Discrete Dynamical Systems. Springer (2007)

6. Basu, S., Ghosh, S.: Fuzzy cellular automata model for discrete dynamical system representing spread of MERS and COVID-19 virus. In: Internet of Medical Things for Smart Healthcare, pp. 267–304. Springer (2020)
7. Ghosh, S.: Evolutions of some one-dimensional homogeneous cellular automata. Complex Syst. **30**, 75–92 (2021)
8. Ghosh, S., Basu S.: Evolution patterns of some one-dimensional non-uniform cellular automata. In: Proceedings of First Asian Symposium on Cellular Automata Technology, pp. 69–77, Springer Nature Singapore, (2022)
9. Ghosh, S., Basu, S.: Study of evolution patterns of some one-dimensional spatial and temporal non-uniform cellular automata. J. Cell. Autom. **17**, 3–24 (2023)
10. Gururanjan, N., Sambassivame, M.: Chaos-logistic map - tent map - corresponding cellular automata. Mapano J. Sci. **9**(2), 28–34 (2010). https://doi.org/10.12723/mjs.17.4.
11. Ibrahimi, M., et al.: Logistic Cellular Automata (2019)
12. Kumar, M., et al.: Intertwining logistic map and cellular automata based color image encryption model. In: IEEE Xplore, International Conference on Computational Techniques in Information and Communication Technologies (ICCTICT), New Delhi, India, pp. 618–623 (2016). https://doi.org/10.1109/ICCTICT.2016.7514653
13. Ning, B., Li, K.P., Gao, Z.Y.: Modeling fixed-block railway signaling system using cellular automata model. Int. J. Mod. Phys. C **16**(11), 1793–1801 (2005). https://doi.org/10.1142/S0129183105008308
14. Stariolo, D.A., Tsallis, C.: Cellular automata controlled logistic map: an unusual type of noise. Int. J. Mod. Phys. C **3**(03), 547–552 (1992). https://doi.org/10.1142/S0129183192000348
15. Strogatz, S.: Non-linear Dynamics and Chaos, Perseus Books Publishing (1994)
16. Wolfram, S.:A New Kind of Science, Wolfram Media (2002)
17. WolframAlpha, logistic map for r (2024). https://www.wolframalpha.com

An Algorithmic Approach of Two-Dimensional Cellular Automata Rules Splicing with Comparative Distance Metric

M. Kushpu[1], D. K. Sheena Christy[1(✉)], and D. G. Thomas[2]

[1] Department of Mathematics, Faculty of Engineering and Technology, SRM Institute of Science and Technology, Kattankulathur, Chennai 603203, Tamil Nadu, India
{km2925,sheenac}@srmist.edu.in

[2] Department of Mathematics, Madras Christian College, East Tambaram, Chennai 600059, Tamil Nadu, India
dgthomasmcc@yahoo.com

Abstract. This paper explores the characteristics of splicing of linear rules of two-dimensional cellular automata (2D CA) with m states and n neighborhoods. The study of splicing in 2D CA is worth examining since it enhances the average distance between linear rules and improves their distribution across the linear rule space. We present an algorithm for splicing two 2D CA linear rules and investigate several properties of the splicing operation. We also examine the algebraic properties of the proposed splicing operation on two-dimensional cellular automaton rules, which help characterize the underlying properties of splicing. These properties are motivated by the connection between splicing and DNA recombination and provide a foundation for its use in biologically motivated computational models. The distribution and proximity of linear rules within the rule space are analyzed using Hamming distance and Manhattan distance.

Keywords: Splicing · Linear rules · Hamming distance · Manhattan distance · DNA recombination

1 Introduction

John von Neumann pioneered the study of Cellular Automata (CA) to model biological self-reproduction [5,6]. A cellular automaton is a discrete dynamical system consisting of an ordered network of finite-state automata (cells) [1]. The future state of each cell is determined by a local update rule based on the states of its neighboring cells [11]. Numerous studies have examined CA from structural, dynamical, and computational perspectives [2,4,7,8].

Formal language theory has been applied to describe the evolution of one-dimensional CA, where the set of finite symbol blocks represents configurations evolving over time from random initial states [4,9,12,14].

H. Raju et al. (Eds.): ASCAT 2026, CCIS 2801, pp. 194–206, 2026.
https://doi.org/10.1007/978-3-032-18612-6_15

Head introduced the concept of a *splicing system*, a theoretical framework for DNA computing that models the recombination of DNA molecules through cutting and rejoining at specific subsequences [16]. Later, Bartlett [3] presented splicing as a crossover operator in CA, enabling the construction of nonlinear rules via piecewise linear combinations. This approach demonstrated that employing splicing among linear CA rules can asymptotically minimize the maximum distance between rule pairs, producing exact configurations within CA rule space. Although crossover operators for one dimensional cellular automata are well studied, two dimensional rules involve more complex neighborhood interactions and spatial patterns. Our proposed splicing operator accounts for center cell interactions, showing that one dimensional results cannot be directly extended and motivating dedicated two dimensional crossover methods.

Further studies by Sheena Christy et al. explored different splicing models affecting array formation and splicing complexity [17–19], while Li and Packard used the Hamming distance to analyze CA genotype distributions [15]. Integrating such splicing mechanisms into CA enhances their biological realism, capturing the essence of genetic recombination and structural diversity in evolution. In genetic algorithms, crossover operators produce offspring by recombining parts of two parent genotypes, thereby preserving characteristics from both. Similarly, splicing of cellular automata rules combines states of two parent rules, which can be interpreted as genotypes, while the resulting cellular evolution represents the phenotype. This interpretation allows CA rule splicing to be viewed as a biologically inspired, crossover-like mechanism within the cellular automata framework.

This paper focuses on the linear rules among the 65,536 total rules of 2D CA with two states and four neighborhoods. Section 2 recalls the fundamental notions of one-dimensional CA and the splicing of 2D CA linear rules. Section 3 introduces the splicing operation for 2D CA, provides an algorithm for combining two linear rules, and discusses their structural properties and genetic interpretation of splicing operation on 2D CA Rules. Section 4 defines the Hamming and Manhattan distance metrics for 2D CA, illustrating how these metrics describe the geometric structure of the rule space and reveal inter-rule relationships. Section 5 concludes with remarks and directions for future research.

2 Linear Rules of 2D CA

In this section, we give the basic notions of one dimensional CA or simply a CA and define the 2D CA linear rules. The set of all possible rules with cardinality m^{m^n} is defined as the rule space of a cellular automaton with m states and a neighborhood of n cells. Here we consider the 2D CA with $m = 2$ states and $n = 4$ neighborhoods.

Definition 1. A cellular automaton is a quadruple (I, S, Σ, χ) where

- $I \in \mathcal{L}^D$ is the cellular space in D dimensions. A collection of cells is arranged in I's positions to create a regular network.

- S is the set of finite states.
- $\Sigma = (\overrightarrow{u_1}, \overrightarrow{u_2}, ..., \overrightarrow{u_N})$ is the neighborhood vector of N different elements of I that connects nearby cells. The cells that are closest to a cell are typically its neighborhoods. But if we know the neighborhood vector N, then for every $i \in \{1, 2, ..., N\}$, the neighborhoods of a cell at position $\overrightarrow{u} \in I$ are at locations $(\overrightarrow{u} + \overrightarrow{u_i}) \in I$.
- $\chi : S^N \rightarrow S$ is referred to as the automaton's local rule. A cell's subsequent state can be found with $\chi(a_1, a_2, ..., a_N)$, where $a_1, a_2, ..., a_N$ represent the states of its N neighborhoods.

Linear rules (Table 1) are expressed in terms of polynomials in a 1D cellular automata. A well-known example of this is Wolfram's elementary cellular automata, which allow the future state of a cell to be derived from the states of its two adjacent neighborhoods. In this context, a linear rule can be denoted as

$$S_u(y+1) = c_1 S_{u-1}(y) + c_2 S_u(y) + c_3 S_{u+1}(y) \pmod 2$$

where $S_u(y)$ denotes the state of cell u at time y and the coefficients c_1, c_2 and c_3 are taken from the ring $(\mathbb{Z}_2, +, *)$.

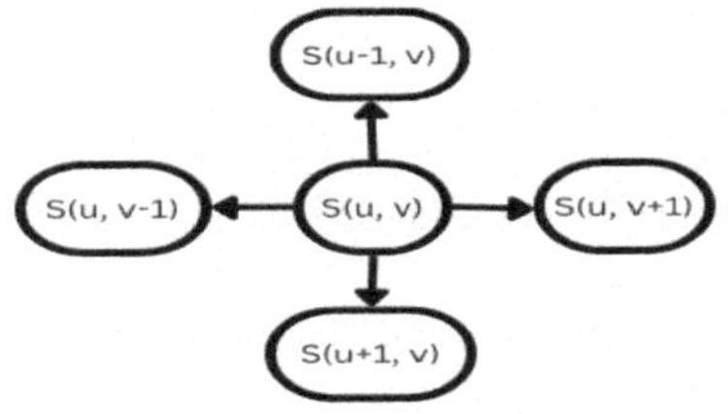

Fig. 1. Next state neighborhood

A **2D cellular automaton** operates on a 2D lattice of cells, each containing a state from a finite set, which concurrently updates its state at discrete time intervals according to the state of its neighbors. The update follows a predetermined local rule.

Definition 2. The next state of the cell $S_{(u,v)}(y)$ in 2D CA is defined by

$$S_{(u,v)}(y+1) = \chi(S_{(u-1,v)}(y), S_{(u,v+1)}(y), S_{(u+1,v)}(y), S_{(u,v-1)}(y))$$

where $S_{(u-1,v)}$ denotes the top neighborhood of $S_{(u,v)}$, $S_{(u,v+1)}$ denotes the right neighborhood of $S_{(u,v)}$, $S_{(u+1,v)}$ denotes the bottom neighborhood of $S_{(u,v)}$ and $S_{(u,v-1)}$ denotes the left neighborhood of $S_{(u,v)}$. Let $S_{(u,v)}(y)$ denote the state of the cell at position (u, v) at time y. "χ " is the rule function that maps the configuration of the four neighborhoods to the next state. This definition specifies how the state of each cell in the 2D cellular automaton grid is updated based on the states of its top, right, bottom, and left neighborhoods, ignoring the cell's center state. The geometrical view of this next state neighborhood is shown in Fig. 1.

Table 1. 2D CA Linear Rules with four neighborhoods and two states

Rule No	Next State Logic Function
0	0
4080	$S_{(u-1,v)}(y) \oplus S_{(u,v+1)}(y)$
13260	$S_{(u-1,v)}(y) \oplus S_{(u+1,v)}(y)$
15420	$S_{(u,v+1)}(y) \oplus S_{(u+1,v)}(y)$
21930	$S_{(u-1,v)}(y) \oplus S_{(u,v-1)}(y)$
23130	$S_{(u,v+1)}(y) \oplus S_{(u,v-1)}(y)$
26214	$S_{(u+1,v)}(y) \oplus S_{(u,v-1)}(y)$
27030	$S_{(u-1,v)}(y) \oplus S_{(u,v+1)}(y) \oplus S_{(u+1,v)}(y) \oplus S_{(u,v-1)}(y)$
38550	$S_{(u,v+1)}(y) \oplus S_{(u+1,v)}(y) \oplus S_{(u,v-1)}(y)$
39270	$S_{(u-1,v)}(y) \oplus S_{(u+1,v)}(y) \oplus S_{(u,v-1)}(y)$
42330	$S_{(u-1,v)}(y) \oplus S_{(u,v+1)}(y) \oplus S_{(u,v-1)}(y)$
43690	$S_{(u,v-1)}(y)$
49980	$S_{(u-1,v)}(y) \oplus S_{(u,v+1)}(y) \oplus S_{(u+1,v)}(y)$
52428	$S_{(u+1,v)}(y)$
61680	$S_{(u-1,v)}(y)$
65280	$S_{(u,v+1)}(y)$

3 Splicing in 2D CA

Splicing, originally proposed in the context of DNA recombination [16], models the biological "cut-and-paste" process in which two DNA molecules are cleaved and rejoined to form hybrid sequences. In a similar way, splicing of cellular-automata (CA) rules combines two local transition rules to generate a new rule that inherits properties from both parents. In 2D CA, such an operation can modify the distribution of rules within the rule space, often leading to a more homogeneous structure and richer dynamical behaviors.We consider two-dimensional binary CA ($m = 2$) with four von Neumann neighbors ($n = 4$). Each rule can be represented by a 16-bit binary string corresponding to all possible neighborhood configurations. The decimal value of this binary string is referred to as the rule number. Splicing improves the similarity of various rules within a collection of linear rules in cellular automata. This process homogenizes the linear rule space, ensuring that the rules become more closely related and exhibit increasingly predictable behavior. This section discusses the splicing of 2D CA rules with m states and n neighborhoods. We have given the algorithm for splicing two 2D CA linear rules with m states and n neighborhoods, tabulated the results, and studied some of the algebraic properties for spliced 2D CA linear rules.

Definition 3. Let R_1, R_2 be any two 2D CA rules, each defined with m states and n neighborhood. The neighborhood combination is represented as a tuple $t = (t_{n-1}, \ldots, t_1, t_0)$, where $t_i \in \{0, 1, \ldots, m-1\}$ is the state of the i-th cell in the neighborhood tuple. The *center cell* refers to the specific position within the neighborhood tuple that determines the new state of the cell being updated.

The state of the center cell, denoted as $t_c(i)$, is determined as follows:

$$t_c(i) = \begin{cases} t_{\frac{n-1}{2}}, & \text{if } n \text{ is odd,} \\ (t_{\frac{n}{2}} + t_{\frac{n}{2}-1}) \bmod 2, & \text{if } n \text{ is even.} \end{cases}$$

The *splicing* of R_1, R_2 is defined as a new 2D CA rule R_s, where the update rule for a cell depends on the state t_c of the center cell in the neighborhood:

$$R_s(t_i) = \begin{cases} R_1(t_i), & \text{if } t_c(i) = 0, \\ R_2(t_i), & \text{if } t_c(i) = 1, \end{cases}$$

Here, R_s is the resultant rule formed by splicing of 2D CA rules, such that the update rule applied to a cell depends on the center cell's state (Fig. 2).

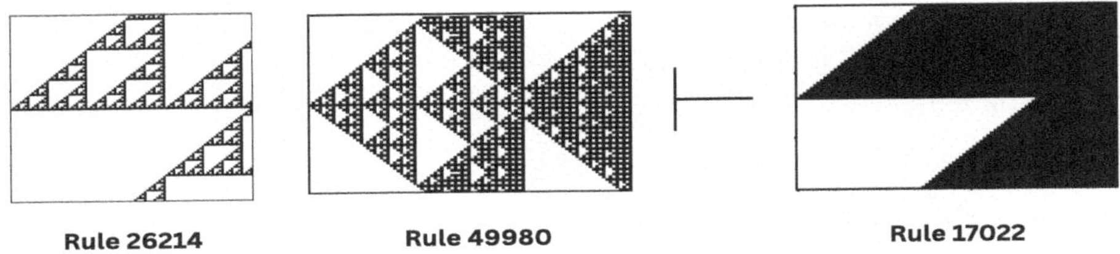

Fig. 2. Resultant rule 17022 obtained through the splicing of rules 26214 and 49980.

3.1 Algorithm for Splicing Two Rules in 2D CA

In this section the algorithm for splicing two 2D CA rules with m-states and n-neighborhoods is discussed. Algorithm 1 deals with the case of 2 - states and 4 - neighborhoods, and in Table 2, it shows the results for the splicing of 2D linear CA rules with 2 - states and 4 -neighborhoods are given. The algorithm can be altered for any arbitrary states and neighborhoods.

3.2 Structural Properties and Genetic Interpretation of Splicing Operation on 2D CA Rules

In this section, we examine the algebraic properties of the proposed splicing operation on two-dimensional cellular automaton rules. These properties help us to characterize the structural behavior of splicing and are motivated by its interpretation as a formal abstraction of DNA recombination. Such an analysis provides a foundation for its use in biologically inspired computational models.

Algorithm 1. Splicing of Two 2D CA Rules

```
function spliceRules(rule1, rule2)
    function convert To 16 Bit Array(rule)
        binary_array ← array of size 16
        binary_representation ← binary representation of rule with exactly 16 bits
        for i = 0 to 15 do
            binary_array[i] ← bit value at position i of binary_representation
        end for
        return binary_array
    end function
    array1 ← convert To 16 Bit Array(rule1)
    array2 ← convert To 16 Bit Array(rule2)
    spliced_array ← array of size 16 initialized to 0
    for n = 0 to 15 do
        Extract neighborhood bits (t3, t2, t1, t0) from n
        central_state ← (t2 + t1) mod 2
        if central_state = 0 then
            spliced_array[n] ← array1[n]
        else
            spliced_array[n] ← array2[n]
        end if
    end for
    spliced_rule ← integer value of spliced_array
    return spliced_rule
end function

function main
    rule1 ← input first rule number
    rule2 ← input second rule number
    spliced_rule ← spliceRules(rule1, rule2)
    Print spliced_rule
end function
```

Table 2. Splicing of 2D CA linear rules with four neighborhoods and two states. In this the header row and header column denotes the linear rules which involves in splicing and the intersection of the header row and header column denotes the resultant rule obtained by splicing

Linear Rules	0	4080	13260	15420	21930	23130	26214	27030	38550	39270	42330	43690	49980	52428	61680	65280
0	0	3120	12300	15420	5160	6168	9252	10260	5140	6180	9240	10280	60	3084	12336	15360
4080	960	4080	13260	16380	6120	7128	10212	11220	6100	7140	10200	11240	1020	4044	13296	16320
13260	960	4080	13260	16380	6120	7128	10212	11220	6100	7140	10200	11240	1020	4044	13296	16320
15420	0	3120	12300	15420	5160	6168	9252	10260	5140	6180	9240	10280	60	3084	12336	15360
21930	16770	19890	29070	32190	21930	22938	26022	27030	21910	22950	26010	27050	16830	19854	29160	32130
23130	16962	20082	29262	32382	22122	23130	26214	27222	22102	23142	26202	27242	17022	20046	29298	32322
26214	16962	20082	29262	32382	22122	23130	26214	27222	22102	23142	26202	27242	17022	20046	29298	32322
27030	16770	19890	29070	32190	21930	22938	26022	27030	21910	22950	26010	27050	16830	19854	29106	32130
38550	33410	36530	45710	48830	38570	39578	42662	43670	38550	39590	42650	43690	33470	36494	45746	48770
39270	33090	36210	45390	48510	38250	39258	42342	43350	38230	39270	42330	43370	33150	36174	45426	48450
42330	33090	36210	45390	48510	38250	39258	42342	43350	38230	39270	42330	43370	33150	36174	45426	48450
43690	33410	36530	45710	48830	38570	39578	42662	43670	38550	39590	42650	43690	33470	36494	45746	48770
49980	49920	53040	62220	65340	55080	56088	59172	60180	55060	56100	59160	60200	49980	53004	62256	65280
52428	49344	52464	61644	64764	54504	55512	58596	59604	54484	55524	58584	59624	49404	52428	61680	64704
61680	49344	52464	61644	64764	54504	55512	58596	59604	54484	55524	58584	59624	49404	52428	61680	64704
65280	49920	53040	62220	65340	55080	56088	59172	60180	55060	56100	59160	60200	49980	53004	62256	65280

1. Closure Property Let $\mathcal{R}$ be the set of all two-dimensional cellular automaton rules with $m = 2$ states and $n = 4$ neighborhood cells. For any $R_1, R_2 \in \mathcal{R}$, define the splicing operation $R_1 * R_2 = R_s$. The set $\mathcal{R}$ is closed under splicing.

However, the subset of linear rules $\mathcal{L} \subset \mathcal{R}$ is not closed under splicing, and the splicing of two linear rules may result in a nonlinear rule is illustrated in Table 2.

2. Non-Commutativity Let $\mathcal{R}$ denote the set of all two-dimensional cellular automaton rules with $m = 2$ states and $n = 4$ neighborhood cells. In general, the splicing operation is not commutative. That is, for two rules $R_1, R_2 \in \mathcal{R}$,

$$R_1 * R_2 \neq R_2 * R_1.$$

This occurs because the splicing process treats the two parent rules differently the choice of which parent rule contributes to the resultant depends on the state of the center cell in the neighborhood. As a result, exchanging the order of the input rules can lead to a different spliced rule is illustrated in Table 2.

3. Associativity Let $\mathcal{R}$ denote the set of all two-dimensional cellular automaton rules with $m = 2$ states and $n = 4$ neighborhood cells. For $R_1, R_2, R_3 \in \mathcal{R}$, the splicing operation is associative, that is,

$$(R_1 * R_2) * R_3 = R_1 * (R_2 * R_3).$$

Example: Let $R_1 = 13260,\ R_2 = 21930,\ R_3 = 38550$

$$(13260 * 21930) * 38550 = 13260 * (21930 * 38550) = 11240.$$

Thus, $*$ is associative on $\mathcal{R}$.

4. Identity Element Let $\mathcal{R}$ denote the set of all two-dimensional cellular automaton rules with $m = 2$ states and $n = 4$ neighborhood cells. A rule $R_I \in \mathcal{R}$ is called an identity element with respect to the splicing operation $*$ if

$$R * R_I = R_I * R = R, \quad \forall\, R \in \mathcal{R}.$$

However, such an identity rule does not exist in the rule set $\mathcal{R}$ under the proposed splicing operation.

Theorem 1. *Let $\mathcal{R}_{m,n}$ denote the set of all two dimensional cellular automaton rules with m states and n neighborhood cells. Each rule $R \in \mathcal{R}_{m,n}$ is represented as a function, $R : \{0, 1, \ldots, m-1\}^n \to \{0, 1, \ldots, m-1\}$, equivalently as a finite string of length $L = m^n$. Define the splicing operator $* : \mathcal{R}_{m,n} \times \mathcal{R}_{m,n} \to \mathcal{R}_{m,n}$ by*

$$R_s(t_i) = \begin{cases} R_1(t_i), & \text{if } t_c(i) = 0, \\ R_2(t_i), & \text{if } t_c(i) = 1, \end{cases}$$

where $t_i \in \{0, 1, \ldots, m-1\}^n$ and $t_c(i) \in \{0, 1\}$ is the center cell dependent control function. Then the operator $*$ defines a valid crossover operator in the sense of genetic algorithms.

Proof. (i) Closure: For every neighborhood configuration t_i, the spliced rule R_s equals either $R_1(t_i)$ or $R_2(t_i)$, both belonging to $\{0, 1, \ldots, m-1\}$. Hence,

$$R_s : \{0, 1, \ldots, m-1\}^n \rightarrow \{0, 1, \ldots, m-1\},$$

and therefore $R_s \in \mathcal{R}_{m,n}$.

(ii) Reversibility: Define the control function $t_c(i) = 0$ if $R_s(t) = R_1(t_i)$ and $t_c(i) = 1$ if $R_s(t) = R_2(t_i)$. The pair $(R_s(t), t_c(i))$ uniquely identifies the parent source of each rule entry. Thus, no information is lost during splicing, and the operation is reversible given the control information.

iii) Recombination: Let

$$T_i = \{\, t \mid t_c(i) = i \,\}, \quad i \in \{0, 1\}.$$

Then the spliced rule R_s is defined by

$$R_s(t) = \begin{cases} R_1(t), & \text{if } t \in T_0, \\ R_2(t), & \text{if } t \in T_1. \end{cases}$$

In other words, R_s inherits the states of R_1 on T_0 and the states of R_2 on T_1. This formulation explicitly shows how the offspring rule combines complementary components from both parents while preserving their respective contributions.

4 Different Metrics in Cellular Automata and Splicing

The linear rules of 2D CA exhibit smooth, regular behavior. Splicing two linear rules produces new ones that inherit characteristics of both parents, reducing divergence within the rule space. As splicing continues, differences among rules diminish, minimizing both average and maximum distances. Eventually, the rule space becomes more uniform, and all linear rules converge toward similar behavior. Thus, continuous splicing drives the 2D CA rule space toward homogeneity, with the maximum inter-rule distance approaching zero.

4.1 Hamming Distance Between Random-Random Rules:

Let A and B be two 2D CA random rules(non linear rules) written as array over $\mathbb{Z}_m$ with neighborhood of size n and there are m^n next-states.

i.e., A $=[a_{11}, a_{12}, a_{13}, ..., a_{1s}]$ and B $= [b_{11}, b_{12}, b_{13}, ..., b_{1s}]$

The Hamming distance $d_H(A, B)$ counts the number of positions at which the two arrays differ:

$$d_H(A, B) = \sum_{i=1}^{s} |a_i - b_i|.$$

Let X_r represent the random variable that is connected to a certain neighborhood, which is given by

$$X_r = \begin{cases} 1, & \text{if } a_{(u,v)} = b_{(u,v)} \\ 0, & \text{if } a_{(u,v)} \neq b_{(u,v)} \end{cases}$$

We can find the expected Hamming distance and variance of random-random rules of 2D CA as given by [3,10].

4.2 Hamming Distance Between Linear-Linear Rules:

Let A and B be two linear rules of a two dimensional cellular automaton (2D CA) with $m = 2$ states and $n = 4$ neighborhoods. Each rule can be represented as a vector of length $s = m^n$, corresponding to all possible neighborhood configurations.

The Hamming distance between the 2D CA linear rules A and B is

$$\mid A, B \mid = s - dm^{n-1}$$

where $d = gcd((a_{11} - b_{11}), ..., (a_{1\,s} - b_{1\,s}), m)$

We can find the expected Hamming distance and variance of linear-linear rules of 2D CA as given by [3].

4.3 Relation Between Splicing Operation and Hamming Distance

Theorem 2. *Let R_1 and R_2 be two 2D Cellular Automata (CA) rules represented as binary strings of length m^n.*

Let $R_s = R_1 * R_2$ where $*$ is the splicing operation.

Then, the Hamming distances satisfy the following equality

$$d_H(R_1, R_s) + d_H(R_2, R_s) = d_H(R_1, R_2),$$

where $d_H(R_1, R_2)$ is the Hamming distance between the binary strings R_1 and R_2.

Proof. Hamming Distance Between R_1 and R_2 is

$$d_H(R_1, R_2) = \sum_{i=0}^{m^n-1} |R_1[i] - R_2[i]|.$$

Hamming Distance Between R_1 and R_2 using the XOR $\oplus$ operation is

$$d_H(R_1, R_2) = R_1[i] \oplus R_2[i].$$

Thus $d_H(R_1, R_s) = \sum_{i=0}^{m^n-1}|(R_1[i] \oplus R_s[i])|$. Similarly, Hamming Distance between R_2 and R_s is

$$d_H(R_2, R_s) = \sum_{i=0}^{m^n-1} |(R_2[i] \oplus R_s[i])|.$$

The XOR operation has the property that: $x \oplus y = 1$ if and only if $x \neq y$. For every bit i, the following holds:

$$|R_1[i] - R_s[i]| + |R_2[i] - R_s[i]| = |R_1[i] - R_2[i]|.$$

Summing over all i from 1 to $m^n - 1$, we obtain:

$$\sum_{i=0}^{m^n-1} |R_1[i] - R_s[i]| + \sum_{i=0}^{m^n-1} |R_2[i] - R_s[i]| = \sum_{i=0}^{m^n-1} |R_1[i] - R_2[i]|.$$

This simplifies to: $d_H(R_1, R_s) + d_H(R_2, R_s) = d_H(R_1, R_2)$.

Theorem 3. *Let $\mathcal{R}$ be the set of 2D CA rules with $m = 2$ states and $n = 4$ neighborhoods and let $R_1, R_2 \in \mathcal{R}$. For each neighborhood index i, let $t_c(i) \in \{0, 1\}$ denote the fixed center cell indicator.*
Define the splice operator $* : \mathcal{R} \times \mathcal{R} \to \mathcal{R}$ by

$$(R_1 * R_2)[i] = \begin{cases} R_1[i], & \textit{if } t_c(i) = 0, \\ R_2[i], & \textit{if } t_c(i) = 1, \end{cases}$$

for every index i. Then

$$\big((R_1 * R_2) * R_1\big) = R_1 \qquad \textit{and} \qquad \big(R_2 * (R_1 * R_2)\big) = R_2.$$

Proof. Fix an arbitrary neighborhood index i. There are two possible cases.

Case 1: $t_c(i) = 0$. From the definition of $*$ we have $(R_1 * R_2)[i] = R_1[i]$. Applying the splicing operator again with R_1 as the second rule gives

$$\big((R_1 * R_2) * R_1\big)[i] = \begin{cases} (R_1 * R_2)[i], & \text{if } t_c(i) = 0, \\ R_1[i], & \text{if } t_c(i) = 1. \end{cases}$$

Since $t_c(i) = 0$, the value of $\big((R_1 * R_2) * R_1\big)[i]$ is determined by the first condition of the definition, hence $\big((R_1 * R_2) * R_1\big)[i] = (R_1 * R_2)[i] = R_1[i]$.

Case 2: $t_c(i) = 1$. From the definition, $(R_1 * R_2)[i] = R_2[i]$. Applying the splice operator with R_1 gives

$$\big((R_1 * R_2), R_1\big)[i] = \begin{cases} (R_1 * R_2)[i], & \text{if } t_c(i) = 0, \\ R_1[i], & \text{if } t_c(i) = 1. \end{cases}$$

Since $t_c(i) = 1$, the value of $S\big(S(R_1, R_2), R_1\big)[i]$ is obtained from the second condition, which gives $\big((R_1 * R_2), R_1\big)[i] = R_1[i]$.

Therefore, for every index i, we have $\big((R_1 * R_2) * R_1\big)[i] = R_1[i]$. Hence, $\big((R_1 * R_2) * R_1\big) = R_1$. By symmetry, the same argument holds for R_2, yielding $\big(R_2 * (R_1 * R_2)\big) = R_2$.

4.4 Manhattan Distance

The Manhattan distance between the rules can be computed by summing the absolute difference between corresponding elements in the rules. This provides a measure how much one rule differs from another. Consider any two 2D CA rules as A and B with $n = 4$ neighborhoods and $m = 2$ states where,

$$A = [a_{11}, a_{12}, ..., a_{1s}] \text{ and } B = [b_{11}, b_{12}, ..., b_{1s}]$$

Then, the Manhattan distance between the two rules A and B is given by i.e., $M_{d(A,B)} = |a_{11} - b_{11}| + |a_{12} - b_{12}| + ... + |a_{1s} - b_{1s}|$

$$M_{d(A,B)} = \sum_{i=0}^{s} |a_{1i} - b_{1i}|$$

The Manhattan distance refers to the method for calculating cell distances. In a 2D grid, the Manhattan distance between two cells (a_1, b_1) and (a_2, b_2) is the sum of their absolute coordinate differences.

$$M_d = |a_1 - b_1| + |a_2 - b_2|$$

Manhattan distance can be utilized in several ways to establish or affect cellular automaton rules. Rather than simply using the nearest cells (as in a Moore or Von Neumann neighborhood), the rule could be based on the states of cells within a fixed Manhattan distance. Consider a neighborhood of radius r that comprises all cells with Manhattan distances less than or equal to r For example:

1. A radius - 1 neighborhood includes cells with Manhattan distance d$\leq$1.
2. A radius - 2 neighborhood would include cells with Manhattan distance d$\leq$2, meaning cells up to two steps are away horizontally or vertically.

This can allow for more distant interactions and influence in the 2D cellular automaton, even the patterns are more complex.The central cell's state could be modified by a weighted sum of its neighborhood states, with the weight determined by the Manhattan distance. Closer cells may have more influence, while more distant cells (within a range) have less influence.

Corollary 1. *$M_{d(A,B)} = d_{H(A,B)}$ for two states A and B.*

Consider two 2D celluar automata rules A and B with 2 states where,

$$A = [a_{11}, a_{12}, ..., a_{1s}] \text{ and } B = [b_{11}, b_{12}, ..., b_{1s}]$$

Then, the Manhattan distance between the two rules A and B is

$$M_{d(A,B)} = |a_{11} - b_{11}| + |a_{12} - b_{12}| + ... + |a_{1s} - b_{1s}|$$

In the case of binary values (0 or 1), the absolute difference is equal to 1 if the bits differ and 0 if they are the same .

The Hamming distance d_H between the two rules A and B is defined as the number of positions at which the corresponding bits differ.

$$d_{H(A,B)} = \begin{cases} 1 \text{ if } a_{(u,v)} \neq b_{(u,v)} \\ 0 \text{ if } a_{(u,v)} = b_{(u,v)} \end{cases}$$

Since both distances count the number of differing bits between the two binary strings, We have $M_{d(A,B)} = d_{H(A,B)}$

Remark 1. For binary cellular automata ($m = 2$), the Hamming distance and Manhattan distance are identical since each differing bit contributes exactly one unit to both measures. However, for multi state cellular automata ($m > 2$), the Manhattan distance $M_d(A, B)$ is always greater than or equal to the Hamming distance $d_H(A, B)$:

$$M_d(A, B) \geq d_H(A, B).$$

Equality holds only when all differing cells differ by one state, i.e., $|a_i - b_i| = 1$ for every i such that $a_i \neq b_i$. Hence, for ternary ($m = 3$) or quaternary ($m = 4$) state automata, the Manhattan distance reflects not only the number of differing positions but also the magnitude of those differences, resulting in $M_d(A, B) \geq d_H(A, B)$.

5 Conclusion

This study consider 2D linear CA rules with $m = 2$ states and $n = 4$ neighborhoods. Different metrics of cellular automata are studied. In particular, the Manhattan distance is defined for measuring how one rule differs from another linear rule and shown that the Manhattan distance is equal to the Hamming distance in the case of two states. By continuously splicing linear rules, the maximum possible distance between the new rule and other rules tends to decrease. Thus, the linear rules are much significant in the rule space of the cellular automata. Crossover operators for one-dimensional cellular automata are extensively researched; however, two dimensional rules introduce further complexities due to neighborhood dependencies and spatial anisotropy. The suggested splicing operator deals with these problems by taking into account interactions between center cells, which shows that simply extending results from one dimension is not enough this shows the potential of the splicing operator for two dimensional cellular automata. Our next work will concentrate on shift-equivalent rules.

References

1. Bhattacharjee, K., Naskar, N., Roy, S., Das, S.: A survey of cellular automata: types, dynamics, non-uniformity and applications. Nat. Comput. **19**, 433–461 (2020). https://doi.org/10.1007/s11047-018-9696-8

2. Sarkar, P.: A brief history of cellular automata. ACM Comput. Surv. **32**(1), 80–107 (2000). https://doi.org/10.1145/349194.349202
3. Bartlett, R., Garzon, M.: Distribution of linear rules in cellular automata rule space. Complex Syst. **6**, 519–532 (1992)
4. Maji, P., Shaw, C., Ganguly, N., Sikdar, B.K., Chaudhuri, P.P.: Theory and application of cellular automata for pattern classification. Fundamenta Informaticae **58**(3–4), 321–354 (2003)
5. Von Neumann, J., Burks, A.W.: Theory of self-reproducing automata. IEEE Trans. Neural Netw. **5**(1), 3–14 (1966). https://doi.org/10.1016/0020-0271(69)90026-6
6. Moore, E.F.: Machine models of self-reproduction. In: Proceedings of Symposia in Applied Mathematics, vol. 14, no. 5, pp. 17–33 (1962). https://doi.org/10.1090/psapm/014/9961
7. Packard, N.H., Wolfram, S.: Two-dimensional cellular automata. J. Stat. Phys. **38**(5), 901–946 (1985). https://doi.org/10.1007/BF01010423
8. Kari, J.: Reversibility of 2D CA is undecidable. Physica D **45**(1–3), 379–385 (1990). https://doi.org/10.1016/0167-2789(90)90195-U
9. Lindenmayer, A.: Cellular automata, formal languages and developmental systems. Stud. Logic Found. Math. **74**, 677–691 (1990). https://doi.org/10.1016/S0049-237X(09)70394-5
10. Clark, A.B., Disney, R.L.: Probability and Random Processes: A First Course with Applications. John Wiley & Sons, New York (1985)
11. Wolfram, S.: Theory and Applications of Cellular Automata. World Scientific (1986)
12. Pettorossi, A.: Automata Theory and Formal Languages: Fundamental Notions, Theorems, and Techniques. Springer, Heidelberg (2022). https://doi.org/10.1007/978-3-031-11965-1
13. Wolfram, S.: Statistical mechanics of cellular automata. Rev. Mod. Phys. **55**(3), 601–603 (1983). https://doi.org/10.1007/BF01009958
14. Wolfram, S.: Computation theory of cellular automata. Commun. Math. Phys. **96**(1), 15–57 (1984). https://doi.org/10.1007/BF01217347
15. Li, W., Packard, N.: The structure of the elementary cellular automata rule space. Complex Syst. **4**(3), 281–97 (1990)
16. Head, T.: Splicing schemes and DNA. In: Lindenmayer Systems: Impacts on Theoretical Computer Science, Computer Graphics, and Developmental Biology, pp. 371–383 (1992). https://doi.org/10.1007/978-3-642-58117-5-23
17. Sheena Christy, D.K., Masilamani, V., Thomas D.G.: Accepting H-array splicing systems and their properties. Sci. Technol. **21**(3), 298–309 (2018). https://doi.org/10.1007/978-3-662-45049-9-51
18. Sheena Christy, D.K., Thomas, D.G.: Accepting hexagonal array splicing system with respect to arrowheads. Int. J. Pure Appl. Math. **113**(6), 91–100 (2017). https://doi.org/10.1007/978-981-10-0451-3-44
19. Sheena Christy, D.K., Masilamani, V., Thomas, D.G.: Iso-array splicing grammar system. In: Proceedings of Seventh International Conference on Bio-Inspired Computing: Theories and Applications (BIC-TA 2012), vol. 1, pp. 157–167. Springer (2013). https://doi.org/10.1007/978-81-322-1038-2-14

Predator-Prey Dynamics in a Three-Species Wa-Tor Model Using Cellular Automata with Memory

Mukund Pareek[1] and Sudhakar Sahoo[2](✉)

[1] Department of Physics and Astrophysics, University of Delhi, New Delhi, India
mukundhp11@gmail.com
[2] Department of Computer Science, Institute of Mathematics and Applications, Bhubaneswar, India
sudhakar@iomaorissa.ac.in

Abstract. Classic predator-prey simulations done using Lotka-Volterra equations, the Wa-Tor or the Cellular Automata (CA) model, are often limited to two-species systems and lack agent memory (they are Markovian). This paper presents a novel simulation model that addresses both limitations. We extend the Wa-Tor framework to a three-species food chain (sharks, fish and plankton) and as our core innovation, integrate it with Cellular Automata with Memory (CAM). This CAM framework allows agents (sharks and fish) to use a history of recent states to make *intelligent*, adaptive decisions based on environmental patterns. To ensure ecological realism, the model also introduces crucial constraints, including a fish starvation mechanic and a sustainable plankton regrowth rule that emulates *ecological memory*. Our simulations show that the final model successfully reproduces persistent, cyclical oscillations and achieves stable long-term coexistence, as predicted by classical theory. We discuss the significant impact of agent memory on system stability and propose simulating a trophic cascade as a robust method for qualitative validation.

Keywords: Predator-Prey Dynamics · Cellular Automata · Wa-Tor Model · Cellular Automata with Memory · Three-Species Model · Agent-Based Modeling · Trophic Cascade

1 Introduction

The dynamics between predator and prey are a cornerstone of ecological research, with foundational approaches ranging from continuous Lotka-Volterra equations [15] to discrete, grid-based models like the Wa-Tor [6] or Cellular Automata (CA) [3,4,8]. However, these classic models often have significant limitations. Firstly, they tend to be ecologically oversimplified; the original Wa-Tor for example, as well as its CA formulation [9], focuses on an isolated two-species food chain, a

This work was done as a project in the Indian Summer School on Cellular Automata 2025.

H. Raju et al. (Eds.): ASCAT 2026, CCIS 2801, pp. 207–223, 2026.
https://doi.org/10.1007/978-3-032-18612-6_16

setup that's rare in real, complex ecosystems. Secondly, they are behaviorally simple. Standard CA is Markovian, meaning its agents are *memoryless* and just react to their immediate surroundings. This fails to capture how real animals use past experience to make decisions.

This work tackles both of these limitations. We move beyond the two-species setup by implementing a linear three-species food chain (Sharks, Fish and Plankton), which is more ecologically significant and one step closer to the complex food chains found in real world than the classic two-species model. As the core innovation of this work, we then integrate this system with the concept of Cellular Automata with Memory (CAM) [11,13]. This allows agents to *remember* recent states, letting them make *intelligent*, adaptive decisions—like sharks targeting areas with a high prevalence of fish, or fish learning to avoid predator hotspots. To the best of our knowledge, this is the first implementation of a three-species Wa-Tor model that explicitly incorporates such memory-based decision-making.

Our model is built upon the classic Wa-Tor framework, retaining core parameters like agent breed times and grid setup from earlier two-species simulations [6,9]. However, adapting this framework for a three-species food chain and adding agent memory required a new set of essential ecological constraints to better capture the dynamics of real-world ecosystems. Therefore, our model integrates three key new mechanics: (1) a fish starvation rule, which is crucial for linking fish survival to plankton consumption; (2) a sustainable plankton regrowth rule, which is *influenced by local history* to emulate the *ecological memory* of real-world resource hotspots; and (3) a balanced memory model for the agents' decision-making. These components work together to create a robust and ecologically-grounded simulation model.

The remainder of this paper is structured as follows. Section 2 provides the theoretical background. It starts by comparing the continuous Lotka-Volterra model to discrete CA and then details the CAM framework and its implementation. Section 3 offers a comprehensive description of our simulation model's design, covering the grid setup, parameter definitions, agent behavioral rules (including the specific CAM-based decision logic), and the resource regeneration mechanism. It concludes with an analysis of the model's computational complexity and practical feasibility. Section 4 presents and analyzes simulation results, interpreting the observed dynamics as distinct outcomes predicted by Volterra's theory and discusses the role of model parameters and agent intelligence. It also addresses the significant challenges associated with validating ecological models against real-world data and outlines our proposed strategy of using trophic cascade simulation for robust qualitative validation. Section 5 summarizes our key findings, acknowledges limitations, and proposes specific directions for future research.

2 Theoretical Background

Our model is a hybrid, combining two distinct elements: classical Lotka-Volterra theory and the grid-based Wa-Tor (CA) framework. This section explains the details behind this theory and this framework, and then details our main innovation: adding agent memory.

2.1 The Theoretical Benchmark: Volterra's 3-Species Food Chain

When people think of Lotka-Volterra equations, they usually picture the classic two-species cycle. But in his 1926 paper, Vito Volterra specifically analyzed a *biological association* of three species [15]. He considered several combinations but paid special attention to the linear three-species food chain: an apex predator (Species 1) eating an intermediate species (Species 2), which in turn eats a basal resource (Species 3).

Food chains like this are very common in real ecosystems. Famous examples include the Sea Otter → Urchin → Kelp marine system and the Wolf → Elk → Aspen terrestrial system. The Shark → Fish → Plankton chain is another classic real-world example. Our choice of these species is arbitrary, however, as our model is designed to be general enough to represent any similar three-species linear food chain.

Volterra modeled this three species interaction using the following system of coupled differential equations:

$$\beta_1 \frac{dN_1}{dt} = (\beta_1 \epsilon_1 + a_{21} N_2 + a_{31} N_3) N_1 \tag{1}$$

$$\beta_2 \frac{dN_2}{dt} = (\beta_2 \epsilon_2 + a_{12} N_1 + a_{32} N_3) N_2 \tag{2}$$

$$\beta_3 \frac{dN_3}{dt} = (\beta_3 \epsilon_3 + a_{13} N_1 + a_{23} N_2) N_3 \tag{3}$$

Here, N_i represents the population of each species (e.g., N_1 = Sharks, N_2 = Fish, N_3 = Plankton). The terms in the equations can be understood in a broad sense:

ϵ_i **(epsilon):** This is the intrinsic growth/decay rate of each species if it were all alone. For predators like sharks (ϵ_1) and fish (ϵ_2), this value is negative, as they would starve. For the basal resource like plankton (ϵ_3), this is positive, as it grows on its own.

a_{ij} **(the *a* terms):** These are the interaction coefficients that define who eats whom. For example, a_{21} is the positive benefit to sharks from eating fish, while a_{12} is the negative harm to fish from being eaten. In this linear chain, a_{13} (sharks eating plankton) is zero.

β_i **(beta):** These are constant weighting factors related to the average biomass (or *weight*) of each species. They essentially balance the equations, ensuring the interaction between a large predator and its much smaller prey is scaled in a way that makes sense.

Volterra's analysis revealed that, depending on the relative values of these parameters, the system could exhibit distinct long-term behaviors, which he categorized into two main cases:

- **Case 1:** This case occurs when the food transfer is insufficient for the carnivores to survive. It assumes that the plant growth rate (k) is constant. Volterra justifies this by noting that in this scenario, the plant population is kept in check by the herbivores and *can not increase indefinitely.* The result is that the carnivorous species (N_1) is depleted, while the herbivores (N_2) and plants (N_3) *tend toward a periodic fluctuation not damped.*
- **Case 2:** This case occurs when the food transfer is sufficient for the carnivores. Here, Volterra states that a constant growth rate for the plants is no longer realistic, as their population *would grow indefinitely.* He therefore suggests a self-limiting rule (logistic growth). This case is then broken down into three subcases:
  - **Subcase a:** The food provided by the plants is not sufficient to maintain the herbivore population. As a result, both the herbivorous species (N_2) and the carnivorous species (N_1) that feed on them die out.
  - **Subcase b:** The plants are sufficient to maintain the herbivores, but there is not sufficient food for the carnivores through the herbivores. The carnivorous species (N_1) is depleted, while the herbivore (N_2) and plant (N_3) species tend toward a stable, damped fluctuation.
  - **Subcase c:** The food is sufficient at all levels, allowing all three species to live. The system tends toward a stable, stationary state, often through *asymptotic and damped variations.*

Volterra's Case 1, which assumes a constant plant growth rate, does not apply to our model. Just as Volterra noted that a simple constant growth rate was unrealistic for his mathematical model, we also found it to be unrealistic for our spatial simulation. However, our concern was not with infinite growth (which our finite grid prevents), but with the unrealistic even spread it produced. We therefore introduced a plankton regrowth rule based on *ecological memory* to better capture the realistic patches and clusters that resources form in nature. This approach, which ensures that the plankton is a self-limiting, renewable resource, aligns our model with Volterra's Case 2 and its three subcases (a, b, and c). Our results will therefore be analyzed against these predictions.

While these continuous models provide fundamental insights, they have a major limitation, as Volterra himself noted: they assume spatially homogeneous, *well-mixed* populations and treat individuals as continuous variables. This neglects the discrete, individual-level interactions and uneven distribution that we see in natural systems. To capture these missing spatial and discrete elements, we therefore turn to a different methodological foundation.

2.2 The Methodological Foundation: Cellular Automata and Wa-Tor

While Volterra's equations provide the theory, our method is built on Cellular Automata (CA), which offers a powerful alternative. Unlike the continuous, *well-mixed* ODEs, CA models are built on a grid where location matters, and they simulate individual organisms (or *agents*) instead of just a total population. This approach naturally captures the key elements that Volterra's continuous model misses. These include *Discreteness*—representing individual agents, discrete events like birth and death, and even *agent age*, which Volterra's equations neglect—and *Locality*, where all interactions are governed by simple rules based only on a cell's immediate neighbors. This allows complex, large-scale patterns to emerge from simple, individual behaviors.

The classic application of this approach is the Wa-Tor simulation, introduced by A. K. Dewdney [6]. Wa-Tor is a simple model that simulates sharks (predators) and fish (prey) on a grid, where each agent follows some local rules. Crucially, Dewdney's model (and its later CA formulation by Jafelice & Silva [9]) used a *toroidal grid.* This *wrap-around* grid connects the top edge to the bottom and the left to the right, thereby simulating a boundless environment. It's a much better approach than a simple grid with *walls* because it completely prevents *edge effects*—unrealistic behaviors that happen when agents get *stuck* at the boundaries.

Our work builds directly on this well-established foundation. We adopt the classic Wa-Tor CA framework on a toroidal grid as our starting point, before adding our own novel extensions.

2.3 A Novel Extension: Cellular Automata with Memory (CAM)

In nature, predator-prey dynamics are not memoryless; they are deeply influenced by environmental history, evolutionary pressures, and learned behaviors [1,5,10]. Studies show that past environmental conditions, such as stressors, toxins, and habitat features, have lasting impacts on predator and prey behaviors, vulnerability, and population dynamics [7,12,14]. This means that current predator-prey behavior is shaped not only by immediate environmental factors but also by the accumulated effects of prior environmental states.

Standard CA is memoryless (Markovian) and hence fails to capture this crucial aspect. Therefore, to increase the ecological realism of our simulation, we introduced Cellular Automata with Memory (CAM), allowing our agents to make decisions based on past events and more realistically mimic natural ecosystems [11,13]. Each cell (y, x) maintains a record of its state over the previous μ timesteps (`memory_depth`). This information allows agents to perceive and react to recent trends in their local environment.

The influence of past states diminishes over time, controlled by the decay factor α (`alpha`). We employ a Balanced Memory weighting scheme, where the weight W_i assigned to the state i steps in the past ($i = 1$ being the most recent memory) is:

$$W_i = \alpha^{i+1} \quad (\text{for } i = 1, 2, ..., \mu) \tag{4}$$

This ensures that even the most recent state ($i = 1$) receives a weight less than 1 (specifically, α), preventing instantaneous events from completely overshadowing persistent historical patterns, unlike a Dominant Memory model ($W_i = \alpha^i$). The memory score for a specific state S at cell (y, x) aggregates this weighted history:

$$\text{Score}_{memory}(y, x, S) = \frac{\sum_{i=1}^{\mu}(H_i(y, x, S) \cdot W_i)}{\sum_{i=1}^{\mu} W_i} \quad (5)$$

where $H_i(y, x, S)$ is an indicator function (1 if the state at (y, x) i steps ago was S, 0 otherwise). This score, ranging from 0 to 1, provides agents with a quantitative measure of the recent prevalence of state S at a potential location, forming the core input for their *intelligent* decision-making processes described below.

3 Simulation Model Design

The simulation was implemented as an object-oriented program in Python 3, utilizing the NumPy library for efficient multi-dimensional array manipulation and Matplotlib for visualization and animation generation. The core logic is encapsulated within a *WaTorSimulation* class. Time proceeds in discrete steps, and agent states across the grid are updated synchronously based on the configuration at the beginning of each step.

3.1 Model Setup and Grid Representation

The simulation environment is represented by a two-dimensional grid (lattice) of cells, with dimensions `width` × `height`. Toroidal boundary conditions are implemented using the modulo operator when calculating neighbor coordinates, effectively connecting opposite edges of the grid (like the surface of a doughnut) and preventing edge effects [6]. Each cell (y, x) can exist in one of four discrete integer states, representing the entities within the ecosystem: `EMPTY` (0), `FISH` (1), `SHARK` (2), or `PLANKTON` (3).

The dynamic state of the simulation is maintained using several parallel NumPy arrays of the same dimensions:

- `self.grid`: Stores the primary state (species type or empty) of each cell at the current timestep.
- `self.fish_age`, `self.shark_age`: Store the age (in timesteps since birth or last reproduction) for each fish or shark, respectively. Initialized to -1 for non-applicable cells. Age increments by 1 at each timestep for living individuals.
- `self.fish_starve`, `self.shark_starve`: Store the hunger level, measured as the number of timesteps since the last successful feeding event, for each fish or shark. Initialized to -1. Reset to 0 upon eating, incremented otherwise.
- `self.cell_history`: A 3D array (`height` × `width` × `memory_depth`) storing the state of each cell for the previous μ timesteps. This array is crucial for the CAM calculations. It is updated at the conclusion of each timestep by shifting older states back and inserting the grid state from the beginning of the just-completed step.

Table 1. Core Simulation Parameters, Experimental Values, and Ecological Constraints.

Parameter	Value	Range	Description & Ecological Role
Grid Size	100^2	$50^2 - 300^2$	Toroidal world dimensions: Must be large enough to allow local clustering patterns to emerge, yet computationally manageable.
Init. Fish (N_2)	2400	20% – 40%	Starting prey population: Adheres to the *Trophic Pyramid* principle- Prey (N_2) must significantly outnumber Predators (N_1) to fuel the food chain.
Init. Sharks (N_1)	400	2% – 10%	Starting predator population: A high initial density ($> 15\%$) leads to *over-predation*, causing immediate prey extinction.
Fish Breed (B_F)	7	3 – 15	Reproduction threshold: Prey must reproduce faster than predators ($B_F < B_S$) to recover population losses from hunting.
Shark Breed (B_S)	8	6 – 15	Reproduction threshold: Predators require more time and energy to breed. If B_S is too low, sharks overpopulate and crash the system.
Fish Starve (S_F)	6	5 – 20	Resilience to hunger: Cap on survival without food. Prevents the prey population from exploding infinitely *(Metabolic Limit)*.
Shark Starve (S_S)	10	6 – 20	Resilience to hunger: Predators must be resilient enough to survive the search for food ($S_S \geq B_S$) without starving mid-hunt.
Memory Depth (μ)	4	1 – 10	Steps of history remembered: Reflects the limited cognitive capacity of simple organisms. Higher μ yields diminishing returns vs. computational cost.
Alpha (α)	0.6	$[0.1, 0.9]$	Memory decay rate: Ensures agents react to *recent* trends. $\alpha \to 1$ causes historical bias; $\alpha \to 0$ ignores history.
Plankton Regrowth (P)	0.1	$[0.01, 0.30]$	Basal resource renewal rate. Determines the environment's *Carrying Capacity.* If too high, food becomes infinite, negating the starvation pressure.

The simulation is initialized by setting all cells to `PLANKTON`. Then, `initial_fish` and `initial_sharks` are placed at distinct, randomly chosen locations, overwriting the plankton. To avoid synchronized initial breeding pulses, newly placed fish and sharks are assigned random initial ages uniformly distributed between 0 and their respective `breed_time` -1 [6]. Their initial starve timers are set to 0. Core simulation parameters and their ecological roles are defined in

Table 1. Agent interactions occur within their Moore neighborhood âĂŞ the eight cells immediately surrounding a central cell (horizontally, vertically, and diagonally). Figure 1 provides a visual representation of a typical grid state during a simulation run.

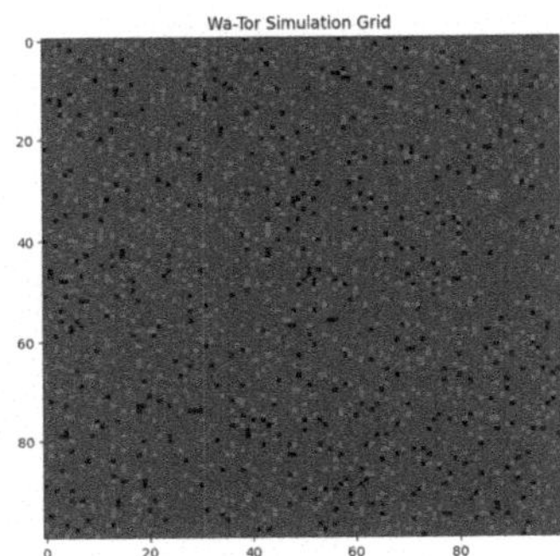

Fig. 1. Snapshot of the 100x100 simulation grid at t = 10,000 (see Fig. 3). Colors: Black (`EMPTY`), Green (`FISH`), Red (`SHARK`), Blue (`PLANKTON`). (Color figure online)

3.2 Agent Decision Logic with Memory

The core update loop iterates through each cell of the grid. To ensure synchronous updates (all decisions based on state at time t), temporary copies (`new_grid`, `new_fish_age`, etc.) are made. Agent actions modify these temporary grids. A boolean `moved` array prevents agents from acting multiple times within a single timestep.

Shark Logic: Sharks exhibit prioritized behavior; hunt first, then move.

1. **Scan for Prey:** Identify adjacent cells containing `FISH` in the original grid (`self.grid`).
2. **Hunting Action (if prey present):**
   - Calculate the *fishiness* memory score (Eq. 5) for each adjacent fish location using `self.cell_history`.
   - Select the location (ny, nx) with the maximum score. If multiple locations share the max score, choose one randomly.
   - Update the temporary grid: set `new_grid[ny, nx] = SHARK`, `new_shark_age[ny, nx] = self.shark_age[y, x] + 1`, `new_shark_starve[ny, nx] = 0`. Now mark `moved[ny, nx] = True`.
3. **Movement Action (if no prey present):**
   - Identify adjacent `EMPTY` or `PLANKTON` cells in the original grid.
   - If such cells exist, calculate the *sharkiness* memory score for each potential move location.
   - Select the location (ny, nx) with the minimum score (avoiding competition). If multiple locations share the min score, choose one randomly.

- Update the temporary grid: set `new_grid[ny, nx] = SHARK`, `new_shark_age[ny, nx] = self.shark_age[y, x] + 1`, `new_shark_starve[ny, nx] = self.shark_starve[y, x] + 1`. Mark `moved[ny, nx] = True`.

4. **Stay Put Action:** If no prey and no valid empty/plankton cells are available, the shark remains in cell (y, x). Update its state in the temporary grid: `new_shark_age[y, x] = self.shark_age[y, x] + 1`, `new_shark_starve[y, x] = self.shark_starve[y, x] + 1`. Mark `moved[y, x] = True`. (Note: Staying put doesn't reset age/hunger).

Post-Action Checks (applied to shark at its location in the *temporary* grid):

- **Starvation Check:** If `new_shark_starve` at the shark's location $\geq$ `shark_starve_time`, the shark dies: set `new_grid` at that location to `EMPTY`, reset corresponding age/starve info to -1.
- **Breeding Check:** If the shark did not starve and its `new_shark_age` $\geq$ `shark_breed_time`, it attempts to reproduce. Check the shark's original location (y, x) in the `new_grid`. If `new_grid[y, x]` is currently `EMPTY` (meaning the parent moved and didn't die there, or another agent didn't occupy it), place a new shark: set `new_grid[y, x] = SHARK`, `new_shark_age[y, x] = 0`, `new_shark_starve[y, x] = 0`. Crucially, also reset the parent's age at its new location: `new_shark_age[ny, nx] = 0` (and reset starve timer: `new_shark_starve[ny, nx] = 0`).
- **Vacate Original Cell:** If the shark moved and did *not* breed (or breeding failed because the original cell was occupied), ensure the original cell (y, x) is set to `EMPTY` in the `new_grid` and its associated shark age/starve info is reset to -1.

Fish Logic: Risk-Reward System; Fish seek plankton while avoiding sharks.

1. **Scan for Moves:** Identify adjacent `EMPTY` or `PLANKTON` cells in the original grid.
2. **Evaluate Moves (if possible):** For each potential move location (ny, nx), calculate the risk-reward score using Eq. 6:

$$\text{Score} = \text{Score}_{memory}(\text{Plankton}) - \text{Score}_{memory}(\text{Shark}) \tag{6}$$

3. **Movement Action:** Select the location (ny, nx) with the maximum score. If multiple locations share the max score, choose one randomly. Move there in the temporary grid: set `new_grid[ny, nx] = FISH`, `new_fish_age[ny, nx] = self.fish_age[y, x] + 1`. Mark `moved[ny, nx] = True`.
4. **Update Hunger:** Check the original state of the destination cell (ny, nx) in `self.grid`. If it was `PLANKTON`, reset hunger: `new_fish_starve[ny, nx] = 0`. If it was `EMPTY`, increment hunger: `new_fish_starve[ny, nx] = self.fish_starve[y, x] + 1`.

5. **Stay Put Action:** If no valid moves exist, the fish remains in cell (y, x). Update
its state in the temporary grid: `new_fish_age[y, x] = self.fish_age[y, x] + 1`, `new_fish_starve[y, x] = self.fish_starve[y, x] + 1`. Mark `moved[y, x] = True`.

Post-Action Checks (applied to fish at its location in the *temporary* grid):

- **Starvation Check:** If `new_fish_starve` $\geq$ `fish_starve_time`, the fish dies: set `new_grid` to `EMPTY`, reset age/starve info.
- **Breeding Check:** If not starved and `new_fish_age` $\geq$ `fish_breed_time`, reproduce similarly to sharks (place newborn in original cell if empty, reset ages).
- **Vacate Original Cell:** If the fish moved and did not breed successfully, set original cell (y, x) to `EMPTY` in `new_grid` and reset fish age/starve info.

3.3 Ecosystem Sustainability: Plankton Regrowth

In the real world, plankton regrowth isn't random. Real ecosystems have *memory.* A spot that was productive in the past is very likely to be productive in the future. For plankton, the perfect real-world example is an ocean upwelling zone. These are fixed spots where ocean currents constantly bring deep, nutrient-rich water up to the surface. Because these locations always have a fresh supply of food (nutrients), they become *historically productive* zones [2]. Our *Plankton Regrowth Rule* below is a simple way to mirror this *ecological memory.*

After all agent actions are recorded on the temporary grids, the plankton regrowth rule is applied to every cell marked as `EMPTY` in `new_grid`.

1. Calculate the plankton memory score (Eq. 5) for the empty cell (y, x).
2. Calculate the probability of regrowth using Eq. 7:

$$P(\text{regrowth}) = P_{base} + (1 - P_{base}) \times \text{Score}_{memory}(y, x, \text{Plankton}) \tag{7}$$

   where P_{base} is the global parameter `plankton_regrowth_chance`.
3. Generate a uniform random number $r \in [0, 1)$. If $r < P(\text{regrowth})$, update the temporary grid: `new_grid[y, x] = PLANKTON`.

This step ensures that the basal resource is renewable, with regrowth favored in historically productive locations. Numerical examples of this calculation are shown in Table 2.

3.4 Timestep Completion and Data Logging

Finally, the simulation state is advanced:

1. The memory history array (`self.cell_history`) is updated: states shift back one step, and the state from `self.grid` (the grid at the *start* of the current timestep) is inserted as the most recent memory ($i = 1$).

Table 2. Numerical examples illustrating how local plankton history ($\mu = 4$, $\alpha = 0.6$) determines the Memory Score (Eq. 5) and the corresponding Plankton Regrowth Probability (Eq. 7). Binary history is ordered from most recent to oldest timestep (1 = Plankton present, 0 = Empty).

Time t	Cell (y, x)	History	Memory Score $S(t)$	Regrowth Prob. $P(t)$
50	(12,45)	`1 1 1 0`	0.9018	0.9116
300	(34,88)	`1 1 0 1`	0.8346	0.8511
4500	(56,12)	`1 0 0 1`	0.5589	0.6030
10,000	(89,23)	`0 0 0 0`	0.0000	0.1000

2. The contents of all temporary arrays (`new_grid`, `new_fish_age`, etc.) are copied back to overwrite the main simulation arrays (`self.grid`, `self.fish_age`, etc.).
3. Global population counts (`np.sum(self.grid == STATE)`) for sharks, fish, and plankton are calculated and appended to time-series lists (`self.shark_counts`, etc.).

This completes one timestep, and the process repeats for `max_steps`.

3.5 Computational Complexity and Practical Feasibility

The computational cost of the proposed simulation is primarily governed by the spatial grid size and the memory depth introduced by the Cellular Automata with Memory (CAM) framework. At each timestep, the model performs a synchronous update over the entire lattice, consisting of N cells. Since all agent interactions are local and restricted to a fixed Moore neighborhood, the baseline computational cost per timestep scales linearly with the number of cells, i.e., $O(N)$, consistent with classical Wa-Tor and cellular automata models.

- **Time Complexity:** The CAM extension introduces an additional, controlled computational cost. Specifically, evaluating the memory score requires aggregating a short history of past states over a memory window of length μ. This results in an effective per-timestep complexity of $O(N \times \mu)$. Importantly, μ is treated as a small, fixed parameter in this work (Table 1), ensuring that the inclusion of memory does not change how the computational cost grows with system size. Over a full simulation run of T timesteps, the total time complexity is therefore $O(T \cdot N \cdot \mu)$.
- **Memory Usage:** Memory storage requirements also scale linearly with grid size and memory depth, as each cell maintains a compact record of its recent states. In our standard configuration (100×100 grid with $\mu = 4$), this corresponds to storing only 40,000 integer values, resulting in a negligible memory footprint relative to modern hardware capabilities. Even substantially larger grids remain well within practical memory limits.

- **Optimization and Visualization:** Visualization and animation are treated as post-processing steps and are excluded from the core simulation loop. This separation ensures that graphical rendering does not interfere with how the simulation itself evolves over time or restrict how long the simulation can be run. To support efficient execution of the core update rules, the simulation is implemented in Python using the `NumPy` library, which allows state updates and memory calculations to be carried out efficiently using optimized array operations. As a result, the model remains stable and computationally efficient for long simulation runs (up to 10,000 timesteps on a 100×100 grid or even more), while enabling richer, history-dependent dynamics than memoryless formulations.

Overall, these considerations demonstrate that the proposed CAM-enhanced Wa-Tor model is not only ecologically expressive but also computationally efficient and scalable.

4 Results, Discussion and Validation

We ran our simulation model (using the parameters in Table 1) for 10,000 timesteps to analyze its long-term stability. The resulting population dynamics are shown in Fig. 3.

4.1 Analysis of Simulation Dynamics

The graph reveals two distinct phases. The first is a brief, violent *overshoot and collapse* (from $t = 0$ to $t \approx 300$). This dynamic is a direct result of our model's *initial memoryless phase.* At $t = 0$, the agents have no past data to draw from, meaning all memory scores are zero. This forces their movement choices to be comparatively random. In this initial phase, the simulation heavily favours the sharks. A shark moving randomly in a grid dense with fish can still hunt effectively by pure chance. A fish, however, moving with this same randomness, has no *intelligent* way to avoid danger. This imbalance allows the sharks to become highly efficient hunters, causing their population to boom and the fish population to crash catastrophically.

Visual Evidence of Crash ($t < 300$): The mechanism driving this crash is visually apparent when comparing it to the initial state. At $t = 0$ (Fig. 2a), agents are placed in the grid with a uniform random distribution and zero memory. By $t \approx 20$ (Fig. 2b), the shark population booms while fish are nearly depleted. Visual inspection reveals that despite their high abundance, sharks do not form clusters but remain scattered across the grid. Similarly, the few remaining fish show no signs of forming *defensive groups* (or *schooling*). This confirms that during this transient phase, agent movements are largely uncoordinated due to insufficient memory history. The sharks are successful not because they are smart, but because the random spread of prey makes *blind* hunting highly effective, driving the rapid population spike and subsequent collapse described above.

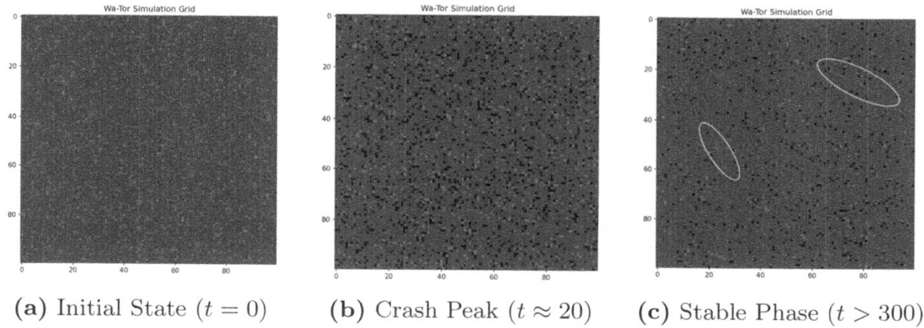

(**a**) Initial State ($t = 0$) (**b**) Crash Peak ($t \approx 20$) (**c**) Stable Phase ($t > 300$)

Fig. 2. Evolution of Spatial Dynamics. (**a**) Agents begin with uniform random distribution. (**b**) Memoryless hunting leads to shark overshoot and fish depletion without clustering. (**c**) CAM logic leads to stable diagonal fish wavefronts (highlighted by yellow ovals) and shark spatial separation. (Color Key: Red = Shark, Green = Fish, Blue = Plankton, Black = Empty). (Color figure online)

The second phase (from $t \approx 300$ to $t = 10,000$), shows the model's real success: recovery and long-term stable coexistence. By this point, the CAM logic is fully engaged. Fish can now use their *intelligence* to calculate meaningful *sharkiness* scores and actively move to safer areas. This balances the sharks' hunting intelligence, creating a dynamic equilibrium or *truce*. This new stability is the result of all three of our new mechanics working together: the fish's defensive intelligence, the shark's hunting intelligence, and the stable foundation provided by our *ecological memory* i.e. the plankton regrowth rule.

Because of this new balance, the system never again experiences a catastrophic crash like the one in the initial phase. It settles into a stationary regime with bounded oscillations (*stable oscillatory phase*).

Emergent Spatial Patterns: Visually, the grid undergoes a clear transformation during this stable oscillatory phase. As highlighted by the *yellow ovals* in Fig. 2c, fish no longer move randomly but organize into *distinct diagonal wavefronts* and localized schools. This structured behavior contrasts with the ragged, shapeless conglomerations Dewdney observed in the classic memoryless Wa-Tor model [6].

These patterns emerge naturally from the fish decision logic (Eq. 6). As individual fish seek locations with high plankton memory and low shark memory, they repeatedly select similar low-risk paths. Over time, this leads to the formation of diagonal chains and clustered groups, as multiple agents converge on the same favorable regions. The resulting wavefronts therefore reflect a collective outcome of local, memory-driven decisions rather than any imposed global structure.

In contrast, sharks consistently maintain spatial separation, typically remaining a few cells apart. This behavior follows directly from the shark decision logic described in Sect. 3.2, which directs predators toward locations with minimal shark memory when prey is unavailable, thereby reducing intra-species competition.

Across all phases, plankton remains the dominant background, visually confirming that the plankton regrowth rule (Eq. 7) sustains the large basal resource required to support higher trophic levels. This hierarchy is consistent with Volterra's theoretical expectations and reinforces the ecological realism of the model.

Further analysis of the data corresponding to Fig. 3 showed that the *dominant oscillation period* was roughly 150–180 timesteps. Fish peaks reliably lead shark peaks by 40–80 steps, and plankton minima coincided with fish maxima. This confirms the clear cause-and-effect that the prey (fish) population drives the predator (shark) population. Looking at the average population counts after the initial crash ($t > 300$), the system stabilizes with Sharks typically in the ∼180–270 range, Fish in the ∼750–1100 range, and Plankton around ∼8200–8700. These numbers clearly obey the expected hierarchy $N_3 > N_2 > N_1$ (Plankton > Fish > Sharks), which reflects the realistic energy dissipation at each trophic level, where the large biomass at the bottom is required to support the smaller biomass at the top.

Over 10,000 timesteps, no extinctions occur, indicating robust coexistence under this parameterization. This result is a clear validation of our model. It demonstrates that our agent-based simulation achieves the exact outcome Volterra predicted in his *Subcase c*: stable, three-species coexistence. This correspondence is evident from a comparison of Fig. 3 with the theoretical curves for Volterra's Case 2 subcases shown in Fig. 4. The inclusion of both fish starvation and our plankton regrowth rule effectively enforces the same logistic limitation that Volterra had to add manually to make his Case 2 realistic.

However, our model also reveals a crucial distinction from the idealized theory. Volterra's equations predicted that the system would settle into a perfectly flat stationary state. In contrast, our agent-based model settles into a persis-

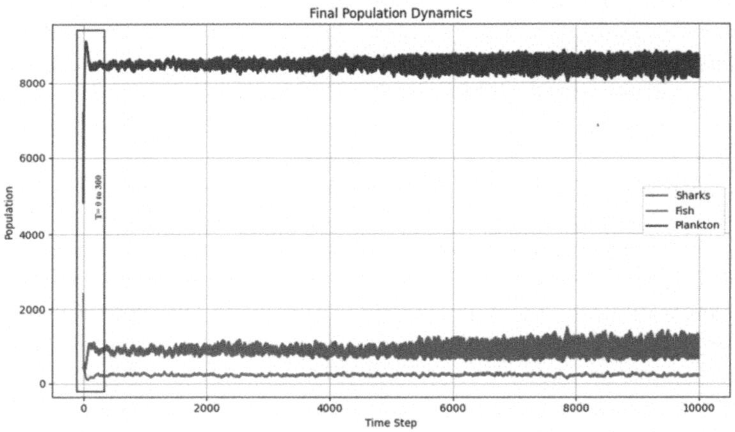

Fig. 3. Population dynamics of our Simulation Model. (Parameters: width = 100, height = 100, initial_fish = 2400, initial_sharks = 400, fish_breed = 7, shark_breed = 8, shark_starve = 10, fish_starve = 6, alpha = 0.6, memDepth = 4, plankton_regrowth = 0.1, t = 10,000 timesteps.) (Color figure online)

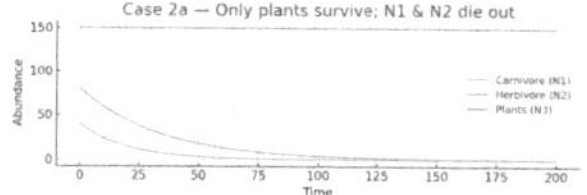

a) **Case 2a:** Only plants survive, N1 & N2 die out.

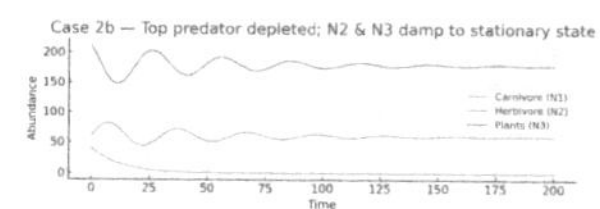

b) **Case 2b:** Top predator is depleted; N2 & N3 damp.

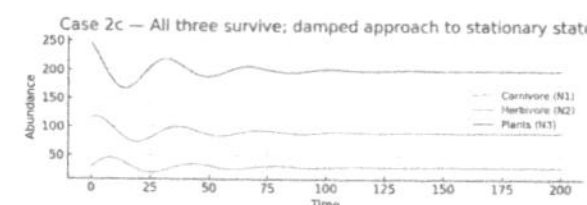

c) **Case 2c:** All three survive; damped approach.

Fig. 4. Volterra's Three-Species Cases (adapted from his 1926 analysis). Our simulation results (Fig. 3) most closely correspond to Subcase 2c. (Color figure online)

tent, cyclical fluctuation (the *fuzzy* lines in Fig. 3). This *fuzziness* is the natural outcome of having thousands of individual agents making decisions on a grid. This is the key difference from Volterra's model, which just assumes that all the species are perfectly *well-mixed* together. It suggests that our model is a robust, practical version of the classic theory. It successfully captures the same fundamental *Subcase c* coexistence, but in a way that is closer to real-world ecosystems because it actually includes patches of high and low resource abundance.

4.2 Discussion: Memory as an Ecological Stabilizer

The classic, memoryless Wa-Tor model often crashes because its agents are *dumb*. Our model is stable precisely because we introduced critical ecological constraints (like fish starvation, plankton regrowth rule) and, most importantly, agent memory.

The agent memory (CAM) itself acts as a stabilizing force. It creates a self-correcting system, smoothing out the wild population swings that would otherwise cause a collapse. We found that our balanced memory scheme ($\alpha = 0.6, \mu = 4$) was optimal; a higher α led to sudden local collapses, while a lower α caused sluggish adaptation and extinction risk. This moderate memory allows for rapid fish recovery without causing the predator population to crash, leading to the stable, bounded oscillations we observed (fish populations typically in the range 750–1100, sharks in the range 180–270, and plankton in the range 8200–8700).

This also directly addresses a puzzle posed by Dewdney in his original 1984 article [6]. He noted that the *geometry* of the populations (e.g., sharks congregating at the edge of a fish school) was the critical factor for stability, which he solved by simply slowing the shark breeding time. He then posed a question about what *intelligent sharks and fish* might do. Our work provides a formal answer: by giving the agents *intelligence* via CAM, we've created a new, more realistic path to stability, one that is driven by adaptive behavior, not just by arbitrary breeding rules.

4.3 Validation Strategy

A direct, quantitative validation of our model is challenging. Finding reliable, long-term, high-resolution population data for a complete three-species food chain is extremely rare. Therefore, we must rely on qualitative validation to test if our model behaves like a real ecosystem.

Our proposed test is to simulate a *trophic cascade.* A future experiment would involve running the stable simulation and then manually removing all apex predators (the sharks). If our model is realistic, we would hypothesize that the *released* fish population (N_2) would boom, which in turn would cause the plankton population (N_3) to crash from over-eating. Successfully reproducing this well-documented phenomenon would provide strong qualitative evidence that our model's logic is sound.

5 Conclusion and Future Work

In this work, we designed and built a new three-species predator-prey model by extending the classic Wa-Tor framework. Our key innovation was giving the agents *intelligence* using Cellular Automata with Memory (CAM).

We showed that our agent model can successfully reproduce the stable coexistence predicted by Volterra's theory, solving the instability problems that simpler, memoryless models often face.

Our work is a useful contribution to fields like computational ecology, complex systems, and artificial life, as the CAM framework we used provides a flexible platform to continue exploring how individual behavior, spatial patterns, and overall ecosystem stability all connect and influence one another.

While our model is a significant step forward, it still has limitations, which point to clear directions for future research.

- **Behavioral Simplicity:** Our agents' *intelligence* is limited: they react to memory, but they don't learn or evolve. A future model could incorporate genetic algorithms (GAs), allowing agents to evolve their own parameters (like memory depth or breed time) to find the optimal strategy for survival.
- **Ecological Simplicity:** Our model is still a linear food chain, but real ecosystems are complex webs. A clear next step is to introduce a fourth species (e.g., a *competitor* fish that also eats plankton) to see how our CAM-driven agents adapt to this new complexity.
- **Parameter Sensitivity:** We found one combination of parameters that leads to stable coexistence, but there are millions of possible sets. This raises the same question Dewdney had [6]: is there a general rule to predict which parameters lead to stability? A major part of our future work will be to perform *systematic parameter sweeps*, perhaps using GAs, to map out the different stability regimes. This would allow us to discover what other dynamics are possible and see which parameters correspond to Volterra's other subcases (a and b).

References

1. Baghel, R.S.: Memory and delay-driven dynamics in a tritrophic food chain model with allee effect and nonlinear predation. Nonlinear Science (2025). https://doi.org/10.1016/j.nls.2025.100069

2. Balaguer, L., Escudero, A., Martin-Duque, J.F., Mola, I., Aronson, J.: The historical reference in restoration ecology: Re-defining a cornerstone concept. Biol. Cons. **176**, 12–20 (2014)
3. Boccara, N., Roblin, O., Roger, M.: Automata network predator-prey model with pursuit and evasion. Phys. Rev. E **50**(6), 4531–4541 (1994)
4. Cattaneo, G., Dennunzio, A., Farina, F.: A full cellular automaton to simulate predator prey systems. In: Lecture Notes in Computer Science. Springer (2006)
5. Debnath, S., Majumdar, P., Sarkar, S., Ghosh, U.: Memory effect on prey-predator dynamics: exploring the role of fear effect, additional food and anti-predator behaviour of prey. J. Comput. Sci. **66** (2023). https://doi.org/10.1016/j.jocs.2022.101929
6. Dewdney, A.K.: Computer recreations: sharks and fish wage an ecological war on a toroidal planet. Sci. Am. **251**(6), 14–22 (1984)
7. Gestoso, I., Arenas, F., Olabarria, C.: Feeding behaviour of an intertidal snail: Does past environmental stress affect predator choices and prey vulnerability? J. Sea Res. **97**, 66–74 (2015)
8. Human, D.J.: Modelling predator prey interactions with cellular automata. Research assignment, Department of Logistics, Stellenbosch University (2014)
9. Jafelice, R.L.C., Silva, R.: Modelling population dynamics using cellular automata. In: VII WSEAS International Conference on Mathematical Biology and Ecology. WSEAS (2011)
10. Jin, D., Yang, R.: Hopf bifurcation in a predator-prey model with memory effect and intra-species competition in predator. J. Appl. Anal. Comput. **13**(3), 1321–1335 (2023). https://doi.org/10.11948/20220127
11. Martinez, G.J.: A note on memory cellular automata. Complex Syst. **31**(1) (2022)
12. Mella, V.S.A., Cooper, C.E., Davies, S.J.J.F.: Effects of historically familiar and novel predator odors on the physiology of an introduced prey. Curr. Zool. **62**(1), 53–59 (2016). https://doi.org/10.1093/cz/zov005
13. Sanz, R.: Cellular automata with memory: an extension of ca for complex modelling. In: Das, S., Roy, S., Bhattacharjee, K. (eds.) The Mathematical Artist. Springer (2022)
14. Smee, D.L.: Environmental context influences the outcomes of predator-prey interactions and degree of top-down control. Nat. Educ. Knowl. **3**(10), 58 (2010)
15. Volterra, V.: Variations and fluctuations of the number of individuals in animal species living together. Journal du Conseil International pour l'Exploration de la Mer **3**, 1–51 (1926)

CA-CNN Deep Learning Model for Image Classification

M J Elizabeth[1(✉)] and Raju Hazari[2]

[1] Department of Computer Science, Mary Matha Arts and Science College, Mananthavady, India
elizabethmj@marymathacollege.ac.in

[2] Department of Computer Science and Engineering, National Institute of Technology, Calicut, Kerala, India
hazariraju0201@gmail.com

Abstract. Image classification is a fundamental task in various computer vision applications such as medical diagnosis, remote sensing, and object recognition. Traditional convolutional neural networks (CNNs) have shown remarkable performance in feature extraction and pattern recognition. However, they often face challenges in capturing local interactions and spatial dependencies efficiently. To address this limitation, we propose a Cellular Automata-based Convolutional Neural Network (CA-CNN) model that integrates the dynamic evolution rules of Cellular Automata (CA) into CNN layers. The CA component enhances spatial feature learning and local dependency modelling while maintaining computational efficiency. Experimental analysis on benchmark image datasets demonstrates that the proposed CA-CNN outperforms conventional CNN models in terms of various performance analysis metrics.

Keywords: Cellular Automata (CA) · Image Classification · Deep Learning · Machine Learning · Convolutional Neural Network

1 Introduction

Among various architectures, the CNN model is commonly seen for image classification by enabling automatic feature extraction and pattern recognition. CNNs learn hierarchical representations, transforming raw pixels into complex abstract features through a convolution and pooling process and through non-linear activation. However, traditional CNNs process data in a static manner. In contrast, CA offers a powerful framework for modelling local interactions and emergent patterns. It consists of a two-dimensional grid of cells that evolve over discrete time steps based on local rules. This iterative evolution captures spatial dependencies and dynamic behaviours within the data. It helps in providing deeper insight into the internal behaviour of the network.

Traditional CNNs have achieved good results in extracting spatial features from images; however, they cannot often capture localized interactions and evolving spatial dynamics within the data. To address this limitation, a CA-CNN deep

H. Raju et al. (Eds.): ASCAT 2026, CCIS 2801, pp. 224–237, 2026.
https://doi.org/10.1007/978-3-032-18612-6_17

learning model is proposed to analyze image data. CA operates through iterative updates of cell states using a kernel as a local rule. It allows the system to simulate feature evolution across layers to capture local correlations. In the CNN model, the CA layers are embedded after convolutional feature extraction. The CA layer refines feature maps across multiple iterations before classification. That is, the feature maps can be iteratively updated, allowing the network to simulate spatial dynamics to enhance the learning process. Visualization of the intermediate CA states shows how image representations evolve over successive CA steps, revealing the internal dynamics of feature transformation, which offers interpretable deep learning and bridges the gap between conventional convolutional processing and dynamic spatial modelling.

The proposed model is evaluated using the CIFAR-10 and MNIST datasets, which are widely adopted as standard benchmark datasets. The authors in [11] utilized these datasets to investigate federated learning models for analyzing hyperparameter interdependencies. Similarly, in [2], the authors examined the performance of a CNN model on these datasets and achieved an accuracy of 88%. The vulnerability of CNN models was further explored in [15], where ResNet architectures such as ResNet18, ResNet34, and ResNet50 were employed. The highest accuracy of 82.88% was reported using ResNet34. Moreover, several other applications, including the ImageSplit technique [14] and 3D modelling, have also employed these datasets. In the present study, we focus on a classification task using these benchmark datasets. The key contributions of this work are given below:

1. The integration of a learnable two-dimensional CA layer within a CNN framework.
2. The ability of the CA layer to model local neighbourhood interactions through iterative evolution.
3. The interpretability gained by visualizing intermediate CA states.

The paper is organized as follows. Working of the CA layer in traditional CNN is described in Sect. 2. CA-CNN architecture and implementation details are explained in Sect. 3. The results and error rate details are explained in Sect. 4. Finally, Sect. 5 concludes the proposed CA-CNN for the image classification task.

2 Cellular Automata (CA) Layer in CNN

The fundamental studies employing CA for text classification tasks and the formulation of update rules for elementary cellular automata are discussed in [3–7]. For image classification, we employ two-dimensional cellular automata (2D CA) integrated with a CNN to develop a hybrid deep learning model.

An image is represented in three dimensions: width (W), height (H), and channels (C). Thus, the input feature map at time t can be expressed as $X_t \in \mathbb{R}^{H \times W \times C}$. The CA update rule at time $t+1$ is defined as:

$$X_{t+1} = X_t + \alpha \cdot \tanh(K * X_t) \tag{1}$$

Here,

- K is a learnable 3×3 convolution kernel representing neighbourhood interaction weights.
- $*$ denotes the 2D convolution operator.
- $\tanh(\cdot)$ introduces non-linearity and bounds the update magnitude.
- α is a learnable scalar parameter controlling the update rate.

Equation 1 performs a residual update on the feature map for a specified number of iterations, T, thereby simulating the iterative evolution of CA:

$$X_T = X_0 + \sum_{t=0}^{T-1} \alpha \cdot \tanh(K * X_t) \quad (2)$$

This process can be interpreted as a form of nonlinear diffusion, where the pixel states evolve based on the influence of their local neighbourhoods. The nonlinear function tanh controls how much each pixel's value changes based on its state and its neighbours.

For example, consider a 5×5 grayscale patch (part of an image):

$$X_0 = \begin{bmatrix} 0 & 0 & 0 & 0 & 0 \\ 0 & 0.2 & 0.5 & 0.4 & 0 \\ 0 & 0.6 & 0.9 & 0.7 & 0 \\ 0 & 0.3 & 0.6 & 0.2 & 0 \\ 0 & 0 & 0 & 0 & 0 \end{bmatrix}$$

Let the learned neighbourhood kernel be:

$$K_{3x3} = \begin{bmatrix} 0.05 & 0.10 & 0.05 \\ 0.10 & -0.60 & 0.10 \\ 0.05 & 0.10 & 0.05 \end{bmatrix}$$

This K is learned automatically during training through backpropagation. The CA rule is implemented as a learnable convolutional kernel that updates the cell state. The K acts as a learned local rule in which the centre pixel gets a negative weight (—0.60), discouraging over-brightness. Here, all neighbours get a positive weight, which encourages smooth spreading.

The update coefficient $\alpha = 0.8$ controls how much of the new information is added to the existing state. If α is too small (e.g., 0.001), the system changes very slowly. The automaton converges slowly and may not learn meaningful transformations. If α is too large (e.g., >1), updates become unstable. The system overshoots and oscillates. A moderate $\alpha = 0.8$ provides smooth but significant evolution. It maintains numerical stability and allows visible pattern growth and strong gradient flow. It acts like a diffusion coefficient. It balances the stability and responsiveness of the system.

For the central pixel at position [2,2], the update is computed as:

$$Neighbourhood = \begin{bmatrix} 0.2 & 0.5 & 0.4 \\ 0.6 & 0.9 & 0.7 \\ 0.3 & 0.6 & 0.2 \end{bmatrix}$$

$$\begin{aligned}\text{Update pre-activation} &= (0.05 \times 0.2) + (0.1 \times 0.5) + (0.05 \times 0.4) + (0.1 \times 0.6) + (-0.6 \times 0.9) \\ &\quad + (0.1 \times 0.7) + (0.05 \times 0.3) + (0.1 \times 0.6) + (0.05 \times 0.2) \\ &= 0.01 + 0.05 + 0.02 + 0.06 - 0.54 + 0.07 + 0.015 + 0.06 + 0.01 \\ &= -0.245\end{aligned}$$

$$\Delta X_{2,2} = \tanh(K * X_0)_{2,2} = \tanh(-0.245) = -0.240$$

Hence, the new state is:

$$X_{1,2,2} = X_{0,2,2} + \alpha \cdot \Delta X_{2,2} = 0.9 + 0.8(-0.240) = 0.9 - 0.192 = 0.708 \approx 0.71$$

Repeating this operation for all pixels yields the next state as:

$$X_1 = \begin{bmatrix} 0.00 & 0.06 & 0.14 & 0.10 & 0.00 \\ 0.03 & 0.26 & 0.47 & 0.33 & 0.04 \\ 0.03 & 0.51 & \mathbf{0.71} & 0.56 & 0.08 \\ 0.01 & 0.28 & 0.49 & 0.23 & 0.02 \\ 0.00 & 0.03 & 0.07 & 0.05 & 0.00 \end{bmatrix}$$

After four iterations, the state stabilizes:

$$X_4 = \begin{bmatrix} 0.01 & 0.09 & 0.17 & 0.12 & 0.01 \\ 0.04 & 0.29 & 0.48 & 0.34 & 0.05 \\ 0.03 & 0.52 & 0.69 & 0.54 & 0.09 \\ 0.02 & 0.29 & 0.47 & 0.22 & 0.03 \\ 0.00 & 0.04 & 0.07 & 0.05 & 0.00 \end{bmatrix}$$

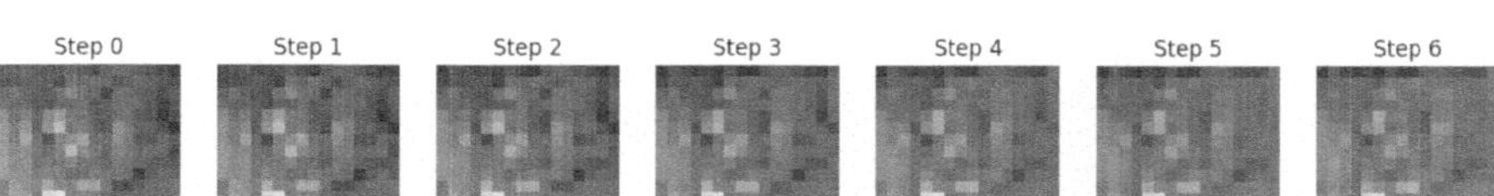

Fig. 1. CA evolution with step size=6 using 16x16 image.

The CA rule behaves like a nonlinear reaction–diffusion process. That is, bright pixels diffuse to nearby regions (smoothing effect) and sharp discontinuities are reduced while structural edges remain intact. The feature map converges to a stable equilibrium state. The evolution of channel 0 is shown in Fig. 1.

3 Proposed CA-CNN Model Implementation

In this proposed model, CA behaviour is embedded through a learnable convolutional kernel that governs the local interaction among neighbouring cells. The CA layer integrates local neighbourhood-based state updates into a Convolutional Neural Network (CNN). Unlike standard convolution layers that extract spatial features in one pass, the CA layer evolves the feature map over several iterative steps based on learned local rules. The CA layer functions as an intermediate operation within the CNN pipeline:

$$F' = \mathrm{CA}(F)$$

where F denotes the feature map obtained from a CNN layer. This integration enables spatial correlations to propagate adaptively, thereby enhancing gradient stability and improving the interpretability of learned representations. That is, in our CA–CNN hybrid model, each iteration allows pixels to interact locally (diffusion). That is the feature values in the image evolve through local interactions. The nonlinear activation (tanh) controls it and ensures complex pattern formation and stability.

The proposed CA-CNN model has four different types of layers shown in Fig. 2. The image data is used as input, and its class labels are given as output. The convolution layer is used to extract local spatial features from the input images. The pooling layer reduces its dimensionality by retaining essential information. CA evolution layer performs iterative CA–based feature refinement and interaction, unlike standard convolution layers that perform only one-pass filtering. Finally, a fully connected layer integrates the learned features for final classification. The iterative evolution of a feature map is given in Eq. 2, used to enhance feature interaction and spatial learning.

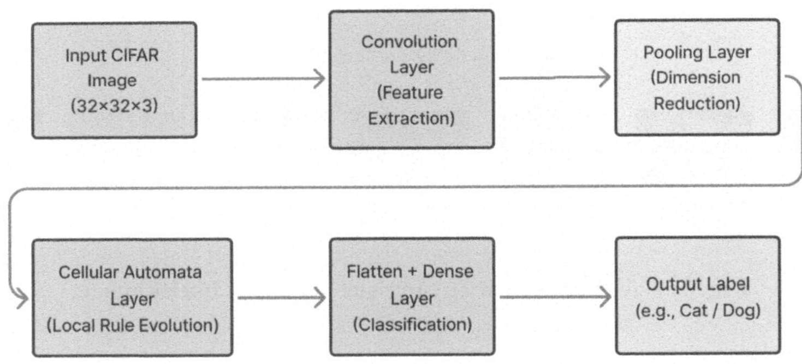

Fig. 2. Block diagram of proposed CA-CNN Model.

The implemented CA-CNN hybrid deep learning model is shown in Fig. 3. The model accepts RGB images of size 32 × 32 × 3 as input. The first and second convolutional layers are used to extract low-level spatial features such as edges and textures. Each convolution operation is followed by Batch Normalization, which accelerates convergence and stabilizes training. The pooling layer reduces the spatial dimensions from 32 × 32 to 16 × 16. The core layer of the proposed

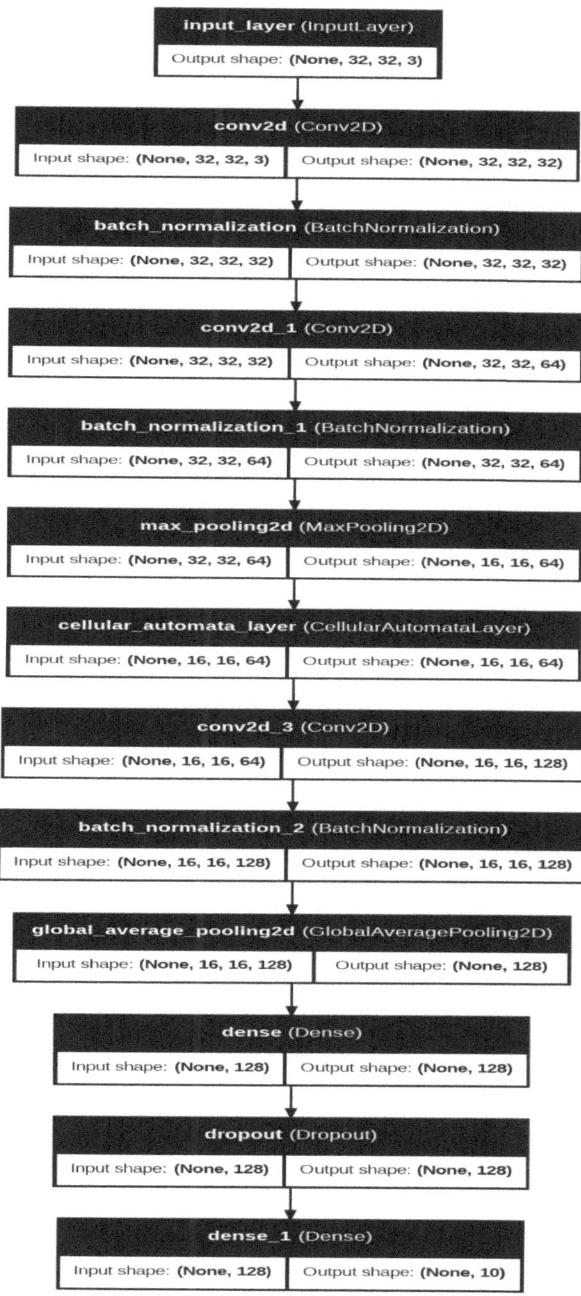

Fig. 3. Proposed CA-CNN Model.

architecture is the CA layer, which performs iterative feature refinement based on local neighbourhood interactions. The CA layer is positioned after two convolutional layers, followed by a pooling operation to ensure that it operates on feature maps that are both semantically meaningful and spatially structured. Early convolutional layers primarily capture low-level features such as edges and textures; inserting the CA layer at this stage may disrupt these primitive representations through excessive smoothing. Conversely, placing the CA layer after deeper layers significantly reduces spatial resolution, limiting the effectiveness of neighbourhood-based interactions. So the selected placement provides a balance between preserving discriminative spatial information and enabling effective CA-based feature evolution. The third convolutional layer (Conv2D_3) increases the number of filters to 128 and captures high-level semantic features. The global average pooling layer converts the 3D feature maps (16 × 16 × 128) into a 1D vector by averaging each feature map, thereby preventing overfitting. The dense layer with 128 neurons learns nonlinear combinations of the extracted features. A final Dense layer with 10 neurons and softmax activation produces the class probabilities corresponding to the 10 categories in the CIFAR-10 or MNIST datasets. So the CA–CNN model combines local interaction dynamics from CA with the hierarchical feature extraction power of CNNs.

4 Prediction Process

In this section, we explored the experiment results, performance analysis, and comparative study with other state-of-the-art models.

4.1 Datasets Description

- *CIFAR-10 dataset* [10]: The CIFAR-10 dataset contains 60,000 colour images of size 32 × 32 pixels. They are categorized into 10 classes with 6,000 images per class. It is divided into 50,000 training images and 10,000 test images. The dataset is organized into five training sets and one test set, each containing 10,000 images. The test set includes exactly 1,000 randomly selected images from each class. The training sets hold the remaining images in random order. Altogether, the training batches comprise exactly 5,000 images per class. The sample images of the dataset are shown in Fig. 4.
- *MNIST dataset* [12]: The MNIST dataset consists of 70,000 grayscale images of handwritten digits ranging from 0 to 9. The training set is of 60,000 samples, and a test set of 10,000 samples. Each image is of size 28 × 28 pixels. The dataset is used as a standard benchmark for evaluating image classification and pattern recognition algorithms. The first 16 images in the MNIST training set are displayed in Fig. 5.

4.2 Results and Performance Analysis

The performance of the proposed approach is evaluated using accuracy, precision, recall, and F1-score metrics. The analysis is conducted on the CIFAR-10 and

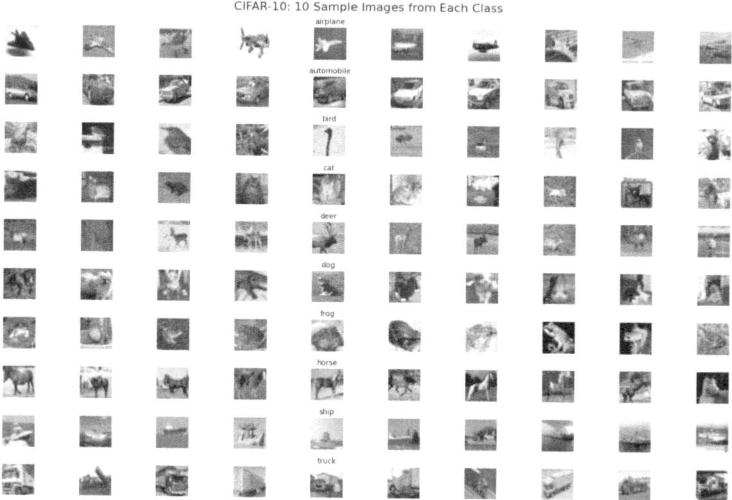

Fig. 4. CIFAR-10 dataset.

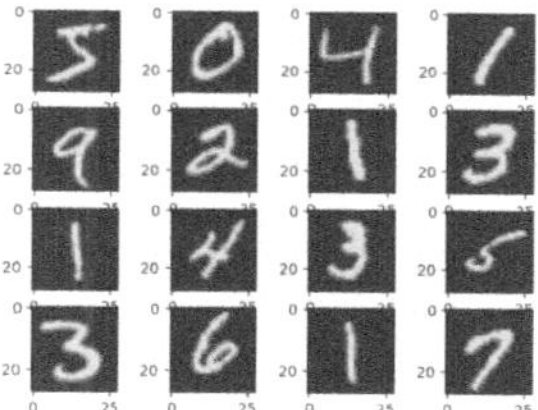

Fig. 5. MNIST dataset.

MNIST datasets, with 80% of the data allocated for training and 20% for testing. The results corresponding to the CIFAR-10 dataset are presented below.

The classification report, presented in Table 1, summarizes the performance of the proposed model on the test data. The normalized confusion matrix in Fig. 6 shows the summary of prediction results of test data. At the top, 0.90 means 90% of Class 0 images were correctly classified as Class 0. The darker the colour, the higher the proportion of correctly predicted samples. The per-class accuracy and error rate of test data are shown in Fig. 7.

The training and Validation accuracy and loss for the CIFAR-10 dataset are shown in Fig. 8. The model learns very well from training data. The validation accuracy fluctuates around 0.80–0.81 throughout training, without significant improvement. The proposed model performed well during the training and validation process over 100 epochs. The accuracy on the training data steadily increases and reaches 96.95%, while the validation accuracy stabilizes around

Table 1. Classification Report on CIFAR-10 Dataset

Class	Precision	Recall	F1-Score	Support
0	0.75	0.90	0.82	1000
1	0.91	0.89	0.90	1000
2	0.69	0.78	0.73	1000
3	0.76	0.58	0.66	1000
4	0.77	0.81	0.79	1000
5	0.76	0.74	0.75	1000
6	0.82	0.87	0.84	1000
7	0.78	0.89	0.83	1000
8	0.95	0.81	0.88	1000
9	0.93	0.79	0.86	1000
Accuracy			**0.81**	**10000**
Macro Avg	0.81	0.81	0.80	10000
Weighted Avg	0.81	0.81	0.80	10000

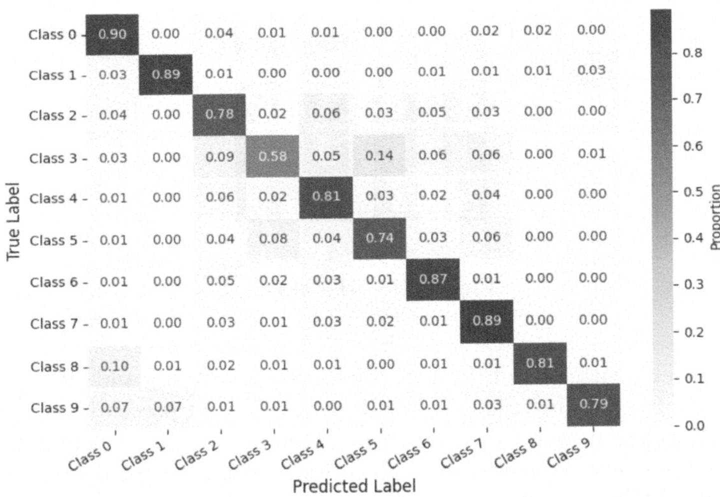

Fig. 6. Confusion Matrix of CIFAR-10 dataset.

80.69%. This noticeable gap between training and validation accuracy indicates overfitting.

At last, we tested the model on grayscale images of the MNIST dataset. The evolution steps are shown in Fig. 9.

The experiment results show, the CA–CNN model achieved a test accuracy of 98.26% compared to 97.13% for the baseline CNN, confirming the benefit of

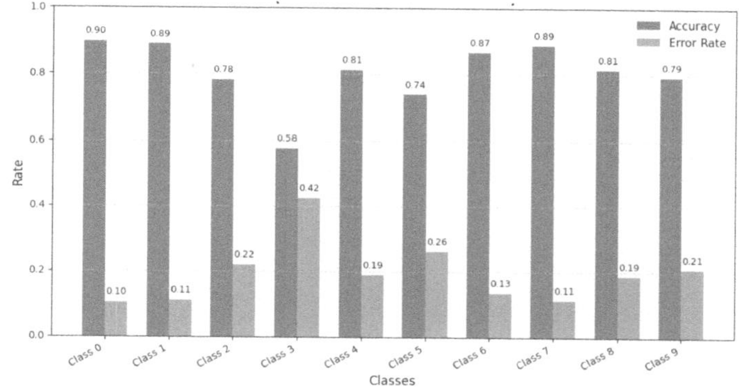

Fig. 7. Accuracy and error rate of CIFAR-10.

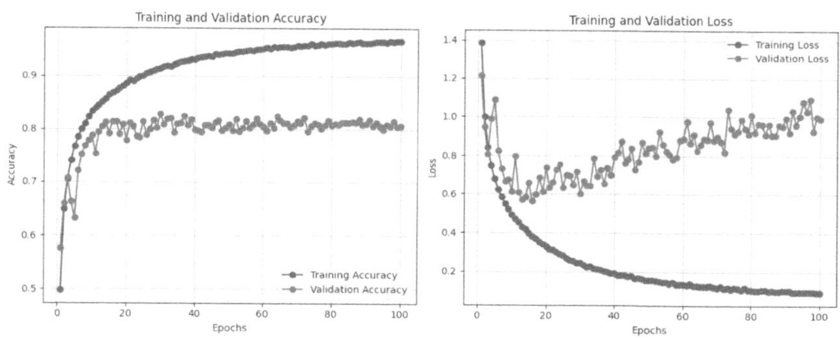

Fig. 8. Training and Validation accuracy and loss of CIFAR-10 dataset.

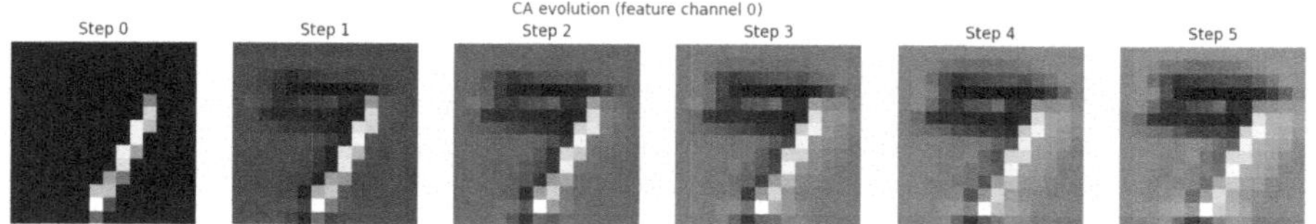

Fig. 9. MNIST Image Evolution Process.

iterative neighbourhood evolution. The accuracy and error rate are shown in Fig. 10.

4.3 Comparative Analysis

In [13], the authors implemented a CNN model and analyzed its training and validation accuracy and losses. The highest training and validation accuracy obtained are 94.64 and 87.34, respectively, using 6 convolution layers and 45

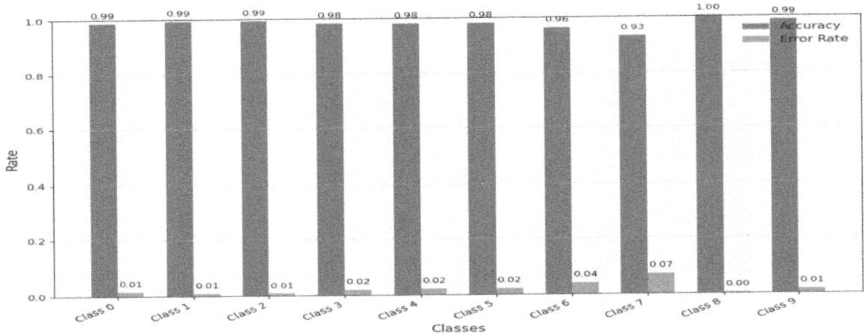

Fig. 10. Per-Class Accuracy and Error Rate using MNIST dataset.

epochs with a 3x3 kernel. The proposed model's training and validation accuracy are 99.44% and 97.67%, respectively, on the MNIST dataset. The comparative analysis with other state-of-the-art models using the CIFAR-10 and MNIST datasets is given in the Table 2. The CA-CNN model achieved 80.69% accuracy on test data, a bit lower than standard CNNs for CIFAR-10. This may happen due to the overfitting problem when we use colored images. The model focuses more on neighbourhood dynamics than deep hierarchical features. For MNIST, the CA-CNN achieved an accuracy of 98.26%, which is comparable to advanced CNN variants, demonstrating its efficiency in recognizing structured handwritten patterns. On CIFAR-10, although the accuracy is slightly lower (80.69%), the model maintains consistent precision and recall values, indicating stable classification behaviour across categories.

Table 2. Comparative Study of results obtained with state-of-art models on testing data of CIFAR-10 and MNIST (Performance comparison of Standard CNN and CA-CNN)

Dataset	Methods with Reference	Accuracy	Precision	Recall	F1-Score
CIFAR-10	Huang (2024) – Efficient CNN Model for CIFAR-10 [8]	0.9114	-	-	-
	Shin & Kim (2024) – Comparison of CNN Models (ResNet variant) [15]	0.8029	-	-	-
	Jordan (2024) – 94% on CIFAR-10 in 3.29 s on a Single GPU [9]	0.9497	-	-	-
	CA-CNN (Proposed)	**0.8069**	**0.8123**	**0.8069**	**0.8051**
MNIST	Ullah et al. (2025) [17] – Handwritten Digit Recognition: An Ensemble-Based Approach (CNN-SVM)	0.9930	0.9930	0.9930	0.9920
	Toyobo et al. (2025) – Performance Comparison of CNN Models [16]	0.9931	0.9900	0.9900	0.9900
	Ahlawat et al. (2023) – Improved Handwritten Digit Recognition Using CNN [1]	0.9987	-	-	-
	CA-CNN (Proposed)	**0.9826**	**0.9801**	**0.9792**	**0.9793**

4.4 Discussion

Effect of CA Iteration Depth: In the current implementation, the number of CA evolution steps is fixed to maintain numerical stability and avoid excessive smoothing of feature maps. Empirical observations indicate that increasing the number of CA iterations intensifies neighbourhood diffusion, which may lead to over-smoothing and loss of discriminative details, particularly in colour images. Conversely, using fewer iterations weakens the influence of the CA layer and reduces its contribution to spatial feature refinement. This behaviour highlights a trade-off between stability and representational richness, suggesting that the CA iteration depth plays a critical role in controlling the impact of neighbourhood evolution on classification performance.

Computational Complexity: The proposed CA layer introduces minimal additional computational overhead, as it employs a single learnable 3×3 convolution kernel with shared parameters across multiple evolution steps. Unlike deeper architectures that increase complexity by stacking numerous convolutional layers, the CA layer reuses the same parameters iteratively, resulting in a modest increase in computation without a proportional growth in model parameters. From an efficiency perspective, the CA-CNN model remains lightweight compared to deep residual networks such as ResNet, which rely on a large number of layers and parameters. The iterative CA updates operate on intermediate feature maps with reduced spatial resolution, further limiting computational cost. Consequently, the proposed model offers a favorable balance between expressive power and training efficiency.

5 Conclusion and Future Work

This work presented a Cellular Automata–based Convolutional Neural Network (CA-CNN) for image classification, integrating iterative neighborhood evolution into the conventional CNN pipeline. The CA layer enhances local spatial interactions and provides an interpretable mechanism for feature evolution, particularly benefiting structured image datasets such as MNIST. Despite these advantages, the study also reveals certain limitations. The CA-based diffusion process may smooth fine-grained details and sharp edges, which can affect performance on complex color images and edge-sensitive applications. In addition, the number of CA evolution steps and the placement of the CA layer introduce trade-offs between stability, smoothing, and representational richness. These factors must be carefully tuned to avoid over-smoothing and loss of discriminative features.

As future work, the proposed CA-CNN framework will be evaluated on additional application domains, including medical imaging and other edge-sensitive tasks, using adaptive or edge-aware CA update rules. A systematic analysis of the impact of CA iteration depth and dynamic control strategies will also be explored. Furthermore, hardware-level performance analysis, including training time, memory usage, and computational overhead, will be conducted to assess the scalability and efficiency of the model. Finally, future research will focus on developing a formal mathematical analysis of CA convergence and stability within deep learning frameworks, aiming to provide theoretical guarantees for the iterative evolution process. These directions will further strengthen the applicability, interpretability, and robustness of CA-based deep learning models.

References

1. Ahlawat, S., Choudhary, A., Nayyar, A., Singh, S., Yoon, B.: Improved handwritten digit recognition using convolutional neural networks (CNN). Sensors (Basel, Switzerland) **20**, 3344 (2020). https://doi.org/10.3390/S20123344, https://pmc.ncbi.nlm.nih.gov/articles/PMC7349603/
2. Divya, S., Adepu, B., Kamakshi, P.: Image enhancement and classification of CIFAR-10 using convolutional neural networks. In: Proceedings - 4th International Conference on Smart Systems and Inventive Technology, ICSSIT 2022. Institute of Electrical and Electronics Engineers Inc. (2022)
3. Elizabeth, M.J., Chaudhuri, P.P., Hazari, R.: Cellular automata-based sentiment analysis for bipolar classification of reviews. Complex Systems (2024). https://doi.org/10.25088/ComplexSystems.34.1.29
4. Elizabeth, M.J., Hazari, R.: Cellular automata-based sentiment analysis using deep learning methods. Complex Systems (2024). https://doi.org/10.25088/ComplexSystems.33.3.353.
5. Elizabeth, M.J., Kommineni, A.K., Hazari, R.: Sentiment analysis for code-mixed data using cellular automata with deep learning models. Lecture Notes Comput. Sci. **14978 LNCS**, 163–176 (2024). https://doi.org/10.1007/978-3-031-71552-5_14
6. Elizabeth, M.J., Panda, A.K., Chaudhuri, P.P., Hazari, R.: Cellular automata-based sentiment analysis. In: Das, S., Martinez, G.J. (eds.) ASCAT 2023. AISC, vol. 1443, pp. 53–64. Springer, Singapore (2023). https://doi.org/10.1007/978-981-99-0688-8_5
7. Elizabeth, M.J., Parsotambhai, S.M., Hazari, R.: Cellular automata enhanced machine learning model for toxic text classification. In: International Conference on Cellular Automata for Research and Industry, pp. 346–355. Springer (2022). https://doi.org/10.1007/978-3-031-14926-9_31.
8. Huang, Y.: Using neural networks to build an efficient classification model for classifying images in the CIFAR-10 dataset (2025)
9. Jordan, K.: 94% on cifar-10 in 3.29 seconds on a single GPU. ArXiv abs/2404.00498 (2024). https://api.semanticscholar.org/CorpusID:268819769
10. Krizhevsky, A.: Learning multiple layers of features from tiny images. University of Toronto, Technical report (2009)
11. Kundroo, M., Kim, T.: Demystifying impact of key hyper-parameters in federated learning: a case study on CIFAR-10 and FashionMNIST. IEEE Access **12**, 120570–120583 (2024). https://doi.org/10.1109/ACCESS.2024.3450894
12. LeCun, Y., Cortes, C., Burges, C.J.C.: The MNIST database of handwritten digits (1998). http://yann.lecun.com/exdb/mnist/
13. Pant, Y., Shah, G., Ojha, R., Thapa, R., Bhatta, B.: Comparison of CNN architecture of image classification using CIFAR10 datasets. Int. J. Eng. Technol. **1**(1), 37–52 (2023)
14. Radhakrishnan, A., Gopi, A., Reghuvaran, C., James, A.: Variability-aware Memristive crossbars with ImageSplit neural architecture. IEEE Trans. Nanotechnol. **23**, 274–280 (2024). https://doi.org/10.1109/TNANO.2024.3375125

15. Shin, S., Shin, D., Kim, J.: Implementation and comparison of CNN models on CIFAR-10 dataset, pp. 1–4. Institute of Electrical and Electronics Engineers (IEEE) (2025)
16. Toyobo, O.J., Olabiyisi, S.O., Ismaila, W.O., Oyedele, A.O.: Performance comparison of convolutional neural network and long short-term memory for the classification of handwritten digits. International Journal of Research and Innovation in Applied Science **X**, 565–572 (2025). https://doi.org/10.51584/IJRIAS.2025.10060041
17. Ullah, S.S., Gang, L., Riaz, M., Ashfaq, A., Khan, S., Khan, S.: Handwritten digit recognition: An ensemble-based approach for superior performance. arXiv preprint arXiv:2503.06104 (2025)

A Study of Non-uniform Cellular Automata with Fully Asynchronous Updating : First Results on the Non-convergent Cases

Sukanya Mukherjee(✉), Mayukh Chatterjee, and Ahana Majumdar

Department of Computer Science and Engineering, Institute of Engineering and Management, Kolkata, University of Engineering and Management, Kolkata, India
sukanyaiem@gmail.com

Abstract. This work investigates finite non-uniform asynchronous cellular automata (NUACAs) under *periodic* boundary condition, following *fully* asynchronous update scheme. Based on the nature of *recurrent* configurations, these NUACAs are categorized into *convergent*, *strongly non-convergent* and *weakly non-convergent*. Furthermore, the convergent ACA rules are *ranked* from 1 to 8 based on the proportion of *active* RMTs. It is shown that NUACAs are strongly non-convergent even if they are composed by solely convergent rules. Moreover, this work identifies the *minimal* proportion of *ranks* 1 and 2 rules necessary to generate strongly non-convergent NUACAs.

1 Introduction

Cellular automata (CAs) are used to model natural phenomena by applying simple *rules* to discrete units called *cells*. These cells are typically arranged either as a linear array or as a grid. The former is known as a 1-dimensional cellular automaton. When a cellular automaton (CA) has a finite number of cells, it is called a *finite* CA. Each cell holds a value at every time step, referred to as its *state*. If the states take values from $\{0, 1\}$, the CA is called a *binary* CA. The collective states of all cells constitute a *configuration*.

A CA evolves over time and during this evolution, each configuration in a finite CA leads to itself or other configuration. Configuration transitions are governed by the *rules* applied to the cells and the *neighborhood* size of the CA. For instance, in a binary CA with 3-neighborhood (here, neighborhood size is 3), the next state of a cell depends on the applied rule and the current states of its neighboring cells. Traditionally, all cells update their states simultaneously at each time step, referred to as *synchronous* update. This makes the system deterministic, ensuring that every configuration has exactly one *next* configuration.

If the update mechanism is relaxed such that cells are not required to update simultaneously, the CA becomes *asynchronous*. Here, cell updates are

This work is carried out as a project in the Indian Summer School on Cellular Automata 2025.

H. Raju et al. (Eds.): ASCAT 2026, CCIS 2801, pp. 238–252, 2026.
https://doi.org/10.1007/978-3-032-18612-6_18

independent, leading to a class known as *asynchronous* CA (ACA). Studies [4,13] show that asynchronous CAs, due to their independent dynamics, may model natural systems more effectively. Unlike synchronous CAs, ACAs are non-deterministic, as a configuration may evolve into multiple possible next configurations.

Asynchronous cellular automata have been extensively studied from a theoretical perspective. Concepts such as *convergent* ACAs [8,16,18], *recurrent* ACAs [7,16,17], reversibility in ACAs [15] and *clouds* in ACAs [14] have been explored. Their relevance to technological applications has also been discussed [16]. Existing studies, however, focus exclusively on *uniform* ACAs, where all cells follow the same rule, despite the inherent *non-uniformity* in their update mechanisms. The behavior of CAs that incorporate non-uniform rule selection under asynchronous update remains unexplored.

Asynchronous cellular automata have been extensively studied theoretically. Concepts such as *convergent* ACAs [8,16,18], *recurrent* ACAs [7,16,17], reversibility [15], and *clouds* [14] have been explored, along with their technological relevance [16]. Existing studies, however, focus on *uniform* ACAs, where all cells follow the same rule, despite inherent *non-uniformity* in update mechanisms. The behavior of CAs with non-uniform rule selection under asynchronous update remains unexplored. In contrast, non-uniform CAs under synchronous update have been extensively studied and are known to exhibit notable theoretical properties, such as convergence, reversibility and number conservation [3,12], which can be mapped to natural systems. Moreover, they have contributed significantly to technological advancements [3,10–12]. These observations motivate the study of *non-uniform asynchronous cellular automata* (NUACAs).

ACAs can be categorized based on their update schemes into two types: (i) *fully asynchronous* and (ii) α-asynchronous. In a fully asynchronous CA, exactly one cell is updated uniformly at random at each time step. In contrast, under α-asynchronous update, each cell is updated independently with probability α. This work focuses on fully asynchronous CAs, which we refer to as *fully* ACAs.

All 256 3-neighborhood binary rules have been analyzed under fully asynchronous update in [16]. Among these, 146 rules were identified as forming convergent uniform ACAs, where configurations are attracted exclusively to *fixed points*. A configuration is a fixed point attractor if its next configuration is itself. The notion of *recurrent* ACAs was also introduced in [16]. In a recurrent ACA, every configuration is *recurrent*, meaning it returns to itself after one or more time steps. Fixed points are trivially recurrent, but recurrent configurations may also have also belong to cycles of length greater than one.

This raises a natural question: if we combine rules from the same class (either convergent or recurrent), does the resulting non-uniform ACA retain the same behavior? The following example shows this is not always the case. The ACA $(50, 243, 220, 92)$, constructed from convergent rules, does not possess a fixed point and is therefore *non-convergent*. This leads to the question of what governs the emergence of convergent or recurrent behavior in non-uniform ACAs. Specifically, we investigate whether convergence is influenced by the proportion of *passive* RMTs in the constituent rules or by the presence of rules sharing the same *rank*, where rank is determined by the number of passive RMTs. This work focuses on NUACAs constructed exclusively from convergent rules.

Section 2 introduces terminology related to asynchronous cellular automata. Section 3 investigates the properties of non-uniform cellular automata under fully asynchronous update. Section 4 presents properties of non-convergence in ACAs, even when constructed using convergent rules.

2 Preliminaries

This work considers a 1-dimensional cellular automaton (CA) with n cells, where the leftmost and rightmost cells are indexed as 0 and $n-1$ respectively. Here, n is called the *size* of CA. Each cell holds a value at a given time step, called its *state*. We restrict our study to 2-state or *binary* CAs (cell states take values from $\{0,1\}$). A *configuration* is an ordered collection of states of cells 0 to $n-1$.

Let x denote a configuration of an n-cell CA at time step t, given by $x = (x_0^t x_1^t \cdots x_{n-1}^t)$, where x_i^t represents the state of cell i at time t. The *next state* of a cell i depends on the *rule* applied to that cell and the current states of its neighbors. This work uses 3-neighborhood CA, where the neighborhood of cell i consists of its left neighbor $(i-1)$, the cell itself (i) and its right neighbor $(i+1)$. Accordingly, a rule is defined as $\mathcal{R} : \{0,1\}^3 \rightarrow \{0,1\}$.

A rule can be represented either as an 8-bit binary string or as its decimal equivalent in the range 0 to 255. Table 1 illustrates this representation. For example, when the present state of a cell and its neighbors is 010 and the applied rule is 76, the next state is 1.

Table 1. Binary 3-neighborhood rules 5, 76 and 236

Present state	111	110	101	100	011	010	001	000	Rule
(RMT)	(7)	(6)	(5)	(4)	(3)	(2)	(1)	(0)	
(i) Next state	0	0	0	0	0	1	0	1	5
(ii) Next state	0	1	0	1	0	1	0	0	76
(iii) Next state	1	1	1	0	1	1	0	0	236

The 3-bit representation of the present state of a cell and its neighbors is called a *rule min term* (RMT). An RMT can also be expressed as the decimal equivalent of the 3-bit binary pattern. For instance, the next state of RMT 010 (decimal 2) under rule 76 (binary 01010100) is 1.

The RMTs of adjacent cells are related as follows. Let $r = x_{i-1}x_i x_{i+1}$ denote the RMT of cell i, where $x_{i-1}, x_i, x_{i+1} \in \{0,1\}$. Then, the possible RMTs at cell $i+1$ are $x_i x_{i+1}0$ and $x_i x_{i+1}1$, while the possible RMTs at cell $i-1$ are $0x_{i-1}x_i$ and $1x_{i-1}x_i$. Thus, a configuration can equivalently be viewed as an *RMT sequence*, i.e., a sequence of RMTs corresponding to cells 0 through $n-1$.

For the terminal cells 0 and $n-1$, neighboring cells are absent. Therefore, boundary conditions are required. This work adopts *periodic* boundary conditions, where the state of cell $n-1$ is treated as the left neighbor of cell 0 and the state of cell 0 as the right neighbor of cell $n-1$. For example, the configuration 1000 corresponds to the RMT sequence $(010, 100, 000, 001) = (2, 4, 0, 1)$.

An RMT $r = x_{i-1}x_ix_{i+1}$ of a rule $\mathcal{R}$ is called *passive* (or inactive) if $\mathcal{R}(x_{i-1}x_i x_{i+1}) = x_i$; otherwise, it is called *active*. For instance, for rule 4, RMT 4 (100) is passive, while RMT 6 (110) is active (see Table 1). These terms are widely used in prior work [1,2,5,6]. If all RMTs in a sequence are passive, the corresponding configuration is a *fixed point attractor*. In a fully asynchronous ACA, the *next* configuration of x is obtained by selecting a cell i uniformly at random and applying the corresponding rule $\mathcal{R}i$. Consequently, $x_j^{t+1} = x_j^t$ for all $j \neq i$ and $x_i^{t+1} = \mathcal{R}i(xi-1^t, x_i^t, xi+1^t)$. A configuration x is a *fixed point attractor* of a fully asynchronous CA if updating any cell leaves it unchanged.

An n-cell CA is said to be *non-uniform* if $\mathcal{R}_i \neq \mathcal{R}_j$ for at least one pair of cells $i \neq j$. The sequence of rules applied to cells 0 through $n-1$ is represented by a *rule vector*. For example, $(246, 36, 144, 166)$ denotes a 4-cell CA where rule 246 is applied to cell 0 and rule 166 to cell 3.

In a fully asynchronous ACA, a configuration may have up to n distinct next configurations, each corresponding to the update of a different cell. For example, in Fig. 2, the configuration 1100 has four next configurations, 0100, 1000, 1110 and 1101, obtained by updating cells 0, 1, 2 and 3, respectively. A configuration is said to be *reachable* from itself or another if it can be obtained through one or more such updates. The sequence of cell indices used to obtain a reachable configuration from a given configuration is called an *update pattern*. For instance, as shown in Fig. 2, the configuration 1100 is reachable from 0101 using update patterns $(0, 3)$ and $(1, 2, 0, 3, 3, 3)$.

A configuration x is *recurrent* if, for every configuration y reachable from x, x is also reachable from y. All configurations reachable from a recurrent configuration are themselves recurrent, forming a *set of recurrent configurations*. For a fixed-point attractor, this set is singleton; conversely, if the set has more than one member then none of the member is fixed point.

3 Non-uniform Asynchronous Cellular Automata

Under fully asynchronous update, the evolution of a cellular automaton is most naturally described in terms of its recurrent configurations. Among recurrent configurations, fixed points play a special role: a fixed-point configuration remains unchanged under the update of any cell.

Based on the relationship between recurrent configurations and fixed points, ACAs can be classified into three distinct classes. In the first class, every recurrent configuration is a fixed point. In the second class, no recurrent configuration is a fixed point. In the third class, some recurrent configurations are fixed points, while others are not. This classification captures all possible asymptotic behaviors of fully asynchronous cellular automata.

As discussed earlier, any reachable configuration of a cellular automaton can equivalently be represented as a sequence of rule minterms (RMTs). For a fixed-point configuration, all RMTs in the corresponding RMT sequence are passive, since updating any cell leaves the configuration unchanged.

The first class of ACAs, in which every recurrent configuration is a fixed point, corresponds to systems whose dynamics eventually stabilize at fixed points. This behavior motivated the notion of *convergence* introduced in [18].

Definition 1. *A cellular automaton under fully asynchronous update is* convergent *if every configuration can reach at least one fixed-point configuration.*

The asynchronous evolution of an ACA can be modeled as a finite Markov chain [18], where configurations are states and single-cell updates are transitions. Fixed-point configurations correspond to absorbing states [9]. Consequently, an ACA is convergent iff all its recurrent configurations are fixed points, that is, if it belongs to the first class.

Prior work considered only *uniform* ACAs under fully asynchronous update and identified 146 local rules that generate convergent ACAs of any finite size [18]. A natural question then arises: if these convergent rules are combined to form an n-cell non-uniform ACA, is the resulting ACA necessarily convergent?

The answer is negative. Consider the following non-uniform ACAs under periodic boundary condition. The 4-cell ACA $(225, 106, 5, 246)$ is convergent and admits a unique fixed point, placing it in the first class. The ACA $(130, 40, 170, 169)$ admits no fixed-point configuration and therefore belongs to the second class. The ACA $(246, 36, 144, 166)$ admits a fixed point, but that fixed point is not reachable from all configurations; this ACA therefore belongs to the third class. Notably, in all three cases, each local rule is individually convergent and has exactly four active RMTs, yet the global behaviors differ significantly. This demonstrates that convergence of a non-uniform ACA cannot be inferred solely from the convergence properties of its constituent rules.

We now formally define *strongly non-convergent* ACAs, which correspond to the second class in the above classification.

Definition 2. *A cellular automaton under fully asynchronous update is* strongly non-convergent *if, starting from any configuration x, no sequence of cell updates can lead to a fixed-point configuration.*

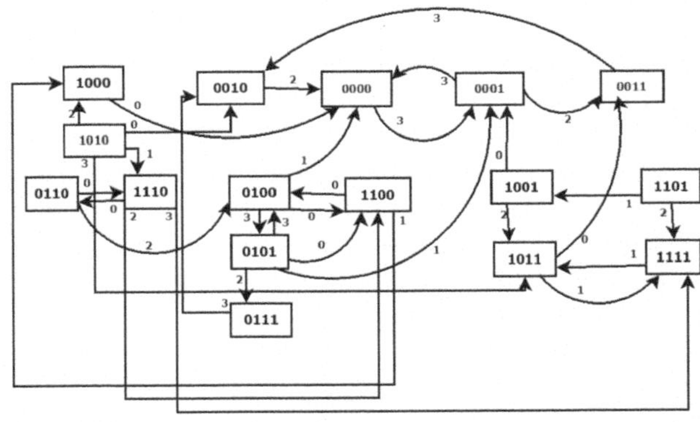

Fig. 1. Partial Transition Diagram of $(130, 40, 170, 169)$.

The ACA $(130, 40, 170, 169)$ is an example of a strongly non-convergent ACA. As shown in Fig. 1, from every configuration the dynamics eventually reach a set of recurrent configurations, none of which is a fixed point. In fact, this ACA admits no fixed-point configuration at all.

Although Definition 2 allows arbitrary update patterns, it is sufficient to consider only single-cell updates, as formalized in the following lemma.

Lemma 1. *A cellular automaton of size n under fully asynchronous update is strongly non-convergent if and only if, for every configuration x, there exists at least one cell i such that updating cell i produces a configuration distinct from x.*

Proof. Suppose a configuration x exists such that updating any cell leaves x unchanged. Then x is a fixed-point configuration and an update pattern of length one leads from x to a fixed point. Hence, the ACA is not strongly non-convergent.

Conversely, suppose that for every configuration x there exists at least one cell i such that updating cell i changes the configuration. Then no configuration is a fixed point. Consequently, no update pattern of any length can lead any configuration to a fixed point. Hence, the ACA is strongly non-convergent according to Definition 2. □

Corollary 1. *An n-cell ACA is strongly non-convergent if and only if every RMT sequence contains at least one active RMT.*

Proof. By Lemma 1, for every configuration there exists at least one cell update that changes the configuration. This implies that, in the corresponding RMT sequence, at least one RMT must be active. Hence, no RMT sequence can consist entirely of passive RMTs. □

In addition to the convergent and strongly non-convergent ACAs, we have the third class, which we refer to as *weakly non-convergent.*

Definition 3. *A cellular automaton under fully asynchronous update is said to be weakly non-convergent if and only if it admits both fixed-point and non-fixed-point recurrent configurations.*

The ACA $(246, 36, 144, 166)$ is weakly non-convergent. As shown in Fig. 2, it has a single fixed-point recurrent configuration (0000) and two additional pairs of recurrent configurations: one pair has configurations 1011 and 1111; another pair consists of 1000 and 1001. The members of each pair communicate among themselves. Here, the fixed point is reachable from 11 configurations.

These examples show that having an equal proportion of active and passive RMTs does not determine convergence behavior. All three non-uniform ACAs discussed above are constructed from rules with identical numbers of passive RMTs, yet they exhibit fundamentally different global dynamics. This motivates a systematic investigation of non-uniform ACAs constructed from rules with varying proportions of passive RMTs. In this work, we restrict attention to candidate rules drawn from the 146 convergent rules identified in [18], in order to isolate the effect of rule heterogeneity on convergence behavior.

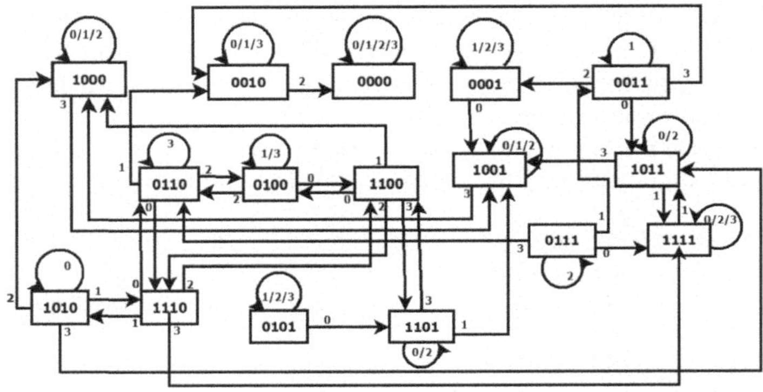

Fig. 2. Transition Diagram of (246, 36, 144, 166).

Table 2. Ranking of 146 convergent rules based on active RMTs

Rules	# Rules	#Active RMTs	Rank
204	1	0	8
76, 140, 220, 196, 200, 205, 206, 236	8	1	7
12, 68, 72, 77, 78, 92, 132, 136, 141, 172, 192, 197, 202, 207, 216, 221, 222, 228, 232, 237, 238, 252	22	2	6
4, 8, 13, 44, 64, 69, 74, 79, 88, 93, 94, 100, 104, 128, 133, 138, 152, 164, 168, 173, 174, 188, 194, 203, 208, 217, 218, 223, 224, 229, 230, 233, 234, 239, 244, 248, 253, 254	38	3	5
0, 5, 10, 24, 36, 40, 66, 80, 90, 95, 96, 106, 120, 130, 144, 154, 160, 165, 166, 169, 170, 175, 180, 184, 189, 190, 210, 219, 225, 226, 231, 235, 240, 245, 246, 249, 250, 255	38	4	4
2, 16, 26, 32, 42, 56, 82, 98, 112, 122, 146, 161, 162, 167, 171, 176, 181, 182, 185, 186, 191, 227, 241, 242, 247, 251	26	5	3
18, 34, 48, 58, 114, 163, 177, 178, 183, 187, 243	11	6	2
50, 179	2	7	1

4 Non-uniform ACA Using Convergent Rules

Building on observations from the above section, we now examine non-uniform ACAs constructed from convergent rules with differing proportions of active RMTs. To facilitate a systematic study, the 146 convergent rules are first classified according to the number of active RMTs they contain. This classification allows us to analyze how the distribution of active RMTs across candidate rules influences the emergence of convergent, strongly non-convergent and weakly non-convergent behavior in non-uniform ACAs under fully asynchronous update. Table 2 shows the ranking of the 146 convergent rules. Rules with a higher number of active RMTs are assigned smaller ranks (for example, rules 50 and 179 have the least rank). The convergent rules are thus ranked from 1 to 8 and each row of Table 2 contains rules of the same rank. We investigate the behavior of non-uniform ACAs under fully asynchronous update using these convergent rules, assuming periodic boundary conditions, and hereafter denoted NUACA.

4.1 Balanced NUACAs

We first consider NUACAs which are composed of rules of the same rank. In this case, all rules have the same proportion of active RMTs.

Definition 4. *An n-cell non-uniform ACA is called* balanced *if all its candidate rules have the same proportion of active RMTs.*

Proposition 1 *An n-cell balanced NUACA composed of rank 1 rules is strongly non-convergent for arbitrary n.*

Proof. Rank 1 convergent rules consist of rules 50 and 179, each having exactly one passive RMT: RMT 0 for rule 50 and RMT 7 for rule 179. Since the ACA is non-uniform and balanced, both rules appear in the rule set. Hence, there exists at least one index i such that $\mathcal{R}_i = 50$ and at least one of $\mathcal{R}_{i-1}$ or $\mathcal{R}_{i+1}$ is 179.

Consider the RMT sequences corresponding to configurations 0^n and 1^n, namely $(0, 0, \ldots, 0)$ and $(7, 7, \ldots, 7)$. For a uniform ACA, these sequences consist entirely of passive RMTs under rules 50 and 179, respectively. However, in the non-uniform case, at least one RMT of $(0, 0, \ldots, 0)$ (resp. $(7, 7, \ldots, 7)$) is active. Therefore, no RMT sequence can be composed solely of passive RMTs. By Corollary 1, the NUACA is strongly non-convergent. □

If we mix odd rules (those with value 1 at RMT 0) and even rules (value 0 at RMT 0) to create a non-uniform ACA, the configuration 0^n (for arbitrary n) can never be a fixed point, because the corresponding RMT sequence $(0, 0, \ldots, 0)$ contains active RMTs. Similarly, strongly non-convergent ACAs are not obtained if all candidate rules are greater than 127, since then the RMT sequence $(7, 7, \ldots, 7)$ generates the fixed point 1^n (since RMT 7 is passive for all candidate rules in this range).

For balanced NUACAs designed by rank 2 rules (Table 2), strongly non-convergence does not occur if all rules are either odd (RMT 7 passive) or even (RMT 0 passive). Let $\mathtt{R}_o^2$ and $\mathtt{R}_e^2$ denote the sets of odd and even rank 2 rules, respectively. Mixing them, however, reveals intrinsic properties that can be exploited to guarantee strong non-convergence.

Proposition 2 *An n-cell balanced NUACA designed by rank 2 rules is strongly non-convergent for arbitrary n if there exists some $0 \leq i < n - 1$ such that any one of the following conditions holds:*

1. $\mathcal{R}_i = 34$ *and* $\mathcal{R}_{i+1} \in \mathtt{R}_o^2 \setminus \{177\}$
2. $\mathcal{R}_i \in \{18, 48\}$ *and* $\mathcal{R}_{i+1} \in \mathtt{R}_o^2 \setminus \{177, 183, 187\}$
3. $\mathcal{R}_i = 114$ *and* $\mathcal{R}_{i+1} \in \mathtt{R}_o^2 \setminus \{177, 163\}$

Proof. Fix an index i such that $\mathcal{R}_i \in \mathtt{R}_e^2$ and $\mathcal{R}_{i+1} \in \mathtt{R}_o^2$. Since RMT 0 is passive for every even rank 2 rule, the RMT sequence from cell 0 through i consists entirely of passive RMTs. To prevent this sequence from forming a fixed point, the cell $i+1$ must possess active RMT. Therefore, both RMTs 0 and 1 are active for $\mathcal{R}_{i+1}$. Among odd rank 2 rules, the use of rule 177 is restricted at cell $i+1$ as RMT 1 is passive here.

Each even rank 2 rule has a second passive RMT $r \neq 0$. To prevent any sequence of all passive RMTs, cell $i+1$ rule must assign active transitions to RMTs $2r$ mod 8 and $2r+1$ mod 8. Excluding the stated exceptional rules ensures this. Thus, under any of these conditions, no RMT sequence consists entirely of passive RMTs; by Corollary 1, the NUACA is strongly non-convergent. □

Proposition 3 *An n-cell balanced NUACA composed of rank 2 rules is strongly non-convergent for arbitrary n there exists some $0 \leq i < n-1$ such that any one of the following conditions holds any one of the following conditions holds:*

1. $\mathcal{R}_i = 177$ *and* $\mathcal{R}_{i+1} \in \mathtt{R}_e^2 \setminus \{58, 114, 178\}$
2. $\mathcal{R}_i \in \{183, 243\}$ *and* $\mathcal{R}_{i+1} \in \mathtt{R}_e^2 \setminus \{18, 34, 114, 178\}$
3. $\mathcal{R}_i = 187$ *and* $\mathcal{R}_{i+1} \in \mathtt{R}_e^2 \setminus \{114, 178\}$.

Proof. The argument follows by left–right symmetry of the neighborhood, and applying reasoning identical to that of Proposition 2. □

Rules 58 and 178 are excluded from Proposition 2 and rule 163 from Proposition 3, as they may generate fixed points. These propositions show that deciding the behavior of an arbitrary NUACA from rank 2 rules is challenging; only specific restrictions guarantee strongly non-convergence. Intuitively, rules with more active RMTs tend to generate more strongly non-convergent NUACAs. Motivated by these observations, we now consider mixing convergent rules of different ranks to design ACAs.

4.2 Single-Rule Perturbation

A NUACA is *unbalanced* if at least one rule contains a different number of passive RMTs compared to the other candidate rules. In our first strategy, the unbalanced NUACAs are generated by introducing a single-rule perturbation: for an n-cell NUACA, $n-1$ cells contain rules of the same rank and a single cell contains a rule of a different rank. For the same rank, we consider rules of ranks 1 to 4 from Table 2; the remaining rule is selected from rank 7.

The population size of such NUACAs of size n is given by $n \times k_1 \times k_2^{n-1}$, where k_1 and k_2 are the counts of convergent rules of rank 7 and rank p, respectively, with $p \in \{1, 2, 3, 4\}$. Specifically, $k_1 = 8$ and k_2 take values 2, 11, 26 and 38 for ranks 1, 2, 3 and 4 respectively. For the experiments, n ranges from 4 to 7 (Fig. 3 to Fig. 4c). Additionally, we have also done the study for a few larger values of n as shown in Figs. 4d, 4e and 4f. For each n, we select 500 NUACAs uniformly at random, except for $n = 4$ and $p = 1$, where the total population (256 NUACAs) is less than 500 and hence fully considered.

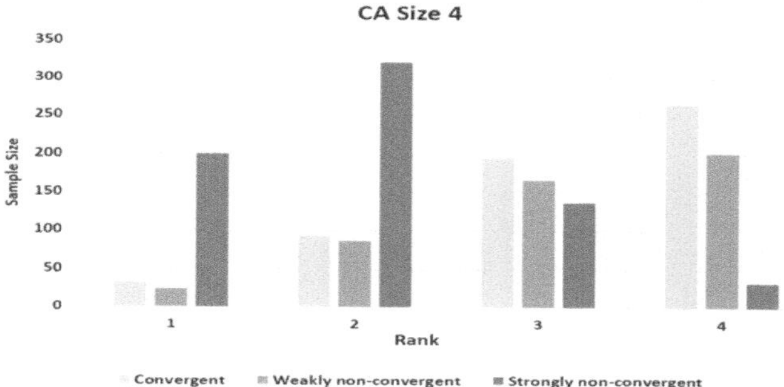

Fig. 3. Distribution of convergent, strongly non-convergent and weakly non-convergent CAs for single-rule perturbation strategy for CA size n=4. For each CA, $n-1$ rules are of rank p and one is of rank 7 (p varies from 1 to 4). 256 samples are selected for $p=1$, 500 samples for every other p. Results show that the proportion of strongly non-convergent CAs increases with decreasing p.

Figure 3 shows the results for $n=4$. When $n-1$ cells contain rank 1 rules, active RMTs dominate and nearly all ACAs are strongly non-convergent. Using rank 2 rules for $n-1$ cells produces similar behavior. As the rank of these $n-1$ cells increases to 3, the number of strongly non-convergent ACAs decreases, while rank 4 rules significantly increase the number of convergent NUACAs.

Figures 4a–4c show results for $n=5$ to 7. In 5-cell NUACAs, each sample size is 500 for $n-1$ cells with fixed-rank rules (1 to 4). Strongly non-convergent ACAs decrease as the rank of these cells increases. Generally, strongly non-convergent NUACAs outnumber weakly non-convergent and convergent ones, except when $n-1$ cells have rank 4 rules. Similar trends hold for $n=6$ and 7. For CA sizes 9, 12, and 14, almost all 500 samples are strongly non-convergent (see Figs. 4d–4f). Observations: (i) when the ranks of the $n-1$ cells are 4, increasing CA size decreases convergent and strongly non-convergent NUACAs and increases weakly non-convergent ones (for CA sizes 5, 6, 7, 9, 12 and 14); (ii) as the rank of $n-1$ cells decreases, strongly non-convergent NUACAs increase while convergent and weakly non-convergent ones decrease.

4.3 Mixing of Rules with Higher Proportion of Active RMTs

The second observation from the previous section motivates us to remove the restriction of fixed-rank rules when generating non-convergent NUACAs. In the *second* strategy, we construct ACAs using convergent rules of different ranks from 1 to 4. Figure 5 shows the behavior of 7000 non-uniform ACAs, with 1000 samples for each CA size of $\{4, 5, 6, 7, 9, 12, 14\}$. The experimental results indicate that almost all ACAs are strongly non-convergent.

A possible explanation follows. Consider two consecutive cells i and $i+1$ containing rank 1 rules. Then the ACA is strongly non-convergent, as the RMT at cell $i+1$ is active when cell i has a passive RMT. Even if rank 1 and rank 2

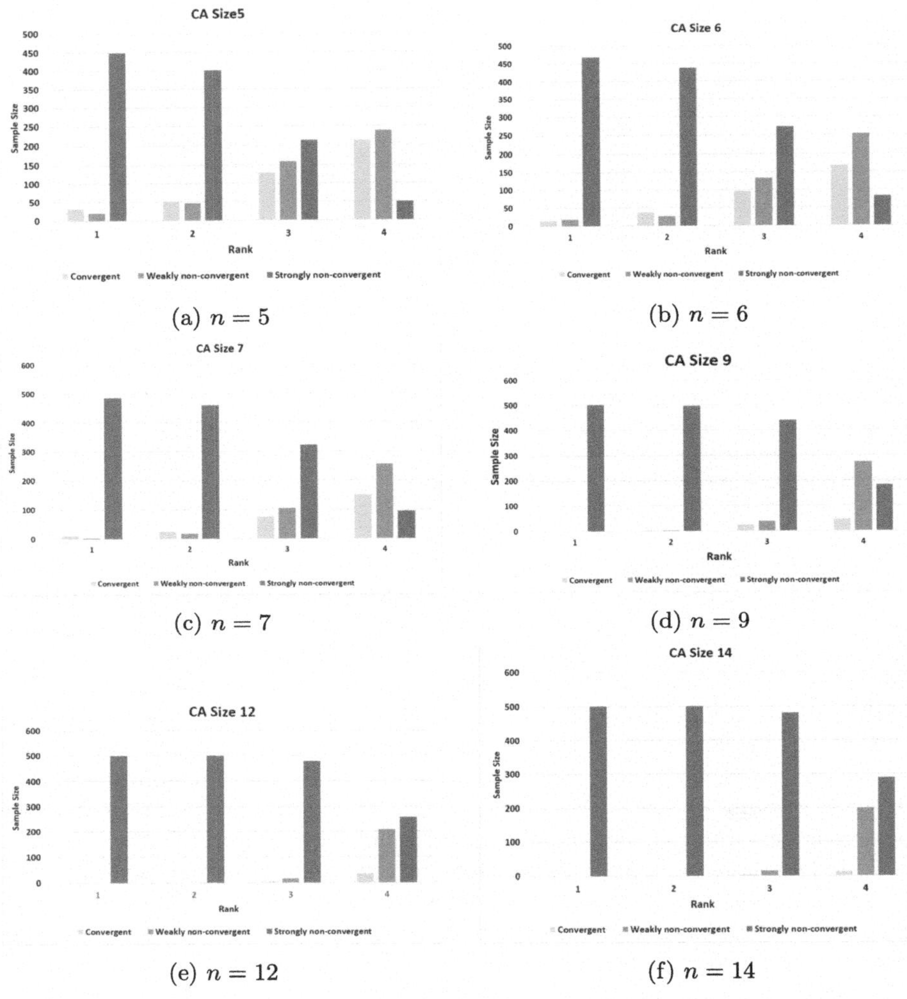

(a) $n = 5$ (b) $n = 6$ (c) $n = 7$ (d) $n = 9$ (e) $n = 12$ (f) $n = 14$

Fig. 4. Distribution of convergent, strongly non-convergent and weakly non-convergent CAs for single-rule perturbation strategy for varied CA sizes n. As before, for each CA, $n - 1$ rules are of rank p and one is of rank 7 (p varies from 1 to 4). 500 samples are used for each n,p combination. Results show that the proportion of non-convergent CAs increases with decreasing p and increasing n.

rules are used strategically in consecutive cells, strongly non-convergence is guaranteed. For example, if $\mathcal{R}_i$ is rule 50 (a rank 1 convergent rule from Table 2), the presence of any rule from $\{163, 183, 187, 243\}$ (rank 2 convergent rules) in the next cell ensures strongly non-convergence. Rule 50 has only one passive RMT (000), while the next cell contains active RMT 000 or 001. Hence, even minimal presence of lower-rank rules guarantees strongly non-convergent ACAs.

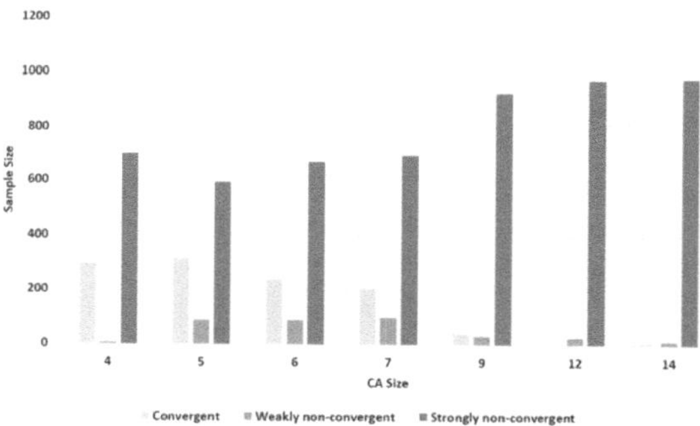

Fig. 5. Experimental results of ACAs constructed from convergent rules of ranks 1 to 4. 1000 samples are considered for each CA size. Results show that the proportion of convergent CAs reduce drastically with n. Similar trend is observed even for weakly non-convergent CAs.

Lemma 2. *An n-cell NUACA is strongly non-convergent if at least one of the following conditions holds for an arbitrary n:*

1. *There exists an $0 \leq i < n-1$ such that if $\mathcal{R}_i$ has only one passive RMT xyz and $\mathcal{R}_{i+1}$ has at least two active RMTs $yz0$ and $yz1$.*
2. *There exists an $0 < i \leq n-1$ such that if $\mathcal{R}_i$ has only one passive RMT xyz and $\mathcal{R}_{i-1}$ has at least two active RMTs $0xy$ and $1xy$.*

Proof. Consider an n-cell CA and let $r = xyz$ be the only passive RMT for the rule at cell i. Assume that in some RMT sequence, all RMTs from cell 0 to $i-1$ are passive, so cells 0 to i contain all passive RMTs. If cell $i+1$ contains both $yz0$ and $yz1$ as active RMTs, then no fixed point can be generated.

Similarly, if cells i to $n-1$ contain passive RMTs and cell i has a rule with only one passive RMT xyz, then the previous cell $i-1$ must have at least two active RMTs $0xy$ and $1xy$ to prevent any fixed point. □

Lemma 2 confirms that in an n-cell ACA with $n-2$ rank 8 rules, the presence of one rank 1 rule and its one adjacent rule with at least two active RMTs suffices to guarantee strong non-convergence, provided the following condition holds: if the rank 1 rule's passive RMT is xyz, then its next (resp. previous) rule can never have $yz0$ and $yz1$ (resp. $0xy$ and $1xy$) as passive.

Corollary 2. *An n-cell ACA is strongly non-convergent if it contains a rank 2 rule (from Table 2) at cell i and $\mathcal{R}_{i-1}$ has at least two active RMTs $0xy$ and $1xy$, where xyz_1 and xyz_2 are the only two passive RMTs of $\mathcal{R}_i$.*

We have provided a constructive characterization of non-convergence in ACAs. In particular, we show that for certain strategies, strongly non-convergence can be detected by examining only neighboring rules and their active RMTs, rather than transitions of configurations or long update patterns. This

Algorithm 1: Methodology for Studying Convergence in Non-Uniform ACAs

Input: CA size n, set of convergent rules $\mathcal{C}$
Output: Classification of NUACAs as convergent, weakly non-convergent, or strongly non-convergent
Classify all rules in $\mathcal{C}$ into ranks based on the number of active RMTs.
Consider the strategies presented in Sections 4.1, 4.2 and 4.3.
for each strategy **do**
  if Balanced NUACAs **then**
    for each rank $r \in \{1, 2\}$ **do**
      Construct NUACAs using only rules of rank r.
      Analyze convergence behavior using Propositions 1, 2, 3.
    end for
  end if
  if Single-rule perturbation **then**
    for rank $p \in \{1, 2, 3, 4\}$ **do**
      Assign rules of rank p to $n - 1$ cells.
      Replace one cell by a rule of rank 7.
      Randomly sample NUACAs and analyze convergence behavior.
    end for
  end if
  if Mixing multiple ranks **then**
    Construct NUACAs using rules of ranks 1 to 4.
    Randomly sample NUACAs and analyze convergence behavior.
    Identify strongly non-convergent ACAs by applying Lemma 2.
  end if
end for
Report counts of convergent, weakly non-convergent and strongly non-convergent NUACAs.

observation motivates an algorithm checking key rule interactions, providing an efficient method to identify strongly non-convergent NUACAs, summarized in Algorithm 1.

5 Conclusion

This work mainly focuses on strongly non-convergent ACAs. Some intrinsic properties have been investigated in reference to restrict the generation of fixed points. This work has preferred to use convergent rules for the experimentation. The presence of only one rule with *seven* active RMTs generates strongly non-convergent ACA even if almost all rules follow *highest proportion* of passive rule. This work primarily focuses on strongly non-convergent asynchronous cellular automata. We investigate intrinsic local rule properties that restrict the generation of fixed points and thereby lead to non-convergent behavior. For experimentation, we use convergent rules in order to isolate the effect of rule interaction rather than individual rule dynamics. Our results show that the count of strongly non-convergent ACAs increases with lower ranked rules for a given CA size. Additionally, the presence of even a single rule of rank 1 may generate a strongly non-convergent ACA, even when all other rules are of low rank.

References

1. Balbi, P.P., de Mattos, T., Ruivo, E.: Characterisation of the elementary cellular automata with neighbourhood priority based deterministic updates. Commun. Nonlinear Sci. Numer. Simul. **104**, 106018 (2022)
2. Concha-Vega, P., Goles, E., Montealegre, P., Ríos-Wilson, M., Santivañez, J.: Introducing the activity parameter for elementary cellular automata. Int. J. Mod. Phys. C **33**(09), 2250121 (2022)
3. Das, S.: Theory and Applications of Nonlinear Cellular Automata In VLSI Design. Ph.D. thesis, Bengal Engineering and Science University, Shibpur, India (2007)
4. Fatès, N.: Guided tour of asynchronous cellular automata. J. Cell. Autom. **9**(5–6), 387–416 (2014)
5. Fatès, N.: Asynchronous cellular automata. In: Cellular Automata, pp. 73–92. Springer (2018)
6. Fatès, N.: A tutorial on elementary cellular automata with fully asynchronous updating: general properties and convergence dynamics. Nat. Comput. **19**(1), 179–197 (2020)
7. Fatès, N., Sethi, B., Das, S.: On the reversibility of ECAs with fully asynchronous updating: the recurrence point of view. In: Reversibility and Universality: Essays Presented to Kenichi Morita on the Occasion of his 70th Birthday, pp. 313–332. Springer (2018)
8. Fates, N., Thierry, É., Morvan, M., Schabanel, N.: Fully asynchronous behavior of double-quiescent elementary cellular automata. Theoret. Comput. Sci. **362**(1–3), 1–16 (2006)
9. Grinstead, C.M., Snell, J.L.: Introduction to probability, the chance project version dated 4 July 2006 of the second revised edition, 1997. American Mathematical Society, Rhode Island, USA (2003)
10. Mukherjee, S., Bhattacharjee, K., Das, S.: Clustering using cyclic spaces of reversible cellular automata. Complex Syst. **30**(2), 205–237 (2021)
11. Naskar, N.: Characterization and Synthesis of Non-Uniform Cellular Automata with Point State Attractors. Ph.D. thesis, Indian Institute of Engineering Science and Technology, Shibpur, India (2015)
12. Pal Chaudhuri, P., Roy Chowdhury, D., Nandi, S., Chatterjee, S.: Additive Cellular Automata – Theory and Applications, vol. 1. IEEE Computer Society Press, USA, ISBN 0-8186-7717-1 (1997)
13. Patel, E.L., Broomhead, D.: A max-plus model of asynchronous cellular automata. arXiv preprint arXiv:1502.04097 (2015)
14. Roy, S., Das, S.: Clouds in the basins of fully asynchronous cellular automata. Adv. Complex Syst. **25**(08), 2250013 (2022)
15. Roy, S., Fatès, N., Das, S.: Reversibility of elementary cellular automata with fully asynchronous updating: an analysis of the rules with partial recurrence. Theoret. Comput. Sci. **1011**, 114721 (2024)
16. Sethi, B.: Theory and applications of fully asynchronous cellular automata. Ph.D. thesis, Indian Institute of Engineering Science and Technology, Shibpur (2018)

17. Sethi, B., Fatès, N., Das, S.: Reversibility of elementary cellular automata under fully asynchronous update. In: International Conference on Theory and Applications of Models of Computation, pp. 39–49. Springer (2014)
18. Sethi, B., Roy, S., Das, S.: Asynchronous cellular automata and pattern classification. Complexity **21**(S1), 370–386 (2016)

Design of Non-Uniform Elementary Asynchronous Cellular Automata for Pattern Classification

Sukanya Mukherjee[1(✉)] and Joydip Paul[2]

[1] Department of Computer Science and Engineering, Institute of Engineering & Management, Kolkata, University of Engineering and Management, Kolkata, India
sukanyaiem@gmail.com

[2] Indian Institute of Engineering Science and Technology, Shibpur, India
joydipp92@gmail.com

Abstract. This work focuses on the usage and performance of *convergent* non-uniform asynchronous elementary cellular automata (NAECAs) for supervised machine learning tasks like *pattern classification.* This work targets on some theoretical findings like i) how *rule min terms* (RMTs) can be used to generate *strictly* convergent NAECAs, ii) behavior of cellular automata obtained by mixing different groups of (146) convergent AECA rules based on the different *proportions* of *passive* RMTs and iii) how the properties like number of *fixed point attractors* (FPAs), *rate of convergence* change on adding *impurity* to the obtained strictly non-uniform AECA. Based on the result of our proposed *three* strategies, potential candidate NAECA rules for pattern classification are chosen keeping constraints like *average convergence steps*, number of FPAs. The result of our NAECA based classifier is compared with other cellular automata and *machine learning* based pattern classification benchmark algorithms on real datasets.

Keywords: convergent non-uniform asynchronous elementary cellular automata (NAECA) · classification · passive RMTs · fixed point attractors (FPAs) · average convergence steps

1 Introduction

Cellular Automaton (CA) was introduced by John von Neumann to model biological self-reproduction [16]. Early work by Wolfram in the 1980 s [18–20] established the use of ECAs as generators of complex natural patterns and implicitly framed CA evolution as a form of regime-based pattern discrimination, laying the foundation for later CA-based classification models. Owing to their inherent massive parallelism, CAs soon attracted significant attention for modeling

This work is carried out as a project in the Indian Summer School on Cellular Automata 2025.

H. Raju et al. (Eds.): ASCAT 2026, CCIS 2801, pp. 253–269, 2026.
https://doi.org/10.1007/978-3-032-18612-6_19

physical and natural systems. A CA can be defined as a lattice of cells that evolves in discrete time and space according to local rules [3,17]. Stephen Wolfram introduced simple two-state, three-neighborhood cellular automata, known as elementary cellular automata (ECAs), which are capable of generating complex patterns and processes [11]. ECAs have been widely used to model and classify natural patterns such as seashell markings and snowflakes, and various CA-based approaches have been proposed for pattern classification [11].

However, most existing studies on CA-based pattern classification focus on *uniform* ECAs, where the same local rule is applied to all cells and updates are performed either *synchronously* (all cells update simultaneously) or *asynchronously* (at least one cell updates at a time), whose dynamical properties have been studied in the CA literature [8,9,14]. There are published works on synchronous *non-uniform* ECAs, where different cells follow different rules for classification purposes [5]. In contrast, the behavior and potential of asynchronous non-uniform cellular automata remain comparatively unexplored. Motivated by this, we investigate non-uniform asynchronous ECAs for pattern classification. Related asynchronous and stochastic CA-based classification models have been explored in prior studies [2,6,7,15].

For pattern classification using non-uniform asynchronous cellular automata (ACAs), it is essential to identify categorically non-uniform *convergent* ACAs, in which *fixed point attractors* (FPAs) serve as basins. An ACA is said to be convergent if all configurations are eventually attracted only to FPAs. In this work, we consider the *fully asynchronous* update scheme, where at each time step exactly one cell is selected uniformly at random and updated. A configuration is an FPA if its associated *RMT sequence* consists entirely of *passive* RMTs (where transition is *inactive* [1,4,8,9]). Consequently, in an ACA, a fixed point attractor is a configuration that returns to itself under the update of any single cell.

Identifying convergent non-uniform ECAs is nontrivial. In particular, the number of passive RMTs in individual rules does not by itself guarantee convergence. For example, the ACA with rule vector $(232, 93, 190, 200)$ is non-convergent and does not admit any fixed point attractor, even though the constituent rules have $6, 5, 4,$ and 7 passive RMTs, respectively. In contrast, the ACA $(196, 104, 250, 77)$ is convergent, although each rule again has at least four passive RMTs. These examples indicate that convergence in non-uniform ACAs depends on the specific arrangement of *active* RMTs rather than solely on their count.

In asynchronous CA, a *recurrent* configuration is a configuration that is reachable from itself through some finite sequence of single-cell updates. A fixed point attractor is a recurrent configuration that is reachable only from itself. In a convergent ACA, no recurrent configurations other than fixed points can exist. The key challenge is therefore to detect and prevent the formation of recurrent configurations that are not fixed points. To illustrate this phenomenon, consider a 4-cell ACA under periodic boundary condition, where the leftmost and rightmost cells are indexed as 0 and 3, respectively. Suppose the configurations 1100 and 1000 each return to themselves in two update steps when cell 1 is updated, while updates of the remaining cells $(0, 2, 3)$ return the configuration to itself in one step. These configurations are recurrent but are not fixed points and cannot be attracted by any FPA. Hence, the ACA is non-convergent, irrespective of the update sequence.

In this example, the RMTs of the rules applied at cells 0, 2, and 3 are passive. Non-convergence arises from the rule at cell 1, which contains active RMTs of both forms $x0z$ and $x1z$ where $x, z \in \{0, 1\}$. Specifically, RMT 100 in configuration 1000 causes a transition to 1 and RMT 110 in configuration 1100 causes a transition to 0. This bidirectional toggling of the middle bit prevents convergence. This example demonstrates that an n-cell ACA may be non-convergent even if all but one rule maintain only passive RMTs, provided a single rule contains active RMTs of both forms $x0z$ and $x1z$ (for some x and z).

One way to restrict the formation of non-convergent ACAs is to allow active transitions only for one form of RMTs, either $x0z$ or $x1z$. Let $\mathtt{R_0}$ denote the set of ECA rules whose active RMTs are exclusively of the form $x0z$, and let $\mathtt{R_1}$ denote the set of rules whose active RMTs are exclusively of the form $x1z$. There are 30 such ECA rules in total, with $|\mathtt{R_0}| = |\mathtt{R_1}| = \mathtt{15}$.

Consider an n-cell ACA constructed using only rules from $\mathtt{R_0}$. If an RMT of the form $x0z$ is active at cell i, then after updating that cell the resulting RMT at i must be of the form $x1z$, which is passive. Consequently, the same active RMT cannot reappear at that cell in subsequent updates, and no configuration containing an active RMT can recur. An analogous argument holds for ACAs constructed using only rules from $\mathtt{R_1}$. Therefore, any ACA composed exclusively of rules from $\mathtt{R_0}$ or $\mathtt{R_1}$, in arbitrary spatial order, is convergent under fully asynchronous update. We refer to these 30 rules as *strictly convergent* rules.

Further insight is obtained by introducing *impurity* into the rule vector through the inclusion of rules outside this strictly convergent set. Following [14], the 146 convergent ACA rules are grouped into eight classes based on the number of passive RMTs. Although a higher number of passive RMTs does not by itself guarantee convergence, the activepassive RMT distribution remains a useful structural indicator of ACA dynamics. Passive RMTs locally restrict state changes, while active RMTs introduce transitions that may inhibit stabilization.

In particular, even if all but one rule in a non-uniform ACA consist solely of passive RMTs, the presence of a single rule containing all the eight active RMTs is sufficient to prevent convergence by enabling non-fixed-point recurrent configurations. This observation supports the heuristic that an increasing number of active RMTs—especially when drawn from both RMT sets—reduces the likelihood of fixed point formation.

Motivated by these observations, experiments are conducted by mixing rules from different classes to study convergence behavior, fixed point attractor growth, and convergence rates. These analyses guide the selection of candidate rule vectors for constructing the proposed pattern classifier. It is observed that non-uniform asynchronous elementary cellular automata (NAECAs) with a large number of fixed point attractors often yield poor classification performance, while rules with large convergence times increase training cost. Accordingly, the classifier is designed to balance convergence, expressiveness, and computational efficiency. The necessary preliminaries are presented in the following section.

2 Preliminaries of Cellular Automata

A Cellular Automaton (CA) is a discrete mathematical model consisting of a lattice of cells that evolve in discrete time according to local rules. In this work, we consider one-dimensional *finite* CA with n cells where n is *finite*. In an n-cell CA, the leftmost cell and rightmost cells are indexed as 0 and $n-1$ respectively. Every cell contains a *state* value. This work uses 2-state 3-neighborhood CAs such that the states are the members of $\{0, 1\}$ and the change of state of a cell is governed by its three neighbors (left, self and right).

The value of a cell at time t is called its *present state*, and its value at time $t+1$ is the *next state*. Under periodic boundary condition, the lattice is circular, so the left neighbor of cell 0 is the $n-1^{th}$ cell and the right neighbor of cell $n-1$ is cell 0. The next state of cell i at time $t+1$ denoted by S_i^{t+1}, is determined by a *local rule* $f : (0, 1)^3 \rightarrow (0, 1)$ applied to its neighborhood $(S_{i-1}^t, S_i^t, S_{i+1}^t)$:

$$S_i^{t+1} = f(S_{i-1}^t, S_i^t, S_{i+1}^t)$$

Since there are three neighbors, there are $2^3 = 8$ possible neighborhood patterns. Each ECA rule specifies the NS for all eight neighborhoods, and the decimal equivalent of these outputs defines the *rule*, giving a total of $2^8 = 256$ possible rules.

A *configuration* of a CA is the set of states of all cells at a given time. For an n-cell CA, there are 2^n possible configurations. For example, the 4-cell configuration (0101) represents one such state of the lattice.

Table 1. Look-Up Table for Rules 4, 12, and 206

PSs	111	110	101	100	011	010	001	000	Rule
(RMT)	(7)	(6)	(5)	(4)	(3)	(2)	(1)	(0)	
(i) NSs	0	0	0	0	0	1	0	0	4
(ii) NSs	0	0	0	0	1	1	0	0	12
(iii) NSs	1	1	0	0	1	1	1	0	206

Table 2. Relationship between the ith and $(i+1)^{th}$ RMTs

i^{th} RMT	$(i+1)^{th}$ RMT
0	0, 1
1	2, 3
2	4, 5
3	6, 7
4	0, 1
5	2, 3
6	4, 5
7	6, 7

Definition 1. [14]: *The association of the neighborhood x, y, z to the value f(x, y, z), which represents the result of the updating function f, is called Rule Min Term (RMT). Each RMT is associated to a number 4x+2y+z.*

For instance, RMT 101 can alternatively be represented as $r = 4{\cdot}1+2{\cdot}0+1 = 5$. A configuration can be represented as a sequence of RMTs. For the configuration (0110), the corresponding RMT sequence under periodic boundary conditions is $(1, 3, 6, 4)$. Table 2 shows the relationship between RMTs of consecutive cells; for example, if the cell i has RMT is 5 (101), then the RMT at cell $i+1$ can be either 2 (010) or 3 (011).

Definition 2: *A transition is* active *if it changes the state of a cell* ($f(x, y, z) \neq y$)*. Otherwise, it is* inactive.

Definition 3: *An RMT r (xyz), is* passive *if* $f(x, y, z) = y$*. Otherwise, it is* active.

For example, in Rule 4, RMTs $\{0, 1, 2, 4, 5\}$ are passive, while RMTs $\{3, 6, 7\}$ are active (see Table 1).

Traditionally, CAs are *synchronous*, updating all cells simultaneously. Removing this constraint gives *asynchronous cellular automata* (ACAs), which can follow two update schemes:

1. **Fully asynchronous:** At each step, exactly one cell is selected uniformly at random and updated. If $x = (S_0^t, S_1^t, \ldots, S_{n-1}^t)$ is the current configuration, updating cell i yields a new configuration

$$y = (S_0^t, \ldots, f_i(S_{i-1}^t, S_i^t, S_{i+1}^t), \ldots, S_{n-1}^t),$$

allowing up to n possible next configurations. This y is *reachable* from x.

2. α-**asynchronous:** Each cell updates independently with probability α.

If all cells use the same rule, the CA is *uniform* ($f_0 = f_1 = \cdots = f_{n-1}$); otherwise, it is *non-uniform*. The *rule vector* $(\mathcal{R}_0, \ldots, \mathcal{R}_{n-1})$ specifies which rule each cell follows. For example, a 4-cell CA with rule vector $(4, 136, 12, 72)$ applies Rule 4 to cell 0, Rule 136 to cell 1, and so on. A *non-uniform asynchronous* ECA (NAECA) combines non-uniformity and full asynchrony.

Figure 1 shows a partial *transition diagram* of the 4-cell NAECA $(4, 136, 12, 72)$.

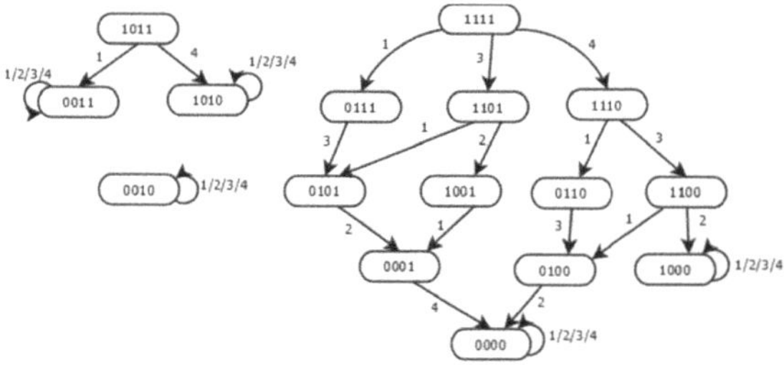

Fig. 1. Partial transition diagram of non-uniform 4-cell ACA $(4, 136, 12, 72)$.

A configuration is a *fixed point attractor (FPA)* if updating any cell leaves the configuration unchanged. An ACA is *convergent* if every initial configuration eventually reaches an FPA. In Fig. 1, (0000), (0010), (0011), (1010), (1000) are FPAs, and all configurations eventually reach one of these FPAs, making the ACA convergent.

During evolution under fully asynchronous updates, the *update sequence* $U = (u_t)_{t\in\mathbb{N}}$ denotes which cell is updated at time t. For example, a sequence from (1111) to (0000) could be $U = (1, 3, 2, 4, \dots)$; (cell indices in the figure are labeled as $i + 1$). A reachable configuration is obtained from itself or another configuration through one or more update sequences.

A configuration x is called *recurrent* if for every configuration y reachable from x, the configuration x is also reachable from y. All configurations reachable from a recurrent configuration are themselves recurrent and collectively form a *set of recurrent configurations*. A fixed point attractor is a recurrent configuration that is reachable only from itself. In convergent ACAs, there is no existence of non fixed point attractor recurrent configurations.

3 Strictly Convergent Non-Uniform ACAs

In uniform ACAs, the convergence of rules is a theoretically established property [14] and not an empirical observation. However, when such rules are combined in a non-uniform asynchronous CA, rule heterogeneity can enable interactions that generate recurrent behavior. To ensure convergence in the non-uniform case, rules must therefore be selected so that no other than recurrent configuration can persist. If there exist a set of at least two non fixed point recurrent configurations, then non-convergence is ensured. Here, we analyze non-convergence in the context.

Property 1 *In a non-uniform asynchronous elementary cellular automaton, for any recurrent configuration that is not a fixed point, every cell i whose state changes during the evolution must have a local rule $\mathcal{R}_i$ containing at least one active RMT from $S_0 = \{0, 1, 4, 5\}$ and at least one active RMT from $S_1 = \{2, 3, 6, 7\}$.*

Proof: Consider a cell i whose state changes at some point during the evolution. Let $s_t(i) \in \{0, 1\}$ denote the state of cell i at time t. Since the configuration is recurrent but not a fixed point, there exists a minimal integer $T > 0$ such that

$$s_{t+T}(i) = s_t(i), \quad \text{and} \quad s_{t+1}(i) \neq s_t(i),$$

for some $t \geq 0$.

Let $\mathcal{R}_i$ denote the local rule applied to cell i, defined over the eight possible RMTs

$$\{0, 1, 2, 3, 4, 5, 6, 7\} \equiv \{000, 001, 010, 011, 100, 101, 110, 111\}.$$

Partition these RMTs into two sets:

$$S_0 = \{0, 1, 4, 5\}, \quad S_1 = \{2, 3, 6, 7\},$$

where the middle bit is 0 for S_0 and 1 for S_1.

Since $s_{t+1}(i) \neq s_t(i)$, the rule $\mathcal{R}_i$ must contain at least one active RMT that changes the middle bit.

- If the middle bit is 0, it can change to 1 only via an active RMT from S_0.
- If the middle bit is 1, it can change to 0 only via an active RMT from S_1.

To return to the original state after T steps, the middle bit must change in both directions (0 to 1 and 1 to 0). Therefore, $\mathcal{R}_i$ must contain at least one active RMT from each of S_0 and S_1.

Hence, the presence of active RMTs from both sets is a necessary condition for the existence of a recurrent configuration that is not a fixed point.

□

Property 2 *Let $\mathcal{R}_i$ be the local rule applied at cell i in an n-cell asynchronous cellular automaton. If, for all $i = 1, \ldots, n$, all active RMTs of $\mathcal{R}_i$ belong exclusively to either*

$$S_0 = \{0, 1, 4, 5\} \quad or \quad S_1 = \{2, 3, 6, 7\},$$

then the ACA converges to a fixed point under fully asynchronous updates.

Proof:

Let x^t denote the configuration at time t, and consider any cell i with local rule $\mathcal{R}_i$.

1. Since all active RMTs of $\mathcal{R}_i$ belong to a single set (either S_0 or S_1), the state of cell i can change only in one direction (state 0 to state 1 or state 1 to state 0).
2. By Property 1, the existence of a non-fixed-point recurrent configuration requires that a cell be able to change state in both directions using RMTs from both sets.
3. As no active RMTs from the opposite set exist in $\mathcal{R}_i$, such bidirectional state changes are impossible.

Therefore, each cell can change its state at most once and cannot revert. Consequently, no recurrent configuration other than fixed points can exist, and the ACA must converge to a fixed point. □

Table 3. Strictly Convergent CA Rules

Active RMT(s) $\in$	Rules
S_0	$R_0=\{205, 206, 207, 220, 221, 222, 223, 236, 237, 239, 252, 253, 254, 255\}$
S_1	$R_1=\{0, 4, 8, 12, 64, 68, 72, 76, 128, 132, 136, 140, 192, 196, 200\}$

Using Property 2, we obtain two sets of rules, each containing 15 rules, as shown in Table 3. Column 1 refers to the set the active RMT(s) belong to and the second column represents the corresponding rules.

Definition 4: *A collection* R *of convergent rules is said to be strictly convergent if every non-uniform ACA formed by rules from any subset of* R *is convergent.*

Corollary 1 $R_0 \cup R_1$ *is strictly convergent.*

Corollary 2 *Any n-cell non-uniform ACA formed with only convergent rules is convergent if at least* $n-1$ *rules are from* $R_0 \cup R_1$.

With a slight abuse of terminology, the term *strictly convergent* non-uniform ACA is used to denote a non-uniform asynchronous cellular automaton whose rule vector consists solely of rules from the strictly convergent rule set.

The average number of FPAs generated by strictly convergent ACAs (of sample size 1000) for different lattice sizes (4 to 30) is shown in Table 4, and the corresponding growth trend is illustrated in Fig. 2.

Table 4. Average number of fixed points for different lattice sizes

Lattice Size	AVG FPA
4	4.646
5	6.668
6	9.868
7	14.314
8	20.959
9	31.099
10	47.383
11	68.220
12	99.953
13	147.622
14	204.941
15	303.580
16	479.495

Lattice Size	AVG FPA
17	698.847
18	964.454
19	1442.730
20	2117.150
21	2916.840
22	4454.400
23	6992.860
24	10445.700
25	15411.300
26	22434.100
27	29883.100
28	46157.000
29	65654.800
30	106142.000

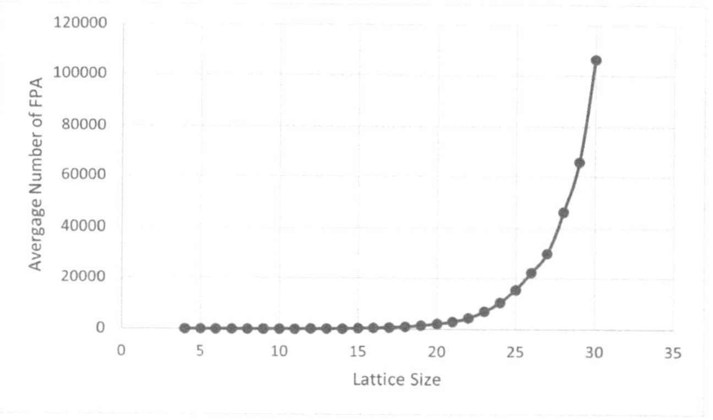

Fig. 2. Relationship between the average number of FPAs and lattice size.

4 Experiments on Non-Uniform ACA and Analysis

In this section, CA refers to non-uniform asynchronous cellular automata (ACA) under periodic boundary conditions. It has been shown that there exist 146 elementary ACA rules that are always convergent under fully asynchronous update schemes. These are proven fact taken from thesis of Biswanath Sethi [14]. These rules are listed in Table 5.

Based on the number of passive RMTs, these rules are grouped into eight classes, as shown in Table 6. Class i contains rules having exactly i passive RMTs. This classification is motivated by the empirical observation that a larger number of passive RMTs generally increases the likelihood of convergence by increasing the number of fixed point attractors (FPAs).

For experimental analysis, hybrid rule vectors are constructed by selecting rules randomly (current time of system is used as seed value) from specified classes of specified lattice sizes. Experiments are conducted for lattice sizes ranging from 4 to 18 in 3 strategies . For each hybridization strategies 10000,1000,500,200 samples are collected for lattice sizes 4 - 8, 10 - 11, 13, 14 - 18 respectively. For each configuration the average numbers of FPAs for convergent CAs($\overline{\text{FPA}}_{\text{conv}} = \frac{1}{|\mathcal{C}_{\text{conv}}|} \sum_{i \in \mathcal{C}_{\text{conv}}} \text{FPA}_i$) , and the number of convergent ACAs are recorded.

Only rules with at least four passive RMTs (Classes 4 to 7) are considered in the hybridization strategies. Class 8, which contains a single rule with all RMTs passive, is excluded, as preliminary observations indicate that its inclusion leads to a rapid growth in the number of FPAs, dominating the dynamics and masking the effects of hybridization.

Table 5. List of 146 Convergent Rules

0, 2, 4, 5, 8	94, 95, 96, 98	165, 166, 167	192, 194, 196	230, 231, 232
10, 12, 13, 16	100, 104, 106	168, 169, 170	197, 200, 202	233, 234, 235
18, 24, 26, 32	112 114, 120	171, 172, 173	203, 204, 205	236, 237, 238
34, 36, 40, 42	122, 128 130	174, 175, 176	206, 207, 208	239, 240, 241
44, 48, 50, 56	132, 133, 136	177, 178, 179	210, 216, 217	242, 243, 244
58, 64, 66, 68	138, 140, 141	180, 181, 182	218, 219, 220	245, 246, 247
69, 72, 74, 76	144, 146, 152	183, 184, 185	221, 222, 223	248, 249, 250
77, 78, 79, 80	154, 160, 161	186, 187, 188	224, 225, 226	251, 252, 253
82, 88, 90, 92	162, 163, 164	189, 190, 191	227, 228, 229	254, 255

4.1 Strategy 1

In this strategy, rules from Classes 6 and 7 are hybridized to form the rule vector randomly. The following empirical observations are obtained:

1. A large proportion of the generated rule vectors (approximately 90%) result in convergent ACAs for the considered lattice sizes.
2. It is clearly seen in the graph 4a that the percentage of forming convergent ACAs decreases as the lattice size increases. The spikes shown the graph are probably due variation of sample size and randomly selecting the rule vectors. More or less it is linear.
3. It is observed that with the increase in the number of Class 7 rules within the rule vector leads to a higher likelihood of convergence.
4. The average number of FPAs increases approximately as $O(n^2)$, where n denotes the lattice size, based on empirical fitting (Fig. 3a Though for higher lattice sizes the trend is not purely visible due to variation of sample size and selecting the rule vectors randomly.

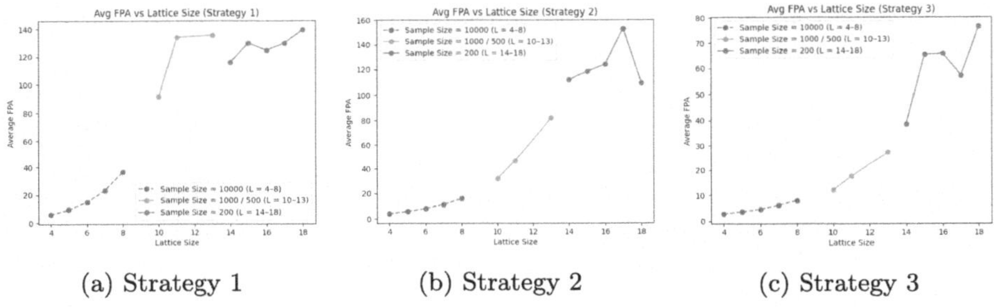

(a) Strategy 1 (b) Strategy 2 (c) Strategy 3

Fig. 3. Average number of fixed-point attractors (Avg FPA) as a function of lattice size for Strategies 13, grouped by sample-size regimes.

Table 6. Classes and their corresponding rules

Class	Number of rules	Rules
8	1	204
7	8	76, 140, 196, 200, 205, 206, 220, 236
6	22	12, 68, 72, 77, 78, 92, 132, 136, 141, 172, 192, 197, 202, 207, 216, 221, 222, 228, 232, 237, 238, 252
5	38	4, 8, 13, 44, 64, 69, 74, 79, 88, 93, 94, 100, 104, 128, 133, 138, 152, 164, 168, 173, 174, 188, 194, 203, 208, 217, 218, 223, 224, 229, 230, 233, 234, 239, 244, 248, 253, 254
4	38	0, 5, 10, 24, 36, 40, 66, 80, 90, 95, 96, 106, 120, 130, 144, 154, 160, 165, 166, 169, 170, 175, 180, 184, 189, 190, 210, 219, 225, 226, 231, 235, 240, 245, 246, 249, 250, 255
3	26	2, 16, 26, 32, 42, 56, 82, 98, 112, 122, 146, 161, 162, 167, 171, 176, 181, 182, 185, 186, 191, 227, 241, 242, 247, 251
2	11	18, 34, 48, 58, 114, 163, 177, 178, 183, 187, 243
1	2	50, 179

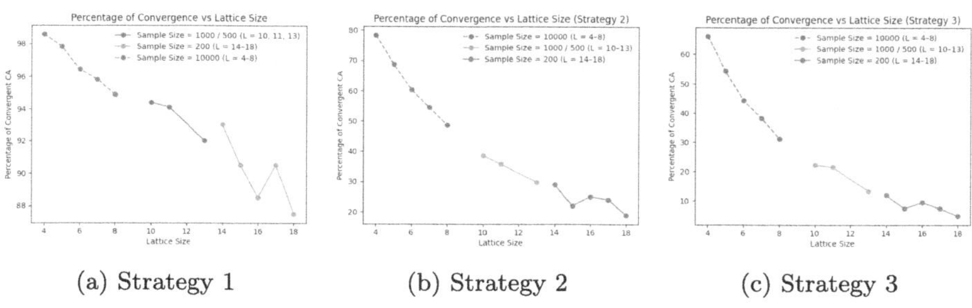

(a) Strategy 1 (b) Strategy 2 (c) Strategy 3

Fig. 4. Percentage of convergent cellular automata as a function of lattice size for Strategies 13 under different sampling regimes.

4.2 Strategy 2

In this strategy, rules from Classes 5, 6, and 7 are hybridized. The following differentiating observations are made:

1. The proportion of convergent ACAs decreases more rapidly with increasing lattice size compared to Strategy 1. 4b
2. The rate of increase in the number of point attractors with respect to lattice size is lower than that observed in Strategy 1, as illustrated in Fig. 3b

4.3 Strategy 3

In this strategy, rules from Classes 4, 5, 6, and 7 are hybridized. The following observations are obtained:

1. The proportion of convergent ACAs decreases more steeply with increasing lattice size compared to Strategies 1 and 2 as shown in Fig. 4c
2. The rate of increase in the number of point attractors with increasing lattice size is lower than that observed in the previous strategies (Fig. 3c)
3. Increasing the number of rules from Classes 4 and 5 within the rule vector leads to a decrease in the probability of convergence.

5 Role of Non-Uniform ACA in Pattern Classification

Asynchronous cellular automata (ACA) possessing multiple fixed point attractors (FPAs) have been shown to be effective for pattern classification tasks [12]. An n-cell cellular automaton with k FPAs can be viewed as a natural k-class classifier [10]. Pattern classification using uniform ACA with multiple FPAs has also been reported in the literature [14]. Motivated by these results, we employ non-uniform ACAs with multiple FPAs for designing supervised pattern classifiers.

Throughout this section, ACA refers to a non-uniform asynchronous cellular automaton under periodic boundary conditions.

5.1 Design of the Pattern Classifier

Pattern classification is a supervised learning problem in which labeled training data are used to learn decision boundaries that generalize to unseen data. We have used 70% of the dataset for training set and rest 30% for test set by using a randomized hold-out strategy implemented in scikit-learn [13]. The training and testing phase of the classifier is described below.

Training Phase : During training, each configuration from the training dataset is iteratively updated using a candidate ACA rule vector until it reaches a fixed point attractor (FPA). The FPA reached by a configuration is assigned the class label of that configuration. If configurations from both classes converge to the same FPA, the class label of the attractor is decided using majority voting.

If a configuration reaches multiple FPAs, the FPA reached first according to the numerically sorted update pattern is considered. This procedure is repeated for all training configurations. For each candidate ACA, the classification accuracy is computed as Accuracy $= \frac{\text{\# correctly classified patterns}}{\text{total number of patterns}} \times 100\%$.. The ACA rule vector that yields the highest training accuracy, along with the corresponding representative attractor sets for each class, is retained. The training procedure is adapted from [14].

Testing Phase : During testing, each unseen configuration from the test dataset is iteratively updated until it reaches an FPA. Using the stored attractor sets obtained during training, the class label of the configuration is predicted based on the class associated with the reached FPA. The classification strategy is illustrated in Fig. 5.

In addition to accuracy, standard performance metrics such as precision, recall, and F_1-score are computed using the confusion matrix shown in Table 7: Accuracy $= \frac{TP+TN}{TP+TN+FP+FN}$, Precision $= \frac{TP}{TP+FP}$, Recall $= \frac{TP}{TP+FN}$, $F_1 = 2 \times \frac{\text{Precision} \times \text{Recall}}{\text{Precision}+\text{Recall}}$.

Table 7. Confusion Matrix

Actual / Predicted	Positive (Class 1)	Negative (Class 2)
Positive (Class 1)	True Positive (TP)	False Negative (FN)
Negative (Class 2)	False Positive (FP)	True Negative (TN)

Let us take an example. Consider Class 1 configurations {0011, 0110, 0010} and Class 2 configurations {1111, 1110, 0111}. Each configuration is iteratively updated using the rule vector (140, 136, 4, 68) until an FPA is reached, as illustrated in Fig. 6

During training, the following transitions are observed: 0011 → 0001, 0110 → 0000, 0010 → 0010, 1111 → 0001, 1110 → 1000, 0111 → 0001.

As a result, the attractor set {1000, 0001} represents Class 2 and {0010, 0000} represents Class 1. Since the attractor 0001 is reached by a majority of configurations from Class 2, it is assigned to Class 2. During testing, any unknown configuration (e.g., 1011) is updated until it reaches 0001 and is therefore classified as Class 2.

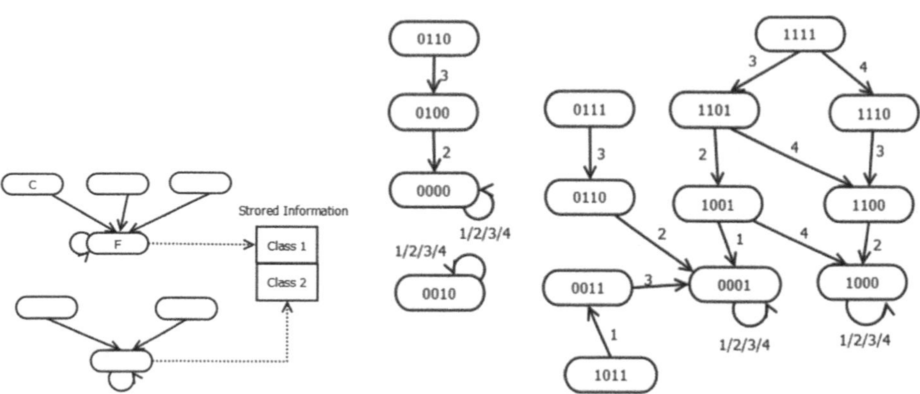

Fig. 5. Pattern classification strategy using ACA.

Fig. 6. ACA-based pattern classifier architecture.

Selection of Candidate Rule Vectors : Not all convergent ACAs are suitable for classification. The following criteria are imposed when selecting candidate rule vectors:

1. The ACA must have at least two FPAs to support binary classification. However, an excessively large number of FPAs is avoided, as it increases memory requirements and degrades generalization.
2. The average convergence time from an arbitrary configuration to an FPA should be small to reduce training and testing time. In this work, the convergence step is restricted to be close to the lattice size n.

Candidate rule vectors are from the strictly convergent ACA rules where at least $n-1$ rules are selected from $\mathtt{R}_0 \cup \mathtt{R}_1$ (See Table 3 of Sect. 3).

Algorithm 1. Computation of Average Convergence Step

Require: Rule vector R of length n, sample size S, maximum steps M
Ensure: Average convergence time T_{avg}

```
Randomly sample min(S, 2^n) configurations
Identify the fixed point set F
total_steps ← 0
for each sampled configuration C do
    steps ← 0
    while C ∉ F and steps < M do
        Asynchronously update one cell of C
        steps ← steps +1
    end while
    total_steps ← total_steps + min(steps, M)
end for
return T_avg = total_steps / S
```

5.2 Time Complexity Analysis

Let n denote the lattice size, R the number of candidate rule vectors selected for training, N_{train} and N_{test} the number of training and testing samples.

Fixed-Point Attractor Computation : For a given rule vector, fixed-point attractors are computed by constructing passive RMTs and traversing a layered directed graph across n lattice positions. In the worst case, the procedure explores an exponential number of admissible configurations, resulting in a time complexity of $\mathcal{O}(n2^n)$ per rule vector.

Average Convergence Step Computation : Checking convergence exhaustively has time complexity $O(2^n)$ and is therefore infeasible for large n. Instead, the number of FPAs is determined using the fixed point graph method with $O(n)$ complexity. [14]. The average convergence step is estimated using Algorithm 1, where the parameters M and S are both set to 1000, resulting in a time complexity of $O(MS)$

Convergence Simulation for a Single Configuration : Convergence from an initial configuration is determined via breadth-first search over the asynchronous state-transition graph. Since each configuration has at most n neighbors and the total state space size is 2^n, the worst-case time complexity for convergence detection is $\mathcal{O}(n2^n)$.

Training Phase Complexity : During training, each of the R candidate rule vectors is evaluated against all N training samples. For each sample, convergence to a fixed-point attractor is simulated. Hence, the overall training phase has a worst-case time complexity of $\mathcal{O}(RN_{train}n2^n)$.

Testing Phase Complexity : In the testing phase, the best-performing rule vector obtained during training is fixed. Each of the M test samples is evolved until convergence, leading to a total testing time complexity of $\mathcal{O}(N_{test}n2^n)$.

Space Complexity : The dominant space requirement arises from storing visited configurations during asynchronous breadth-first search, which requires $\mathcal{O}(2^n)$ memory in the worst case. Additional storage for rule tables and attractor sets incurs only linear overhead in n.

Although the theoretical worst-case time complexity is exponential in the lattice size, the use of sampling strategies and early convergence detection substantially reduces the effective runtime in practice, making the proposed approach feasible for moderate lattice sizes.

5.3 Performance of the Classifier on Datasets

Datasets are collected from the UCI Machine Learning Repository and Kaggle. Each configuration of an n-cell ACA is treated as an n-bit binary string. Table 8 summarizes the dataset characteristics and the performance of the proposed non-uniform ACA-based classifier.

ACA operate on binary strings; therefore, categorical and continuous attributes are converted to binary representations. Categorical features are encoded using thermometer encoding or one-hot encoding, depending on suitability [10]. Continuous attributes are discretized into bins and subsequently encoded.

The proposed classifier is compared with standard classifiers including Decision Tree (C4.5), Random Forest, Naive Bayes, KNN, MLP, Logistic Regression, SVC, and Uniform ACA. The comparative test efficiencies are shown in Table 9. Results marked with * are taken from [14].

Table 8. Performance of Proposed ACA-based Classifier Model

Dataset	Attri- butes	CA Size	Best Rule	Accu- racy (%)	Prec- ison (%)	Recall (%)	F1-score (%)
Personality Dataset	7	14	(196, 141, 206, 196, 236, 172,76, 78, 206, 196, 206, 76, 236, 236)	92.64	93.81	92.22	93.03
Monk Dataset 1	6	11	(236, 238, 92, 68, 172, 136, 252, 238, 92, 92, 72)	86.29	85.71	87.09	86.40
Pima Diabetes Dataset	8	24	(196, 196, 141, 140, 76, 200, 236,205,236, 200, 205, 220, 221, 220,78, 220, 76, 68, 236, 236, 200, 236, 237, 205)	76.62	81.19	82.83	82.00
Heart Disease Dataset	13	18	(196, 76, 76, 236, 205, 220, 77, 236,141, 206, 236, 76, 196, 220, 206, 68, 76, 196)	99.17	100.00	98.52	99.25
Haberman Dataset	3	9	(172, 205, 200, 196, 202, 236, 206,200, 196)	69.35	75.47	86.96	80.81

Table 9. Algorithm Test Efficiency Comparison

Dataset	Algorithm	Test Efficiency (%)
Monk 1	Bayesian*	99.9
	Decision Tree* (C4.5)	100.0
	MLP*	100.0
	Uniform ACA*	81.58
	Non-Uniform ACA	86.3
Pima Diabetes	Decision Tree	75.5
	SVC	77.2
	Random Forest	76.6
	Uniform ACA	76.6
	Non-Uniform ACA	73.3
Personality Dataset	Logistic Regression	93.8
	KNN	96.7
	Uniform ACA	92.6
	Non-Uniform ACA	92.6
Heart Disease	KNN	83.16
	MLP	87.28
	Decision Tree (C4.5)	97.6
	Uniform ACA	98.36
	Non-Uniform ACA	95.6
Haberman Dataset	KNN	68.00
	MLP	69.00
	Decision Tree (C4.5)	70.00
	Uniform ACA*	77.49
	Non-Uniform ACA	69.35

6 Conclusion

This work presents a supervised pattern classification framework based on non-uniform asynchronous elementary cellular automata. By analyzing passive RMTs, strictly convergent ACA rules are identified, ensuring the absence of non fixed point recurrent configuration. Experimental results demonstrate that hybridized convergent rule vectors achieve competitive performance with established machine learning models across multiple benchmark datasets.

References

1. Balbi, P.P., de Mattos, T., Ruivo, E.: Characterisation of the elementary cellular automata with neighbourhood priority based deterministic updates. Commun. Nonlinear Sci. Numer. Simul. **104**, 106018 (2022)
2. Bhattacharjee, K., Paul, S., Das, S.: Affinity classification problem by stochastic cellular automata. Complex Syst. **32**(3), 271–288 (2022)
3. Chaudhuri, P.P., Chowdhury, D.R., Nandi, S., Chattopadhyay, S.: Additive cellular automata: theory and applications, Volume 1, vol. 1. John Wiley and Sons (1997)
4. Concha-Vega, P., Goles, E., Montealegre, P., Ríos-Wilson, M., Santivañez, J.: Introducing the activity parameter for elementary cellular automata. Int. J. Mod. Phys. C **33**(09), 2250121 (2022)
5. Das, S., Sikdar, B.: Non-uniform cellular automata (2026)
6. Fates, N.: A note on the classification of the most simple asynchronous cellular automata. In: International Workshop on Cellular Automata and Discrete Complex Systems, pp. 31–45. Springer (2013)
7. Fatès, N.: Stochastic cellular automata solutions to the density classification problem: when randomness helps computing. Theory Comput. Syst. **53**(2), 223–242 (2013)
8. Fatès, N.: Asynchronous cellular automata. In: Cellular Automata, pp. 73–92. Springer (2018)
9. Fatès, N.: A tutorial on elementary cellular automata with fully asynchronous updating: general properties and convergence dynamics. Nat. Comput. **19**(1), 179–197 (2020)
10. Gallant, S.I.: Neural network learning and expert systems. MIT press (1993)
11. Ganguly, N.: Cellular automata evolution: theory and applications in pattern recognition and classification. India: Doctor Dissertation of CST Dept. BECDU (2003)
12. Maji, P.: Cellular automata evolution for pattern recognition. Ph.D. thesis, PhD thesis, Jadavpur University, Kolkata, India (2005)
13. Pedregosa, F., et al.: Scikit-learn: machine learning in python. J. Mach. Learn. Res. **12**, 2825–2830 (2011)
14. Sethi, B.: Theory and applications of fully asynchronous cellular automata. Ph.D. thesis, Indian Institute of Engineering Science and Technology, Shibpur (2018)
15. Sethi, B., Roy, S., Das, S.: Asynchronous cellular automata and pattern classification. Complexity **21**(S1), 370–386 (2016)
16. Von Neumann, J., et al.: Theory of self-reproducing automata (1966)
17. Wolfram, S.: A new kind of science (wolfram media, inc., 2002). http://www.wolframscience.com/nksonline/toc.html
18. Wolfram, S.: Statistical mechanics of cellular automata. Rev. Mod. Phys. **55**(3), 601 (1983)
19. Wolfram, S.: Cellular automata as models of complexity. Nature **311**(5985), 419–424 (1984)
20. Wolfram, S.: Universality and Complexity in Cellular Automata. Physica D **10**(1–2), 1–35 (1984)

Correlating a Measure of Asynchrony of Block-Sequential Updates with the Dynamics of Elementary Cellular Automata

Daniel de Azevedo Gimigliano(✉) and Pedro Paulo Balbi

Pós-graduação em Engenharia Elétrica e Computação,Faculdade de Computação e Informática, Universidade Presbiteriana Mackenzie,Rua da Consolação 896, Consolação, 01302-907 São Paulo, SP, Brazil
daniel.gimigliano@mackenzista.com.br, pedrob@mackenzie.br

Abstract. The dynamics of cellular automata running on cyclic configurations under block-sequential deterministic updates may drastically depend on the update employed. In order to better understand such a dependence, various analyses have been carried out in the literature. In the same direction, here we explore the potential of a recently proposed metric meant to act as a measure of the *asynchrony degree* of an update; we emphasise that this the first effort trying to probe the potential of the metric as it relates to the dynamics of automata networks. More specifically, we investigate, for all elementary cellular automata, when update schedules with the same degree of asynchrony have similar dynamics, as expressed by their leading to structurally equivalent basins of attraction, that is, those with the same attractors and same configurations in each one. Considering all configuration sizes from 5 to 13, we performed a computational analysis of the dynamics of the rules, trying to correlate them with the asynchrony degree of the update schedules, and grouping the rules into similarity classes of their dynamics. Due to computational limitations, the correlation between the asynchronous degrees and the dynamics sought was not entirely conclusive, but was clearly indicative of the relevance of the metric and the methodology employed.

Keywords: Asynchrony degree · block-sequential asynchronous update · elementary cellular automata · dynamical similarity classes · structrually equivalent basins of attraction

1 Introduction

Cellular automata (CAs) are fully discrete dynamical systems that operate entirely by means of local action among their constituting parts, the cells. At each step, each cell applies the same rule to the states of its neighbours. In the synchronous regime, all cells update simultaneously. In models of natural phenomena, interactions are not simultaneous. Asynchrony in cellular automata can

H. Raju et al. (Eds.): ASCAT 2026, CCIS 2801, pp. 270–282, 2026.
https://doi.org/10.1007/978-3-032-18612-6_20

be implemented using deterministic or stochastic update schedules [8]. Among deterministic approaches, positional schemes stand out, where the update order is defined by the spatial position of the cells, neighbourhood-based schemes, where the order depends on the local state [4], and others [7]. Here we focus on deterministic asynchrony by Block-Sequential Update Schedules (BSUs), which partition cells into ordered blocks: cells in the same block update together and blocks update in sequence [1].

Recent studies have characterised aspects of the dynamics of the elementary cellular automata (ECAs) – one-dimensional, binary with the two next-nearest neighbours of a cell – when subjected to block-sequential asynchronism. More specifically, [9,11] and [6] characterised the dynamics of ECAs in terms of their sensitivity to asynchronism in finite configurations, and [10] extended the analysis to infinite configurations.

In [5] a metric was proposed for the *asynchrony degree* of a block-sequential update schedule, defined according to the amount of nesting of applications of the local transition function in each cell of a configuration. The proposal was made in the context of the need therein of a simple way to discriminate sets of updates in terms of which one would be 'more asynchronous' than others. However, no further analyses has been made since then in terms of trying to figure out how such a measure might correlate (if at all) with the dynamics of a rule under such a schedule. This is exactly the motivation of the current work.

Here we tackle this question by investigating how the asynchrony degree ($\mathcal{T}$) of a BSU might correlate with the basins of attraction of a rule operating under this BSU. To this end we rely on the proposition of the notions of *structural equivalence* between basins of attraction, and of the *partition signature* of groups of BSUs, values of asynchrony degree and patterns of structural equivalence related to basins of attraction. This gives rise to a new view of the elementary space in terms of *Asynchrony Degree based Dynamical Similarity* (ADDS) *classes* related to the dynamics of the rules according to the values of asynchrony degree of the BSUs.

In what follows, Sect. 2 presents the basic definitions related to the work, including block-sequential updates. Section 3 presents the definitions of key notions we proposed that underlie the work. Section 4 describes the computational work carried out that led to a characterisation of the patterns of response to asynchrony of the ECAs, and compares the result with existing classifications in the literature. Section 5 concludes by discussing the results obtained and its limitations, and provides future directions.

2 Basic Definitions

2.1 Cellular Automata

A *cellular automaton* (CA) is a fully discrete dynamical system composed of L cells, each one holding a state from a finite set S. Given a *dimension* $d \in \mathbb{Z}^+$, a *neighbourhood vector* $N = (\boldsymbol{v}_1, \cdots, \boldsymbol{v}_{|N|})$, $\boldsymbol{v}_i \in \mathbb{Z}^d, \forall 1 \leq i \leq |N|$, and a *local transition function* $f : S^{|N|} \to S$, the CA is formally defined as the 4-tuple

(S, N, f, d) [13]. The local rule f is applied uniformly to each cell according to the states in its neighbourhood, inducing a global transition function F on the configuration space.

In defining CAs, it is usually helpful to refer to the neighbourhood extension in terms of its *radius* r, which, in the case of one-dimensional CAs, means that the neighbourhood vector takes the form $N = (-r, \ldots, 0, \ldots, r)$, i.e., that each cell has $2r + 1$ neighbours [13].

Given a radius-r CA with local rule f, such a local rule induces the *global transition function* (or *global rule*) $F : \mathcal{C} \rightarrow \mathcal{C}$ in the set of all configurations, that is, $(F(c))_i = f(c_{i-r}, \cdots, c_i, \cdots, c_{i+r})$.

2.2 Elementary Cellular Automata

Elementary cellular automata. (ECAs) are one-dimensional ($d = 1$) binary ($S = \{0, 1\}$) cellular automata with a neighbourhood radius $r = 1$. The local transition function $f : \{0, 1\}^3 \rightarrow \{0, 1\}$ is applied uniformly to each cell i according to its neighbours, therefore inducing at that cell the global function $(F(c))_i = f(c_{i-1}, c_i, c_{i+1})$. Each rule is described by a binary sequence $(s_7 s_6 s_5 s_4 s_3 s_2 s_1 s_0)$, where s_ℓ is the output value for the neighbourhood represented by ℓ in binary. The elementary space contains $2^{2^3} = 256$ distinct rules, identified by their *Wolfram number*, the decimal value of the binary sequence.

For dynamical behaviour analyses, rules that differ only by symmetries or state can be grouped into *equivalence classes.* Two ECA rules belong to the same *dynamical equivalence class* when they exhibit identical qualitative behaviour under symmetry transformations: conjugation (complementation of states $0 \leftrightarrow 1$), reflection (spatial mirroring of the neighbourhood), or the composition of both. These transformations partition the space of 256 rules into 88 equivalence classes [13], each one represented by the rule with the lowest Wolfram number in each class.

2.3 Block-Sequential Update Schedule

For finite size configurations with length L, a BSU partitions the L cells into *ordered blocks*, with the blocks updating in sequence, while the cells in the same block update together.

Formally, with $1 \leq m \leq L$, a BSU is any *ordered partition* $\sigma = (B_1, B_2, \ldots, B_m)$ of the set $\{1, 2, \ldots, L\}$ of cells [1,9]. When $m = 1$ we have the parallel or synchronous case. The combinatorial structure of these partitions was formalised using the so called *update digraphs*, which capture the update order between the cells and give rise to equivalent BSUs [2]. The *global asynchronous rule* F_σ is the sequential composition of the update operators of each block: $F_\sigma = F_{B_m} \circ \cdots \circ F_{B_1}$ [5]. For configurations of length L, the total number of independent BSUs is given by the *Fubini numbers* [2,5].

Given a pair (rule R, BSU σ) for configurations of length L, the *transition graph* $G(R, \sigma)$ is a digraph whose vertices are the configurations of the state space

$\{0,1\}^L$ and whose edges represent transitions between configurations under the application of the global rule F_σ [6,9].

For cyclic configurations (which is the present case), BSUs obtained by rotation generate equivalent dynamics. Two BSU σ and σ' are *rotationally equivalent* if there exists a shift $j \in \{1,\ldots,L-1\}$ such that $\sigma' = \mathrm{rot}_j(\sigma)$, where $\mathrm{rot}_j : \{1,\ldots,L\} \to \{1,\ldots,L\}$ with $\mathrm{rot}_j(i) = ((i-1+j) \bmod L) + 1$. Such an equivalence stems from the shift invariance of cellular automata [9]: rotations preserve the isomorphism of transition graphs $G(R,\sigma)$ for any rule R [6].

For configurations of length $L = 5$, the total number of BSUs is $F_5 = 541$, the fifth Fubini number [5]. Applying the equivalence by update digraphs [3], this number reduces to 181 equivalence classes. The application of rotational equivalence [9] further reduces the space to 37 representative BSUs.

Figure 1 illustrates the effect of a BSU on the temporal evolution of ECA rule 110, on a same initial configuration size $L = 21$, with time flowing downwards. In panel (a), the update is synchronous: all cells are updated simultaneously at each time step. In panel (b), a fully sequential BSU is employed, where each cell constitutes its own block, updated one at a time from position 1 to position L, i.e., $\sigma = (\{1\},\{2\},\ldots,\{21\})$.

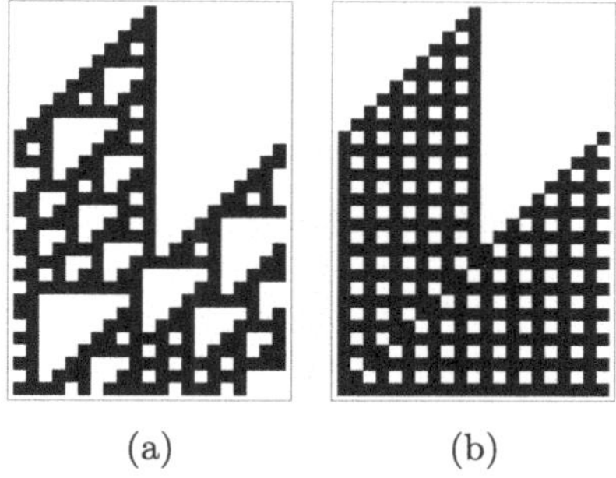

Fig. 1. Temporal evolution of ECA 110 under synchronous update (a) and under a fully sequential BSU (b). Same initial configuration in both cases and time flowing downwards.

2.4 Asynchrony Degree

In the synchronous (parallel) update, every cell in a configuration is subjected to a single and direct application of the local function f. However, in the case of any asynchronous BSU some cells might only be updated after their neighbours, which, in turn might only be updated after their own neighbours, and so on, in a chain of dependence that can be regarded as if, in general, the cells become subjected to a number of indirect applications of f. So, while the cells in the first block of any BSU are always subjected to a single application of f, for the cells in the other blocks, the number of (indirect) applications of f will depend on their position in such a chain of dependence.

In order to account for this fact at the level of each individual cell, we define the *individual asynchrony degree* of cell i, denoted by $d_\sigma(i)$, as a metric related to the number of indirect applications of the local function f that cell i is subjected to in the BSU. Naturally, the minimum value of applications of f is 1, as used in [5]; but since we only want to refer here to the indirect action of f over the cells, we disconsider the direct action of f, which leads to the minimum value of the metric becoming 0. Consequently, all the cells in the first block have $d_\sigma(i) = 0$, while the cells in the subsequent blocks have $0 \leq d_\sigma(i)$, up to a maximum that depends on the configuration size L and the radius r of the local function.

As an illustration, let us consider the following example, where $L = 5$, $r = 1$ and $\sigma = (\{1\}, \{2, 3\}, \{4, 5\})$.

For cell 1 (in block 1), no neighbouring cell has been updated yet; so

$$(F_\sigma(c))_1 = f(c_5, c_1, c_2), \quad d_\sigma(1) = 1 - 1 = 0.$$

For cell 2 (in block 2), since its left-hand neighbour (cell 1) has already been updated, we have that

$$(F_\sigma(c))_2 = f(f(c_5, c_1, c_2), c_2, c_3), \quad d_\sigma(2) = 2 - 1 = 1.$$

For cell 3 (in block 2), which updates simultaneously with cell 2, but has no dependence on cell 1, this leads to

$$(F_\sigma(c))_3 = f(c_2, c_3, c_4), \quad d_\sigma(3) = 1 - 1 = 0.$$

For cell 4 (in block 3), since its left-hand neighbour (cell 3) has already been updated, we get

$$(F_\sigma(c))_4 = f(f(c_2, c_3, c_4), c_4, c_5), \quad d_\sigma(4) = 2 - 1 = 1.$$

Finally, for cell 5 (in block 3), since its right-hand neighbour (cell 1) has been updated,

$$(F_\sigma(c))_5 = f(c_4, c_5, f(c_5, c_1, c_2)), \quad d_\sigma(5) = 2 - 1 = 1.$$

So, based upon the values of the individual asynchrony degrees of the BSU, we define its (total) *asynchrony degree* ($\mathcal{T}$) as

$$\mathcal{T}(\sigma) = \sum_{i=1}^{L} d_\sigma(i).$$

Thus, for the example above, $\mathcal{T}(\sigma) = 0 + 1 + 0 + 1 + 1 = 3$. And notice that for the synchronous case, the corresponding asynchrony degree is 0, a meaningful quantity that clarifies why we disconsider the direct applications of f.

For configurations of length $L = 5$, the 37 representative BSUs (Sect. 2.3) are distributed across 8 distinct $\mathcal{T}$ values: 2, 3, 4, 5, 6, 7, 9, and 11, excluding the synchronous case ($\mathcal{T}$=0).

3 Concepts Underlying the Methodology

3.1 Structural Equivalence Between Basins of Attraction Fields

In the transition graph $G(R, \sigma)$ defined by a pair (rule R, BSU σ) for configurations of length L, each *weakly connected component* represents a *basin of attraction*, which is a set of configurations that converge to the same *attractor* (*fixed point* or *limit cycle*). The *basin of attraction field* (BAF) is the set $\mathcal{B}(R, \sigma, L) = \{A_1, A_2, \ldots, A_{n_b}\}$ of all basins, representing a complete partition of the state space $\{0, 1\}^L$ [12].

In order to illustrate how the transition graph is constructed, consider rule 184, over the configurations of length $L = 3$, and the BSU $\sigma_1 = (\{2\}, \{1, 3\})$. Each of the $2^3 = 8$ possible configurations constitutes a vertex of the graph, and the edges represent transitions under the application of the global rule F_σ.

Since the BSU has two blocks, each transition occurs in two steps. For instance, let us analyse configuration 101: first, block $\{2\}$ is considered, meaning that only cell 2 is updated; since the application of rule 184 to the neighbourhood (1, 0, 1) leads to state 1, this results in the intermediate configuration 111. In the sequence, block $\{1, 3\}$ is considered, entailing cells 1 and 3 to be simultaneously updated; for both cells the neighbourhood is (1, 1, 1) that leads to state 1. Therefore, the final configuration remains 111. Consequently, the edge 101 $\rightarrow$ 111 is added to the graph. By repeating this procedure for all configurations, we obtain the complete graph shown in Fig. 2(a).

The graph has two disconnected components, each corresponding to a basin of attraction, i.e., the set of configurations that converge to the same attractor. In the figure, the fixed point 111 attracts the configurations $\{101, 100, 001\}$, while the fixed point 000 attracts $\{010, 110, 011\}$.

To compare the behaviour of a rule under different BSUs, we define *structural equivalence* between BAFs as follows: two BAFs $\mathcal{B}(R, \sigma_1, L)$ and $\mathcal{B}(R, \sigma_2, L)$ are *structurally equivalent* ($\mathcal{B}(R, \sigma_1, L) \equiv \mathcal{B}(R, \sigma_2, L)$), when they have the same number of basins, with the same attractors and the same configurations associated with each attractor.

Consider now rule 184 in $L = 3$ with the BSU $\sigma_2 = (\{3\}, \{1\}, \{2\})$. Both previous BSUs generate two attractors, but partition the 8 configurations differently, resulting in structurally non-equivalent BAFs, as Fig. 2 illustrates.

3.2 Partition Signature

To investigate the relationship between $\mathcal{T}$ and structural equivalence of BAFs, we introduce two complementary notions. The first, the *partition pattern* ($P_\mathcal{T}$), describes how BSUs with the same asynchrony degree value are grouped, according to the structural equivalence of the BAFs they generate; the sizes of these groups are recorded in descending order. The second, the *partition signature* of a rule R, denoted $\mathcal{S}_L(R) = [(\mathcal{T}_1, P_1), \ldots, (\mathcal{T}_8, P_8)]$, gathers all partition patterns for a given configuration size L, characterising the rule's overall response to asynchrony.

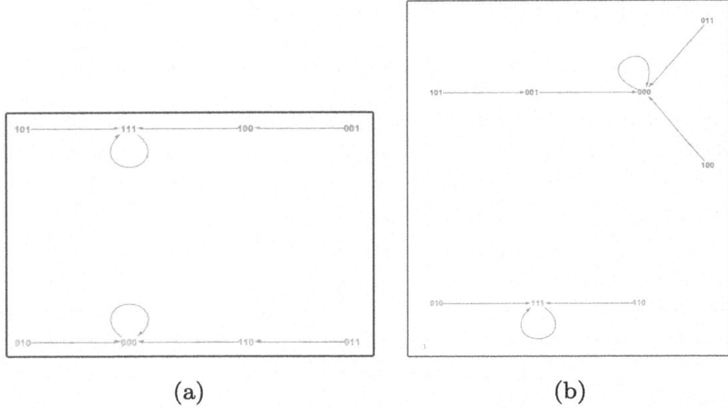

(a) (b)

Fig. 2. Structurally non-equivalent BAFs for rule 184 in $L = 3$. Nodes represent configurations; edges indicate transitions; loops indicate fixed points. **(a)** BSU $\sigma_1 = (\{2\}, \{1, 3\})$: two basins, each with one fixed point and 4 configurations. First basin: $\{(0,1,1), (1,1,0), (0,0,0), (0,1,0)\}$; second basin: $\{(1,0,1), (1,1,1), (1,0,0), (0,0,1)\}$. **(b)** BSU $\sigma_2 = (\{3\}, \{1\}, \{2\})$: two basins with 3 and 5 configurations, respectively. First basin: $\{(1,1,0), (1,1,1), (0,1,0)\}$; second basin: $\{(0,1,1), (0,0,0), (0,0,1), (1,0,0), (1,0,1)\}$.

For example, let us consider $\mathcal{T} = 3$ with 6 representative BSUs, and suppose that 3 of them generate structurally equivalent BAFs, 2 others also generate equivalent BAFs (but distinct from the first group), and 1 generates a unique BAF. In this case, the partition pattern is $P_3 = [3, 2, 1]$. The complete signature $\mathcal{S}_5(R)$ gathers the patterns for all $\mathcal{T}$ values present in $L = 5$: $\mathcal{S}_5(R) = [(2, P_2), (3, P_3), \ldots, (11, P_{11})]$.

The pattern $P_\mathcal{T}$ reflects the rule's sensitivity to asynchrony. When $P_\mathcal{T} = [k]$, this means that all k BSUs with the same $\mathcal{T}$ generate structurally equivalent BAFs, indicating maximum insensitivity to that level of asynchrony. At the opposite extreme, $P_\mathcal{T} = [1, 1, \ldots, 1]$ indicates maximum sensitivity: each BSU produces a distinct BAF. Intermediate patterns, such as the $[3, 2, 1]$ from the previous example, reveal varying sensitivity, meaning that some subsets of BSUs share structure, while others remain unique.

3.3 Asynchrony Degree Based Dynamical Similarity Classes

For a given configuration size L, two rules R_1 and R_2 are said to belong to the same *Asynchrony Degree based Dynamical Similarity class* (ADDS class) when their partition signatures are identical, i.e., $\mathcal{S}_L(R_1) = \mathcal{S}_L(R_2)$. This equivalence relation partitions the rule space according to patterns of response to asynchrony.

As an illustration, suppose that for two rules R_1 and R_2, and for every value of $\mathcal{T}$, the partition patterns $P_\mathcal{T}$ of both are equal to $[n_\mathcal{T}]$, where $n_\mathcal{T}$ denotes the number of representative BSUs with asynchrony degree $\mathcal{T}$. The pattern $[n_\mathcal{T}]$—a single-element sequence—indicates that all $n_{\mathcal{T}}$ BSUs form a single group, that is, all of them generate structurally equivalent BAFs for the rule in question. This behaviour characterises maximum insensitivity to that level of asynchrony.

In contrast, the pattern $[1, 1, \ldots, 1]$ – a sequence of $n_{\mathcal{T}}$ elements, all equal to 1 – indicates that each BSU forms its own group, that is, each one generates a structurally distinct BAF. This behaviour characterises maximum sensitivity.

Therefore, if for every value of $\mathcal{T}$ the partition patterns of R_1 and R_2 coincide, then $\mathcal{S}_L(R_1) = \mathcal{S}_L(R_2)$, and both belong to the same ADDS class, regardless of whether they exhibit insensitivity, sensitivity, or any intermediate behaviour.

4 Results

All the work was carried out computationally, without any prior theoretical formulation to estimate the ADDS classes. For each configuration length $L \in \{5, 6, 7, 8, 9, 10, 11, 12, 13\}$, we generated the complete set of possible update schedules (BSUs) and applied rotational equivalence to eliminate redundant representatives, resulting in 37 representative BSUs for $L = 5$. These 37 representatives were partitioned by the $\mathcal{T}$ values, excluding the synchronous schedule ($\mathcal{T}$=0), resulting in 36 asynchronous schedules distributed across 8 distinct values: $\mathcal{T} \in \{2, 3, 4, 5, 6, 7, 9, 11\}$. The upper limit $L = 13$ was established due to computational limitations.

Starting with $L = 5$, for each of the 88 representative ECA rules we went about generating all corresponding ADDS classes, considering all 36 representative asynchronous schedules. In the process, all pairs (rule R, BSU σ) were obtained and their BAFs pairwise compared, so as to detect the structurally equivalent ones; then by relying on the associated partition signatures, 10 distinct ADDS classes were derived.

Table 1 shows the result. Notice a great discrepancy in the size of the ADDS classes: while class A accounts for 67 of the 88 rules, classes E through J contain only one rule each.

Table 2 presents the complete set of partition signatures for $L = 5$. Each row corresponds to the signature of a specific ADDS class, with columns representing the eight distinct $\mathcal{T}$ values observed for the 36 asynchronous schedules representative of this size. Each cell contains the partition pattern $P_{\mathcal{T}}$, indicating how representative BSUs with that degree of asynchrony are grouped according to the structural equivalence of the associated BAFs.

The partition signatures in Table 2 allows to identify three distinct patterns of response to asynchrony. Class A exhibits maximum sensitivity with pattern $[1, 1, \ldots, 1]$ for all $\mathcal{T}$ values, suggesting that each BSU produces a distinct partition of the state space. Class C exhibits maximum insensitivity with the pattern $[n]$, $n \in \{2, 4, 6, 8\}$, n being the total number of BSUs for each $\mathcal{T}$ value, indicating that BSUs with the same $\mathcal{T}$ value always produce structurally equivalent BAFs. The remaining classes exhibit unstable partition signatures, which vary when the configuration length changes, as detailed in Table 4.

Table 1. Distribution of the 88 ECA rules into 10 ADDS classes for $L = 5$.

Class	Rules	Representatives
A	67	2, 4, 5, 6, 7, 9, 10, 11, 13, 14, 18, 19, 22, 23, 24, 26, 27, 29, 30, 33, 35, 36, 37, 38, 40, 41, 42, 43, 45, 46, 50, 54, 56, 57, 58, 62, 72, 73, 74, 76, 77, 78, 90, 94, 104, 105, 106, 108, 110, 122, 126, 130, 132, 134, 138, 142, 146, 150, 152, 154, 156, 162, 164, 168, 172, 178, 184
B	6	3, 12, 15, 28, 44, 60
C	5	0, 8, 51, 200, 204
D	4	34, 136, 140, 170
E	1	1
F	1	25
G	1	32
H	1	128
I	1	160
J	1	232

In order to investigate the stability of the ADDS classes across different configuration lengths, as mentioned earlier we extended the analysis for all configuration sizes L up to 13. Table 3 summarises the results, showing the number of ADDS classes for each configuration size, the number of representative BSUs that exist for each size, and the time to compute the information about each class.

All computational experiments were carried out using *Wolfram Mathematica* version 14.3; the corresponding Wolfram language code is available upon request. As for the hardware infrastructure, we relied on a dedicated Hetzner EX44 server (Intel Core i5-13500 14-core, 128 GB RAM DDR4 Non-ECC).

The investigation across different configuration lengths revealed that ADDS classes partition into two categories: classes with stable composition across all analysed lengths, and classes with variable composition depending on L.

The stable category comprises four invariant ADDS classes outside Class A: $G_1 = \{3, 12, 15, 28, 44, 60\}$, $G_2 = \{0, 8, 51, 200, 204\}$, $G_3 = \{34, 136, 140, 170\}$, and $G_4 = \{32\}$. Additionally, within Class A, group G_5 constitutes a stable core of 65 rules: $G_5 = \{2, 4, 5, 6, 7, 9, 10, 11, 13, 14, 18, 19, 22, 23, 24, 26, 27, 29, 30, 33, 35, 36, 37, 38, 40, 41, 42, 43, 45, 46, 50, 54, 56, 57, 58, 62, 72, 73, 74, 76, 77, 78, 94, 104, 105, 106, 108, 110, 122, 126, 130, 132, 134, 142, 146, 150, 152, 154, 156, 162, 164, 168, 172, 178, 184\}$. In contrast, seven rules exhibit variable class membership: $G_6 = \{1, 25, 90, 128, 138, 160, 232\}$ transitions between classes or integrates into Class A (G_5) depending on L.

Table 4 details the behaviour of the 7 rules of group G_6 for each configuration size from $L = 5$ to $L = 13$. Absent rules indicate integration into Class A (group

Table 2. Partition signatures by ADDS classes for $L = 5$. Column headers indicate $\mathcal{T}$ values.

Class	2	3	4	5	6	7	9	11
A	[1,1,1,1]	[1,1,1,1,1,1]	[1,1,1,1,1,1,1,1]	[1,1,1,1,1,1]	[1,1,1,1]	[1,1,1,1]	[1,1]	[1,1]
B	[1,1,1,1]	[2,1,1,1,1]	[1,1,1,1,1,1,1,1]	[2,1,1,1,1]	[1,1,1,1]	[1,1,1,1]	[1,1]	[1,1]
C	[4]	[6]	[8]	[6]	[4]	[4]	[2]	[2]
D	[4]	[3,2,1]	[3,3,1,1]	[2,1,1,1,1]	[2,1,1]	[2,2]	[1,1]	[1,1]
E	[1,1,1,1]	[2,1,1,1,1]	[1,1,1,1,1,1,1,1]	[1,1,1,1,1,1]	[1,1,1,1]	[1,1,1,1]	[1,1]	[1,1]
F	[1,1,1,1]	[2,1,1,1,1]	[3,1,1,1,1,1]	[1,1,1,1,1,1]	[2,1,1]	[2,1,1]	[1,1]	[1,1]
G	[2,1,1]	[3,2,1]	[4,2,2]	[3,3]	[2,2]	[4]	[2]	[2]
H	[1,1,1,1]	[2,1,1,1,1]	[2,2,1,1,1,1]	[2,1,1,1,1]	[2,1,1]	[4]	[2]	[1,1]
I	[1,1,1,1]	[1,1,1,1,1,1]	[2,1,1,1,1,1,1]	[2,2,1,1]	[2,1,1]	[2,1,1]	[2]	[1,1]
J	[1,1,1,1]	[1,1,1,1,1,1]	[2,2,1,1,1,1]	[2,2,1,1]	[2,1,1]	[2,2]	[2]	[2]

Table 3. Distribution of the number BSUs and ADDS classes by configuration length, and associated computational time to obtain the classes.

L	BSU	Classes	Time (sec)
5	37	10	0.66
6	104	8	3.55
7	277	7	25.45
8	764	6	143.39
9	2077	7	811.63
10	5720	7	4,494.31
11	15733	6	25,059.85
12	43666	7	331,881.08
13	121381	8	1,958,187.50

G_5) at that size; rules indicated mean that they are present either isolated in an ADDS class or in a pair. Thus, as L varies, rules migrate between classes or integrate into Class A, revealing the dependence of the dynamics on the configuration size.

It is worth comparing the ADDS classes we obtained, with other partitions of the elementary space according to aspects of the dynamics due to block-sequential updates. Accordingly, [6,9,11] analysed the sensitivity of the dynamics of the ECA rules when replacing a BSU by another. More precisely, while [9,11] was interested in maximal sensitivity to the updates – meaning that the dynamics of an ECA changes for every BSU employed – [6] analysed the non-maximal case.

The classification in [6] organized the 19 non-maximally sensitive rules into four classes according to their sensitivity to synchronism: Class I (insensitive), Class II (low-sensitive), Class III (medium-sensitive), and Class IV (almost max-

Table 4. Variable ADDS classes (group G_6) for all configuration sizes analysed.

L	Classes with one rule	Classes with two rules
5	{1}, {25}, {128}, {160}, {232}	-
6	{25}, {128}	{90, 160}
7	{25}, {128}	-
8	{128}	-
9	{25}, {128}	-
10	{25}, {128}	-
11	{25}	-
12	{25}, {128}	-
13	{25}, {128}, {138}	-

sensitive). Comparing our invariant groups with this classification, we clearly observe some correspondences: group $G_2 = \{0, 8, 51, 200, 204\}$ contains rules from Classes I and III, while groups G_1, G_3, and G_4 correspond to Class II. Rules 162 (group G_5), 128 and 160 (group G_6) belong to Class IV.

The classification in [9,11] partitioned the ECA rule space according to their maximum sensitivity to update schedules, identifying 19 rule classes with non-maximal sensitivity for $L \geq 7$. Out of these 19 classes, 18 correspond to the invariant groups G_1, G_2, G_3, G_4 (16 rules) and rules 160 and 162; the remaining class, the one with rule 128, belongs to the variable group G_6 in this work.

5 Concluding Remarks

This work investigated the relationship between the asynchrony degree and the structural equivalence of basins of attraction fields (BAFs) in elementary cellular automata under deterministic block-sequential update schedules. The analysis of the 88 ECA rules across configurations of increasing lengths ($L \in \{5, ..., 13\}$), corresponding to a computational space growing from 37 to over 121,000 representative BSUs, identified hierarchical patterns of response to asynchrony. Five groups were characterised by configuration-length independent partition signatures, while one group exhibited signatures dependent on L.

Comparison of the invariant groups with the classification in [6] suggests a possible correlation between partition signatures and sensitivity to synchronism. The analysis does not produce a fully conclusive correlation due to high computational demands, but it indicates the relevance of the asynchrony degree as a tool for characterising the response to asynchrony in elementary CA rules. The partition signature methodology reveals patterns shared by dynamically distinct rules, offering a complementary perspective to classifications based on quantitative sensitivity.

The structural equivalence criterion (Sect. 3.1) requires exact identity of the sets of configurations in each basin, but ignores internal topology and transient times. Criteria based on branch density or stability to perturbations could reveal distinct grouping patterns. The instability observed in ADDS classes for different configuration lengths – where 7 rules transition between classes depending on L – is yet to be explained. Analyses with configuration sizes beyond $L = 13$ could stabilise the variable classes. The characterisation remains purely empirical, without theoretical justification of the observed correlation with classifications already established in the literature; this is also an open alley for a follow-up research.

The theoretical formalisation of the observed correlation between partition signatures and the classification in [6] may connect the structural equivalence of BAFs with exact measures of sensitivity to synchronism. A promising direction is to explore alternative equivalence criteria, such as incorporating internal basin topology and transient times, to reveal structural patterns not detected by present criterion. Another research direction would consider generalising the notion of partition signature to other classes of cellular automata, verifying whether the identified patterns constitute a phenomenon specific to ECAs or a more comprehensive property.

Acknowledgements. D.G. thanks Instituto Presbiteriano Mackenzie for a graduate grant, and P.P.B. thanks the Brazilian agency CNPq (Conselho Nacional de Desenvolvimento Científico e Tecnológico) for the research grant PQ 303356/2022–7.

References

1. Aracena, J., Goles, E., Moreira, A., Salinas, L.: On the robustness of update schedules in Boolean networks. Biosystems **97**(1), 1–8 (2009). https://doi.org/10.1016/j.biosystems.2009.03.006
2. Aracena, J., Fanchon, E., Montalva, M., Noual, M.: Combinatorics on update digraphs in Boolean networks. Discret. Appl. Math. **159**(6), 401–409 (2011). https://doi.org/10.1016/j.dam.2010.10.010
3. Aracena, J., Demongeot, J., Fanchon, E., Montalva, M.: On the number of different dynamics in Boolean networks with deterministic update schedules. Math. Biosci. **242**(2), 188–194 (2013). https://doi.org/10.1016/j.mbs.2013.01.007
4. Balbi, P.P., de Mattos, T., Ruivo, E.L.P.: Characterisation of the elementary cellular automata with neighbourhood priority based deterministic updates. Commun. Nonlinear Sci. Numer. Simul. **104**, 106018 (2022). https://doi.org/10.1016/j.cnsns.2021.106018
5. Balbi, P.P., Etchebehere, G.S., Ruivo, E.L.P.: A spectral outlook on the elementary cellular automata with cyclic configurations and block-sequential asynchronous updates. In: Adamatzky, A. (ed.) Automata and Complexity: Essays Presented to Eric Goles on the Occasion of His 70th Birthday. Emergence, Complexity and Computation, vol. 42, pp. 93–115. Springer, Cham (2022). https://doi.org/10.1007/978-3-030-92551-2_9
6. Balbi, P.P., Formenti, E., Perrot, K., Riva, S., Ruivo, E.L.P.: Non-maximal sensitivity to synchronism in elementary cellular automata: exact asymptotic measures. Theor. Comput. Sci. **926**, 21–50 (2022). https://doi.org/10.1016/j.tcs.2022.05.024
7. Donoso-Leiva, I., Goles, E., Ríos-Wilson, M., Sené, S.: Impact of (a)synchronism on ECA: towards a new classification. Chaos, Solitons Fractals **199**, 116601 (2025). https://doi.org/10.1016/j.chaos.2025.116601

8. Fatès, N.: A guided tour of asynchronous cellular automata. J. Cell. Automata **9**(5-6), 387–416 (2014)
9. Perrot, K., Montalva-Medel, M., de Oliveira, P.P.B., Ruivo, E.L.P.: Maximum sensitivity to update schedules of elementary cellular automata over periodic configurations. Nat. Comput. **19**(1), 51–90 (2019). https://doi.org/10.1007/s11047-019-09743-9
10. Ruivo, E.L.P., Balbi, P.P., Montalva-Medel, M., Perrot, K.: Maximum sensitivity to update schedules of elementary cellular automata over infinite configurations. Inf. Comput. **274**, 104538 (2020). https://doi.org/10.1016/j.ic.2020.104538
11. Ruivo, E.L.P., Montalva-Medel, M., de Oliveira, P.P.B., Perrot, K.: Characterisation of the elementary cellular automata in terms of their maximum sensitivity to all possible asynchronous updates. Chaos, Solitons & Fractals **113**, 209–220 (2018). https://doi.org/10.1016/j.chaos.2018.06.004
12. Wuensche, A.: Complex and chaotic dynamics, basins of attraction, and memory in discrete networks. Discrete Dynamics Lab (2010). www.ddlab.org
13. Wolfram, S.: A new kind of science. Wolfram Media (2002)

Correction to: Design of Light-Weight Combined Pseudo Random Number Generator Using Cellular Automaton

Bhuvaneswari Arumugam and Kamalika Bhattacharjee

Correction to:
Chapter 5 in: H. Raju et al. (Eds.): *Cellular Automata Technology*, CCIS 2801, https://doi.org/10.1007/978-3-032-18612-6_5

In the originally published version of chapters 5, the author's name is in incorrect order. It should be Bhuvaneswari A. It has been corrected. Also, In the originally published version of chapter 11, the annotated corrections shared by author are not fixed. It has been corrected.

The updated version of this chapter can be found at
https://doi.org/10.1007/978-3-032-18612-6_5

H. Raju et al. (Eds.): ASCAT 2026, CCIS 2801, p. C1, 2026.
https://doi.org/10.1007/978-3-032-18612-6_21

Author Index

H. Raju et al. (Eds.): ASCAT 2026, CCIS 2801, p. 283, 2026.
https://doi.org/10.1007/978-3-032-18612-6

The manufacturer's authorised representative in the EU is Springer Nature Customer Service Centre GmbH, Europaplatz 3, 69115 Heidelberg, Germany. If you have any concerns regarding our products, please contact ProductSafety@springernature.com

Printed and bound by CPI Group (UK) Ltd, Croydon, CR0 4YY
07/07/2026
02160906-0007